AF324127

NEW METHODS FOR CHAOTIC DYNAMICS

WORLD SCIENTIFIC SERIES ON NONLINEAR SCIENCE

Editor: Leon O. Chua
University of California, Berkeley

Series A.   MONOGRAPHS AND TREATISES

WORLD SCIENTIFIC SERIES ON
NONLINEAR SCIENCE

Series A   Vol. 58

Series Editor: Leon O. Chua

NEW METHODS FOR CHAOTIC DYNAMICS

Nikolai Alexandrovich Magnitskii

Sergey Vasilevich Sidorov

Russian Academy of Sciences, Russia

World Scientific

NEW JERSEY · LONDON · SINGAPORE · BEIJING · SHANGHAI · HONG KONG · TAIPEI · CHENNAI

Published by

World Scientific Publishing Co. Pte. Ltd.

5 Toh Tuck Link, Singapore 596224

USA office: 27 Warren Street, Suite 401-402, Hackensack, NJ 07601

UK office: 57 Shelton Street, Covent Garden, London WC2H 9HE

British Library Cataloguing-in-Publication Data
A catalogue record for this book is available from the British Library.

NEW METHODS FOR CHAOTIC DYNAMICS

ISBN 981-256-817-4

Printed in Singapore by World Scientific Printers (S) Pte Ltd

This book is dedicated to Elena and Natalia, our wives. We are greatly indebted to them for their encouragement, attention, patience and support, without which this book would have remained in our dreams only.

Preface

The present book was written on the basis of research carried out in recent years under the direction of one of the authors in the laboratory for nonlinear and chaotic dynamics of the Institute for Systems Analysis of the Russian Academy of Science. In this book the authors state their, in many cases distinct from traditional, point of view on principles of formation, scenarios of occurrence and ways of control of chaotic motion in nonlinear dissipative dynamical systems described by autonomous and non-autonomous ordinary differential equations, diffusion type partial differential equations and differential equations with delay argument.

Systems of nonlinear differential equations are a special case of an extensive family of nonlinear dynamical systems into which also enter various nonlinear algebraic, difference, integral, functional and abstract operational equations. In this connection, until recently the uniform geometrical approach to study nonlinear dynamical systems was represented absolutely natural, allowing to consider from common positions the nonlinear systems described by both discrete mappings, and ordinary and partial differential equations. Intensive application of geometrical approach to the analysis of dynamical systems originated with the well-known work of the American mathematician S. Smale who offered a design of mapping which subsequently received the name of Smale's horseshoe. It has been shown, that stable limit set (attractor) of discrete dynamical system cannot be such a smooth manifold of the whole dimension which are, for example, a stable limit cycle or torus, and everywhere holey, self-similar fractal set of fractional dimension. Besides it has been shown, that the behavior of trajectories of dynamical system for such strange attractor in the terminology of D. Ruelle and F. Takens is complex enough, combining global stability (the trajectory does not leave some area of phase space) with local instability of separate close trajectories, exponentially running up in time, that is characterized by the presence on an attractor both negative, and positive Lyapunov exponents. Other chaotic dynamical systems described by discrete mappings and possessing strange attractors have been further found, for example, logistic map, Henon map, Smale–Williams solenoid, *etc.*

As the analysis of properties of continuous dynamical systems described by ordinary differential equations, can be reduced, as it seemed, to the analysis of properties of some mapping — Poincare mapping, it was observed that in continuous dynamical systems irregular, chaotic behaviour of trajectories became connected to the presence of a strange attractor in the system. However, the proof of this fact directly for the well-known Lorenz system of three ordinary differential equations in which the irregular behavior of trajectories for the first time was revealed, has faced with significant difficulties. For a long time, numerous attempts to prove by the methods of geometrical theory of dynamical systems the presence of a strange attractor in neighbourhoods of a saddle–node or a saddle–focus separatrix loops of the Lorenz system have ended with failure. Moreover, the problem to show, whether the behavior of solutions of the Lorenz system coincides with the dynamics of a geometrical Lorenz attractor was formulated by S. Smale as one of 18 most significant mathematical problems of XXI century. And the recent results of the authors have allowed to confirm definitely, that the geometrical approach developed for discrete mappings even if allowed to obtain a number of brilliant results for them, is not absolutely adequate with reference to the continuous dynamical systems described by the differential equations. Now we can insist absolutely, that the definition of chaotic attractor of continuous dynamical system as a strange attractor, and also such traditional sections of chaotic dynamics as calculation of attractor's dimension, scenarios of transition to chaos, criteria of dynamical chaos and spatio-temporal chaos demand significant updating and revision.

As numerous examples show, neither the presence of a positive Lyapunov exponent, nor the presence of a saddle–node or saddle–focus separatrix loops, nor the presence of a saddle–node or saddle–focus itself is a necessary condition for the existence of chaotic dynamics in a nonlinear dissipative autonomous system of ordinary differential equations. Moreover, irregular attractors of a huge class of three-dimensional nonlinear dissipative autonomous systems of ordinary differential equations, containing all classical chaotic systems, are born as a result of the same cascade of soft bifurcations of stable limit cycles. The start is always the Feigenbaum cascade of period doubling bifurcations. It continues by complete or incomplete Sharkovskii subharmonic cascade of bifurcations of stable limit cycles with an arbitrary period and then continues by complete or incomplete homoclinic cascade of bifurcations, opened and described by authors.

In the present book, the theory of such attractors, named as singular attractors, is proposed. Any singular attractor exists only at a separate accumulation point of values of the bifurcation parameter. It contains unstable cycles of various periods in any its neighbourhood. It is proved, that any singular attractor of three-dimensional nonlinear autonomous dissipative system of ordinary differential equations lies on a two-dimensional (in general, many-sheeted) surface of three-dimensional phase space, that is the closure of a two-dimensional invariant unstable manifold (separatrix surface) of a singular saddle cycle which gives rise to the cascade of period doubling bifurcations. In this connection, fractal dimension of any singular attractor of a three-dimensional system cannot exceed two. Chaotic dynamics in all systems of the considered class of differential equations arises owing to phase shift between trajectories forming separatrix surface of the singular cycle. This leads to an appearance of the continuous one-dimensional mappings having multiple-valued inverse mappings in a two-dimensional rotating plane transversal to the cycle. This is impossible in any Poincare section, transition to which leads to loss of phase. Any singular attractor in any system of the considered class of differential equations cannot have positive Lyapunov exponent and it is not a hyperbolic set.

Thus, in all systems of the class of nonlinear autonomous three-dimensional systems of ordinary differential equations considered in the book, only complete or incomplete subharmonic or homoclinic singular attractors are born during the first stages of transition to chaos. The same scenario of transition to chaos takes place for all known classical three-dimensional autonomous dissipative systems of nonlinear ordinary differential equations, including Lorenz, Rossler, Chua, Magnitskii systems, *etc.* Moreover, as shown in the book, the same universal scenario of transition to chaos is realized in many-dimensional systems of ordinary differential equations, in nonlinear partial differential equations and in differential equations with delay arguments. And as other scenarios of transition to chaos in dissipative systems of nonlinear differential equations, except subharmonic or homoclinic cascades of bifurcations, are not observed yet, the hypothesis is rather believable about universality of the method of appearance of chaotic dynamics in dissipative systems of differential equations described in the book.

The book consists of six chapters. The basic concepts, definitions and theorems of the theory of ordinary differential equations are stated in Chapter 1. Chapter 2 is devoted to the description of the basic bifurcations of singular points, limit cycles, tori and irregular attractors of nonlinear systems of ordinary differential equations. Special attention is given to insufficiently known non-local bifurcations of homoclinic and heteroclinic contours, and also to various cascades of bifurcations of both regular, and irregular attractors. Chapter 3 shows on the basis of the conducted numerical calculations and the large illustrative material, that all classical dissipative nonlinear three-dimensional autonomous systems of ordinary differential equations have the unique universal scenario of transition to chaos through the cascade of period doubling bifurcations, subharmonic and then homoclinic cascades of soft bifurcations of stable limit cycles. This scenario is described by the theory of dynamical chaos in nonlinear systems of ordinary differential equations, developed by one of the authors and stated in Chapter 4.

Chapter 5 shows, that the same scenario of transition to chaos takes place also in many-dimensional autonomous dissipative systems of ordinary differential equations, in infinitely-dimensional systems of partial differential equations of the reaction–diffusion type, and also in the ordinary differential equations with delay argument. In such systems, transition to chaos is carried out through the cascade of period doubling bifurcations, subharmonic and then homoclinic cascades of soft bifurcations of stable two-dimensional tori. Chapter 6 considers both classical and original methods of solution of the basic problem of chaos control, consisting of detection and stabilization of unstable cycles of nonlinear systems of differential equations possessing chaotic dynamics.

The authors would like to thank D.V. Anosov, A.B. Bakushinskii, S.V. Emelyanov, V.A. Il'in, Yu.S. Il'yashenko, Yu.L. Klimontovich, S.K. Korovin, P.S. Krasnoschekov, E.I. Moiseev, Yu.S. Popkov for useful discussions of topics touched in the book, and also "Partner" company, and personally I.I. Vainshtok and A.S. Zaltsman, for the financial support without which the work on the book and its Russian edition would be simply impossible.

The problems raised in the book, methods of their solution and the obtained results go far beyond the frameworks of traditional representations about chaotic attractors of nonlinear dissipative systems of ordinary differential equations. Please send your opinions, remarks and offers to the e-mail address nmag@isa.ru.

N.A. Magnitskii, S.V. Sidorov

Contents

Chapter 1

Systems of Ordinary Differential Equations

A brief review of the basic concepts and theorems of the general character for the theory of ordinary differential equations is given in this chapter. This is necessary to understand the material, stated in the subsequent chapters. More detailed statement of results presented in the given chapter can be found in works which are specified in the list of the cited literature.

1.1 Basic Definitions and Theorems

1.1.1 *Fields of directions and their integral curves*

Consider real finite-dimensional linear space $\mathbb{R}^m$. *Field of directions* in a region M of the space $\mathbb{R}^m$ is the correspondence which compares every point $x \in M$ with a straight line passing through x.

Definition 1.1 *Integral curve of a field of directions* is the curve which in every its point touches the direction of a field in this point.

1.1.2 *Vector fields, differential equations, integral and phase curves*

Vector field F, given in a region M of the space $\mathbb{R}^m$ is the correspondence which compares every point $x \in M$ with the vector F of the space $\mathbb{R}^m$ applied to this point.

System of differential equations, corresponding to a vector field F, is the system

$$\dot{x} = F(x), \ x \in M \subset \mathbb{R}^m, \tag{1.1}$$

1

where the point above the letter means differentiation on t. Region M is called *the phase space* of the system, and direct product $I \times M$ — *the expanded phase space*, where I is an interval of the real axis of time t.

The system (1.1) is also called *an autonomous system* of ordinary differential equations. The system which right part depends as well from t

$$\dot{x} = F(x,t), \ x \in M \subset \mathbb{R}^m, \ t \in I \subset \mathbb{R} \tag{1.2}$$

is called *a non-autonomous system*. A set of systems of equations of a kind

$$\dot{x} = F(x,t,\mu), \ x \in M \subset \mathbb{R}^m, \ \mu \in L \subset \mathbb{R}^k, \ t \in I \subset \mathbb{R}, \tag{1.3}$$

is called *a family* of ordinary differential equations given in the phase space M by vector fields F, depending on coordinates of vectors of system parameters μ, lying in the region L of the space $\mathbb{R}^k$.

Autonomous families (1.3) of ordinary differential equations

$$\dot{x} = F(x,\mu), \quad x \in M \subset \mathbb{R}^m, \quad \mu \in L \subset \mathbb{R}^k,$$

given by families of vector fields $F(x,\mu)$ and representing the greatest interest from the point of view of various applications, will be the basic object of study in the present book.

Definition 1.2 Differentiable mapping $x : I \to M$ of an interval I of a real axis t into phase space is called *the solution of a system of differential Eqs.* (1.2) if for any $\tau \in I$

$$\dot{x}(\tau) = F(x(\tau), \tau).$$

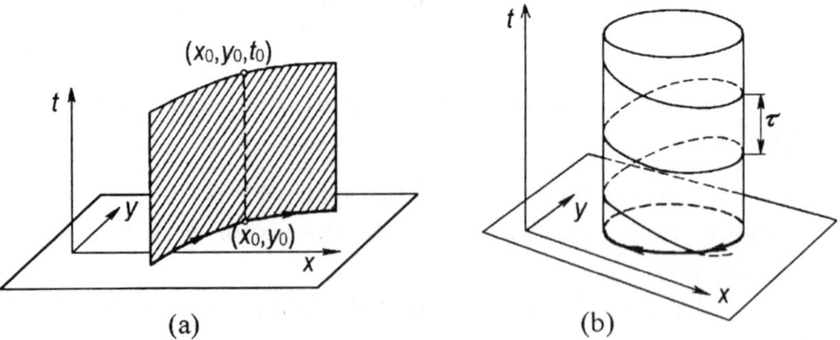

Fig. 1.1　Projections of a segment of an integral curve and a spiral with step τ on a phase plane (x, y) in the form of a segment (a) and the closed curve (b).

Definition 1.3   Graph of a solution of a system of differential equations is called its *integral curve*, and a projection of an integral curve on phase space along the axis t is called *a phase curve*.

Examples of integral and phase curves are represented in Fig. 1.1. Phase curves often are also named *trajectories of solutions* of systems of differential equations.

In the following we shall be interested in unlimited time intervals $I = [0, \infty)$ and $I = (-\infty, \infty)$ only. We shall also suppose, that the initial moment of time is $t_0 = 0$.

1.1.3   *Theorems of existence and uniqueness of solutions*

Vector field with differentiable components is called *a differentiable vector field*. Vector field with components having continuous derivatives of all orders which are not exceeding r is called *a field of class C^r*.

Theorem 1.1   [Coddington and Levinson (1955); Hartman (1964)] *One and only one integral curve of a system of differential equations with real time passes through each point of expanded phase space of the corresponding differentiable vector field.*

The condition of smoothness of the right part in the formulated theorem can be weakened.

Definition 1.4   Mapping $f : M \to N$, $M \in \mathbb{R}^m$, $N \in \mathbb{R}^n$ satisfies to *the Lipshitz condition* if there is such positive constant q, that

$$\|f(x_1) - f(x_2)\| \le q\|x_1 - x_2\|$$

for all $x_1 \in M$, $x_2 \in M$.

Theorem 1.2   [Coddington and Levinson (1955); Hartman (1964)] *Let the right part of a system of differential Eqs. (1.2) be continuous in some region $I \times M$ of the space $\mathbb{R}^{m+1}$ and satisfy in it the Lipshitz condition on x with the same constant q. Then one and only one integral curve of a system of differential Eqs. (1.2) passes through each point of the region $I \times M$ of the expanded phase space.*

If the right part of a system (1.2) is only continuous, then through each point of the expanded phase space there passes at least one integral curve. Thus uniqueness can be broken, that is visible from an example of the equation $\dot{x} = 2x^{1/2}$, having integral curves $(t,0)$ and (t,t^2), passing through the point $(0,0)$.

By virtue of theorems formulated above the given initial condition at the initial moment of time $x(0) = x_0$ uniquely determines the solution of a system of differential equations at any moment $t : x(t) = \varphi^t(x_0)$. The problem of finding a solution of a system of differential equations, satisfying to the given initial condition, is called *the Cauchy problem*. Mapping φ^t is a mapping of phase space M in itself and is called *a phase flow*. Any region G of phase space under influence of a phase flow passes in time t to some other region $G_t = \varphi^t(G)$.

1.1.4   *Differentiable dependence of solutions from initial conditions and parameters, the equations in variations*

Let f be a differentiable mapping of region M of the space $\mathbb{R}^m$ into region N of the space $\mathbb{R}^n$. The main linear part of the mapping f at a point x_0 is called *the derivative of the mapping f* at this point, i.e. the linear operator $A : \mathbb{R}^m \to \mathbb{R}^n$ such, that

$$f(x) - f(x_0) = A(x - x_0) + o(\|x - x_0\|).$$

In coordinates $(x_1, \ldots, x_m)$ and $(y_1, \ldots, y_n)$ mapping f can be represented in the form of a vector function $y = f(x) = (f_1(x), \ldots, f_n(x))^{\mathrm{T}}$. The matrix of the operator A in coordinates (x,y) is the Jacobi matrix of a vector function f, i.e.

$$A = \frac{\partial f}{\partial x}(x_0), \ \ A = \{a_{ij}\}, \ a_{ij} = \frac{\partial f_i}{\partial x_j}(x_0), \quad i = 1, \ldots, n, \quad j = 1, \ldots, m.$$

Theorem 1.3   [Coddington and Levinson (1955); Hartman (1964)] *Let the family of differential Eqs. (1.3) be given by vector fields $F(x, t, \mu)$, continuous in some region of the space $\mathbb{R}^{k+m+1}$ together with the derivatives $\partial F/\partial x$ and $\partial F/\partial \mu$. Then the solution φ of the family (1.3) with the initial condition $\varphi(0, \mu) = x$ is continuously differentiable on x, μ. If dependence of a field F from parameters μ is only continuous, then dependence of the solution on parameters is also continuous.*

The equations for derivatives of the solution under initial conditions and parameters can be easy determine. We shall denote through φ_ξ the solution of a system (1.2) with the initial condition $\varphi_\xi(0) = \xi$. We fix $\xi = x$ and denote

$$X(t) = \left.\frac{\partial \varphi_\xi(t)}{\partial \xi}\right|_{\xi=x}.$$

At every moment t the linear operator $X(t)$ operates from $\mathbb{R}^m$ into $\mathbb{R}^m$. From (1.2) follows, that operator-valued function $X(t)$ satisfies the following *equation in variations*

$$\dot{X}(t) = A(t)X(t), \quad \text{where} \quad A(t) = \frac{\partial F}{\partial x}(\varphi_x(t), t).$$

It is the linear homogeneous non-autonomous differential equation, and $X(0)$ is an identity matrix.

Let us write out now the equation in variations for a derivative of the solution of family (1.3) on parameters. Let $\varphi_{\xi,\alpha}$ be the solution of family (1.3) with the initial condition $\varphi_{\xi,\alpha}(0, \alpha) = \xi$. We fix $\xi = x$, and denote

$$Y(t) = \left.\frac{\partial \varphi_{\xi,\alpha}(t)}{\partial \alpha}\right|_{\xi=x, \alpha=\mu}.$$

At every moment t the linear operator $Y(t)$ operates from $\mathbb{R}^k$ in $\mathbb{R}^m$. From (1.3) follows, that operator-valued function $Y(t)$ satisfies the following equation in variations

$$\dot{Y}(t) = A(t)Y(t) + b(t),$$

where

$$A(t) = \frac{\partial F}{\partial x}(\varphi_{x,\mu}(t), t, \mu), \quad b(t) = \frac{\partial F}{\partial \mu}(\varphi_{x,\mu}(t), t, \mu).$$

This is the linear inhomogeneous non-autonomous differential equation, and $Y(0) = 0$.

1.1.5 *Dissipative and conservative systems of differential equations*

We shall denote through $V(\Omega_t)$ Euclidean volume of a region $\Omega_t = \varphi^t(\Omega_0)$ of the phase space, resulting as shift during time t all points of some initial region Ω_0 along phase curves of autonomous system of differential Eqs. (1.1). Then change of volume $V(\Omega_t)$ satisfies the equation

$$\frac{dV(\Omega_t)}{dt} = \int_V \text{div}\, F(x)dx, \quad \text{where} \quad \text{div}\, F(x) = \sum_{k=1}^{m} \frac{\partial F_k}{\partial x_k}(x)$$

is the sum of diagonal elements of the operator $\partial F/\partial x$, and dx is an element of volume.

Definition 1.5 The system of Eqs. (1.1) is called conservative if the volume of any region of phase space does not change in time, and it is called dissipative if the volume of some region of phase space decreases in time.

Thus, if everywhere in phase space $\mathrm{div}F(x) = 0$ then the system saves volume and is conservative. If there is a region of phase space, in which $\mathrm{div}F(x) < 0$, then the system (1.1) is dissipative in this region.

In the further we shall be interested in families of exclusively dissipative systems of differential equations. The readers, who are interested in chaotic dynamics of conservative and, in particular, Hamiltonian systems, we send to the papers [Arnold (1978a); Arnold *et al.* (1988); Haken (1983); Moser (2001)].

1.1.6   *Numerical methods for solution of systems of ordinary differential equations*

The solution of the Cauchy problem for autonomous nonlinear system of ordinary differential equations of a kind (1.1) only in exclusive cases can be found in an explicit form. Integration of such systems demands, as a rule, approximation of the solution by various linear functions on consecutive small intervals of time (steps) of duration τ. The most widespread method of such approximation is the fourth-order Runge-Kutta method giving at small τ difference equations, which approximate the solution with accuracy $O(\tau^4)$.

Let us denote through y_n the already computed approximate value for solution of system (1.2) at the moment t_n. Then the approximate value of solution of system (1.2) at the moment $t_{n+1} = t_n + \tau$ is under the formula:
$y_{n+1} = y_n + \tau(\sigma_1 k_1 + \sigma_2 k_2 + \sigma_3 k_3 + \sigma_4 k_4)$, where $k_1 = F(t_n, y_n)$, $k_2 = F(t_n + a_2\tau, y_n + b_{21}\tau k_1)$, $k_3 = F(t_n + a_3\tau, y_n + b_{31}\tau k_1 + b_{32}\tau k_2)$, $k_4 = F(t_n + a_4\tau, y_n + b_{41}\tau k_1 + b_{42}\tau k_2 + b_{43}\tau k_3)$. Parameters σ_1, σ_2, σ_3, σ_4, a_2, a_3, a_4 and b_{21}, b_{31}, b_{32}, b_{41}, b_{42}, b_{43} can be determined, generally speaking, by various means. All numerical solutions for systems of ordinary differential equations resulted in the present book, are executed at following values of these parameters: $\sigma_1 = \sigma_4 = 1/6$, $\sigma_2 = \sigma_3 = 1/3$, $a_2 = a_3 = 1/2$, $a_4 = 1$ and $b_{21} = b_{32} = 1/2$, $b_{31} = 0$, $b_{41} = b_{42} = 0$, $b_{43} = 1$.

Note, that estimations of accuracy of the approximate solutions coincide, as a rule, with estimations of accuracy of approximation and essentially depend on a length of a time interval T on which the system of differential equations is solved. For example, in the fourth-order Runge-Kutta method the accuracy of solution $\|x(t_n) - y_n\| = CT \exp(\alpha T)\tau^4$, where α and C are some constants [Samarskii and Gulin (1989)].

1.1.7 *Ill-posedness of numerical methods in solution of systems of ordinary differential equations*

The problem of numerical solution of nonlinear differential equations possessing complex irregular dynamics consists, as a rule, not in solution of equation on any small final interval of time, but in finding of limit sets (attractors) of trajectories at $t \to \infty$. Estimations of accuracy of numerical methods lose sense in such statement. But there arises another important problem consisting of ill-posedness of the solved task. The essence of a problem consists of the following. Any method of numerical solution of differential equation demands approximation in any way of the derivative of the function known only approximately in nodes of some grid. The problem of calculation of derivative from not exactly given function is a classical *ill-posed problem* [Tikhonov and Arsenin (1977)]. Greater errors in calculation of derivative correspond to small errors in calculation of the function, i.e. there is no continuous dependence of derivative on the function. For example, let the error in calculation of the function looks like $m^{-1} \sin m^2 t$. The error in calculation of the derivative of this function thus will be $m \cos m^2 t$. Hence, at $m \to \infty$ the error of function in the uniform metrics tends to zero, while the error of a derivative tends to infinity in the same metrics.

The ill-posedness of calculation of derivative $x'(t_n)$ of a net function $x(t)$ leads to that the step of a grid τ cannot be chosen arbitrary by small, and should be determined by an error in calculation of the function. Really, let values of the function $x(t)$ at points t_{n+1} and t_n be given with accuracy δ, i.e. we know the values y_{n+1} and y_n such, that $|x(t_{n+1}) - y_{n+1}| \leq \delta$, $|x(t_n) - y_n| \leq \delta$. Then an accuracy of replacement of the derivative $x'(t_n)$ by an approximate formula, for example, $(y_{n+1} - y_n)/\tau$, is equal not to $O(\tau)$, but $O(\tau) + 2\delta/\tau$. The error thus consists of two parts: accuracy of the method of approximation of derivative $O(\tau)$ and an irreducible error $2\delta/\tau$, which is connected with calculation of the function. At $\tau \to 0$ an irreducible error, obviously, becomes unrestrictedly large. In this case the choice of a step of a grid $\tau = O(\delta^{1/2})$ is optimal. Such way of a choice of an optimal step is called *the regularization on a step*.

Let us show, that the similar effect takes place also in numerical solution of the Cauchy problem for a system of ordinary differential equations of the kind (1.1), for example, by the fourth-order Runge-Kutta method. Let $y_{n+1} = y_n + \tau f(y_n)$. Denote through $y_n - x(t_n) = \delta_n$ the error in calculation of value of the solution $x(t)$ at a point t_n. Then

$$\delta_{n+1} = \delta_n + \tau f'(x_n + \theta \delta_n)\delta_n - (x_{n+1} - x_n) + \tau f(x_n),$$

where $|\theta| < 1$. As accuracy of approximation of a method is equal to $O(\tau^4)$, then $|\delta_{n+1}| \leq |\delta_n| \cdot |1 + \tau L| + K\tau^4$, where $L = f'(x_n + \theta \delta_n)$. It is evident, that if $\tau \to 0$ then $\delta_{n+1} \to \delta_n$, and the total error can only increase from step to step. At the same time if $-1/\tau < L < 0$ it is possible to achieve significant reduction of an error at a choice of the optimal step minimizing an estimation of an error.

Below in the book numerous examples of concrete systems of nonlinear ordinary differential equations will be demonstrated in which at a small step the chaotic behaviour of trajectories is observed, but at some reasonable step — a stable limit cycle is observed. A very striking example of such system is the well-known Lorenz system of three nonlinear ordinary differential equations [Lorenz (1963)].

1.2 Singular Points and Their Invariant Manifolds

1.2.1 *Singular points of systems of ordinary differential equations*

Singular point (fixed point, equilibrium point, stationary point) of a system of differential Eqs. (1.1) is a singular point of a corresponding vector field $F(x)$.

Definition 1.6 *The singular point* of a vector field is a point of phase space in which the vector of a field vanishes.

Let x_0 be a singular point of a differentiable vector field $F(x)$, being the right part of an autonomous system of differential Eqs. (1.1), and let $\partial F/\partial x$ be a derivative of the mapping F. The system of linear differential equations

$$\dot{y} = Ay, \qquad \text{where} \quad A = \frac{\partial F}{\partial x}(x_0), \quad y = x - x_0$$

is called *linearization of the system* (1.1) at a singular point x_0, the field Ay is called *a linear part* of a field F at the point x_0, and an operator A is called *the operator of this linear part* or *the operator of linearization*.

Definition 1.7 The singular point of a vector field is called *a nondegenerate point*, if the operator of a linear part of a field at this point is nondegenerate.

Definition 1.8 The singular point of a system of differential Eqs. (1.1) is called *a hyperbolic point* if any eigenvalue of the operator of a linear part of a field at this point does not lie on an imaginary axis.

Definition 1.9 Two systems of differential equations (or, that the same — two vector fields) are *topologically equivalent* in a neighbourhoods of their singular points, if there exists *a homeomorphism* (one-to-one and mutually continuous mapping), transforming a singular point of the first system and the trajectories lying in its some neighbourhood, in a singular point and trajectories of the second system with preservation of orientation of trajectories.

Theorem 1.4 (the Grobman–Hartman theorem [Bilov *et al.* (1966); Hartman (1964)]) *Continuously differentiable vector field with a hyperbolic singular point is topologically equivalent to its linear part in some neighbourhood of this point.*

From the formulated theorem in particular follows, that the qualitative behaviour of solutions of autonomous system of differential Eqs. (1.1) in a neighbourhood of a hyperbolic singular point is completely determined by behaviour of solutions of the system of linear differential equations with the constant operator (matrix) of a linear part of a field at this point.

1.2.2 *Stability of singular points and stationary solutions*

A singular point (fixed point, equilibrium point, stationary point) in an autonomous system of differential equations is called a stable (asymptotically stable) point, if the stationary solution of this system identically equal to the point is stable (asymptotically stable).

Definition 1.10 *The stationary solution* of an autonomous system of differential equations (the solution which is identically equal to a singular point) is called *Lyapunov stable* if all solutions of this system with initial conditions from sufficiently small neighbourhood of the singular point are defined on all positive semi-axis of time and uniformly on time converge to the investigated stationary solution when initial condition tends to the indicated singular point.

In other words, the stationary solution x_0 of the system (1.1) is Lyapunov stable if for any $\varepsilon > 0$ there exists $\delta > 0$, such that for all solutions $x(t)$ of the system (1.1) from $\|x(0) - x_0\| < \delta$ it follows $\|x(t) - x_0\| < \varepsilon$ for all $t > 0$.

Definition 1.11   The stationary solution of autonomous system of differential Eqs. (1.1) is called *asymptotically stable* if it is Lyapunov stable and besides all solutions $x(t)$ of the system (1.1) with initial conditions close enough to an investigated singular point $\|x(0) - x_0\| < \delta$ tend to it when $t \to \infty$, i.e. $\|x(t) - x_0\| \to 0$ when $t \to \infty$.

If in conditions of asymptotic stability solutions of the system of Eqs. (1.1) tend to a singular point exponentially, i.e. $\|x(t) - x_0\| < c \exp(-\gamma t)$ with some positive constants c and γ, the stationary solution x_0 of the system (1.1) is called *exponentially asymptotically stable*.

Stability (and asymptotic stability) of stationary solutions (singular points) is a local property of the vector field setting the system of differential equations. Only simple approaching of solutions to the singular point at $t \to \infty$ is not a local property and not sufficient for asymptotic stability.

Theorem 1.5   (the Lyapunov theorem of stability on the first approximation [Malkin (1966)]) *If the operator of linearization A of a differentiable vector field $F(x)$ of the system (1.1) at a singular point has eigenvalues only with a negative real part, then this singular point is asymptotically stable. If one of eigenvalues of the operator A has a positive real part, then the singular point is not Lyapunov stable.*

It follows from *the Lyapunov theorem*, that the stable hyperbolic singular point is always exponentially asymptotically stable.

Definition 1.12   Differentiable function $V(x)$ is called *a Lyapunov function* for a singular point x_0 of a vector field $F(x)$ if it satisfies the following conditions:

- function V is defined in some neighbourhood of a point x_0 and has at this point a strict local minimum;
- a derivative of the function V along a vector field F in some neighbourhood of a point x_0 is nonpositive, i.e.

$$\frac{d}{dt}V(x) = \sum_{k=1}^{m} \frac{\partial V}{\partial x_k} \frac{dx_k}{dt} = \sum_{k=1}^{m} \frac{\partial V}{\partial x_k} F_k(x) \leq 0.$$

Theorem 1.6 [Malkin (1966)] *The singular point of a differentiable vector field for which there exists Lyapunov function, is stable.*

1.2.3 *Invariant manifolds*

The subset G of phase space is called on an invariant set in relation to a phase flow φ^t or *an invariant set* if $\varphi^t(G) = G$ takes place for all admissible t. The subset G of Euclidean space $\mathbb{R}^m$, having in each point a unique tangent hyperplane, is called *manifold*. In this case it is claimed, that G is smoothly enclosed in $\mathbb{R}^m$. If the set G is smoothly enclosed in phase space M of systems of differential equations, then it is claimed, that G is *submanifold* of phase space.

Definition 1.13 *Invariant manifold* of a vector field and of a corresponding system of differential equations is such submanifold of phase space which in each point touches a vector of the field.

Let us consider the linear operator $A : \mathbb{R}^m \to \mathbb{R}^m$. The space $\mathbb{R}^m$ breaks up in the direct sum of three subspaces:

$$\mathbb{R}^m = T^s \otimes T^u \otimes T^c.$$

All three subspaces in the right part of equality are invariant in relation to the operator A. The spectrum of restriction A on T^s lies in opened left half-plane, restriction A on T^u lies in right half-plane and restriction A on T^c lies on an imaginary axis. For the operator A, being the operator of linearization of a vector field F of the Eqs. (1.1) at a hyperbolic singular point, $T^c = \{0\}$.

Theorem 1.7 (the Adamar–Perron theorem [Marsden and McCracken (1976)]) *Let F be a C^r-smooth vector field with a hyperbolic singular point 0 and with a linear part Ax in zero, T^s and T^u are the hyperplanes corresponding to the operator A. Then the system of differential equations has two C^r-smooth invariant in relation to F manifolds W^s and W^u, passing through 0 and touching in zero of the hyperplanes T^s and T^u accordingly. Solutions with initial conditions on W^s (W^u) exponentially tend to zero when $t \to +\infty (t \to -\infty)$.*

The manifold W^s is called *stable*, and the manifold W^u is called *unstable manifold* of a singular point 0.

Theorem 1.8   (the theorem on the central manifold [Marsden and Mc-Cracken (1976); Hassard *et al.* (1981)]) *If in conditions of the previous theorem the operator A has eigenvalues also on an imaginary axis, i.e. $T^c \neq \{0\}$, then the system of differential Eqs. (1.1) has the third C^{r-1}-smooth invariant manifold W^c, passing through 0 and touching in zero of the hyperplane T^c.*

The manifold W^c is called *the central manifold*, and a hyperplane $T^s \otimes T^u$ is called *a hyperplane of hyperbolic variables*. The behaviour of phase curves on the manifold W^c is defined by nonlinear members.

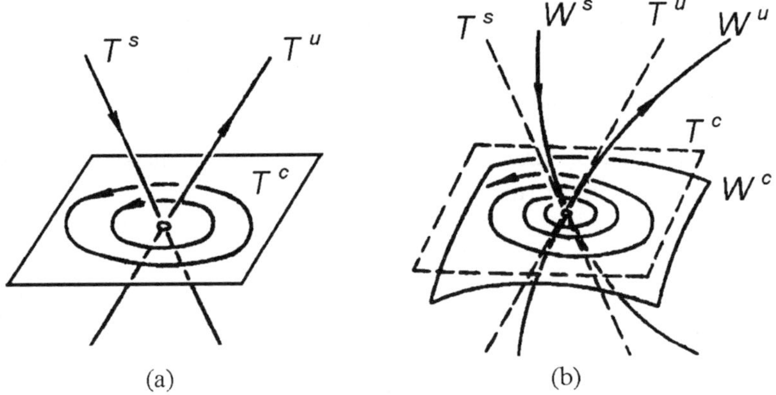

Fig. 1.2   Stable, unstable and central manifolds of linear (a) and nonlinear (b) systems.

Schematically invariant subspaces and manifolds of singular points are represented in Fig. 1.2. The designations accepted for them are clear from the formulated above theorems: s — stable, u — unstable, c — central.

In many cases at study of local topology of a nonlinear vector field and a corresponding system of differential equations in a neighbourhood of a singular point it is important to know only restriction of this field on the central manifold. However, as it will be shown below, this is absolutely insufficient for full understanding of a global topological picture which is determined by hyperbolic variables and bifurcations of singular cycles.

1.2.4 *Singular points of linear vector fields*

Any system of linear differential equations specified by a linear vector field, looks like

$$\dot{y} = Ay. \tag{1.4}$$

The type of a singular point and a character of behaviour of solutions of the system (1.4) are determined by eigenvalues of the linear operator A.

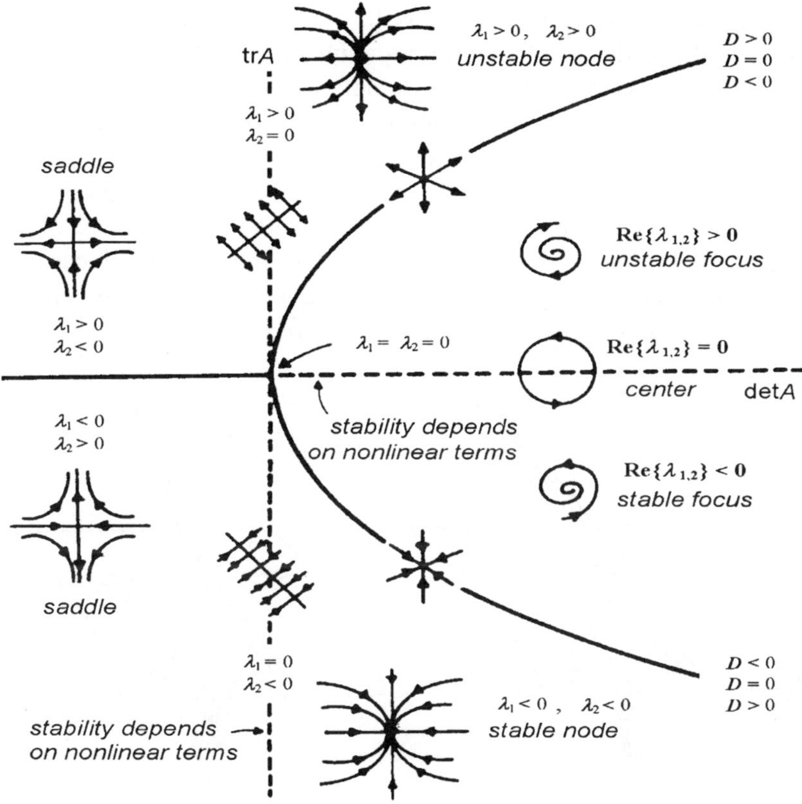

Fig. 1.3   Classification of singular points of linear two-dimensional autonomous systems.

On the real plane a nondegenerate singular point has one of the following four types: *a saddle, a node, a focus* and *a center*. Eigenvalues λ_1 and λ_2

of the linear operator A in this case are defined by the formula

$$\lambda_{1,2} = \frac{1}{2}(\operatorname{tr} A \pm \sqrt{D}),$$

where $D = (\operatorname{tr} A)^2 - 4\det A$, $\operatorname{tr} A$ is *the trace of the matrix* A, i.e. the sum of its diagonal elements, and $\det A$ is *the determinant of the matrix* A. The regions occupied in this case by various types of singular points of the Eq. (1.4) in plane $(\det A, \operatorname{tr} A)$, are presented in Fig. 1.3. In the case of $D = 0$ dicritical node corresponds to a scalar matrix $A = \lambda E$, where E is an identity matrix, and degenerate node corresponds to a matrix which looks like a two-dimensional Jordan cell.

The condition $\det A = 0$ determines a line of degenerate singular points among which it is possible to mark out a degenerate two-dimensional saddle–node, having, as a rule, one nodal and two saddle sectors. Nondegenerate saddle, node and focus are hyperbolic singular points. Therefore, as follows from the Grobman–Hartman theorem, all pictures represented in Fig. 1.3, except center, are keeping safe under small perturbations of linear system (1.4). Besides, saddle is always unstable, and node and focus can be both stable (exponentially asymptotically stable), and unstable depending on signs of real parts of eigenvalues of the matrix A.

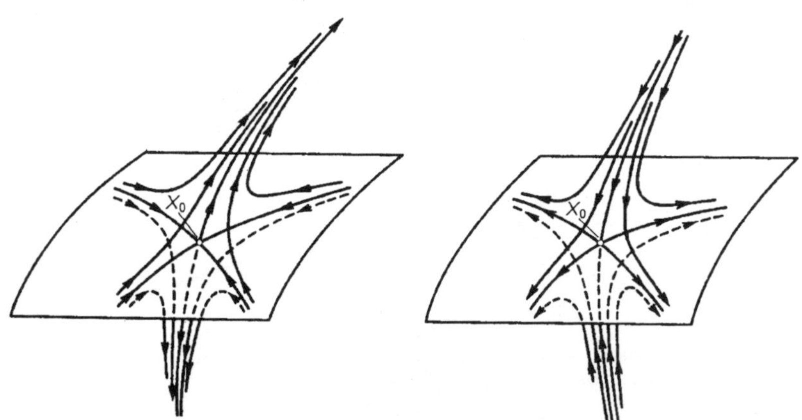

Fig. 1.4   Singular points x_0 of a saddle–node type.

In three-dimensional real space there are more complex hyperbolic singular points, being by combinations of a saddle with a node or focus and named, accordingly, *a saddle–node* (Fig. 1.4) and *a saddle–focus* (Fig. 1.5).

A saddle–node and a saddle–focus are always unstable. They have one-dimensional stable and two-dimensional unstable manifolds (or on the contrary). As we shall see below, in neighbourhoods of such singular points presence of complex irregular dynamics is possible. Therefore these points can play an important role in formation of chaotic attractors of nonlinear systems of ordinary differential equations.

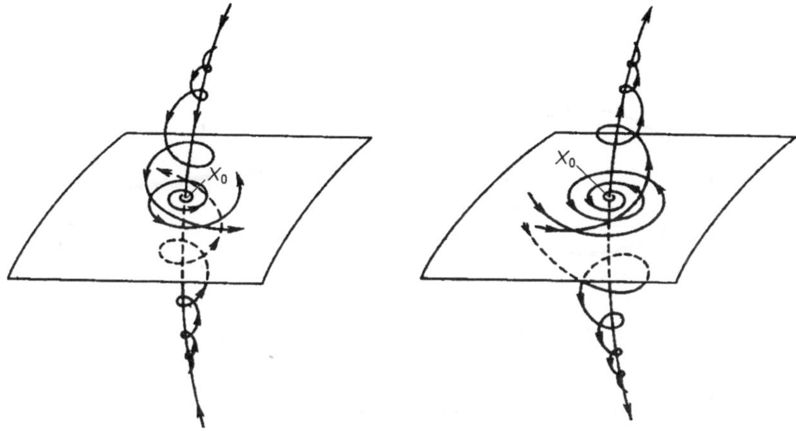

Fig. 1.5   Singular points x_0 of a saddle–focus type.

1.2.5   *Separatrices of singular points, homoclinic and hete-roclinic trajectories, separatrix contours*

The trajectory of autonomous system of ordinary differential equations is called *a separatrix of a singular point* if it tends to this singular point or when $t \to +\infty$, or when $t \to -\infty$.

One-dimensional invariant stable (unstable) manifold of a singular point consist of the singular point and two entering in it (starting with it) separatrices. Many-dimensional invariant manifolds of singular points sometimes are also named *separatrix surfaces*.

The trajectory (a phase curve), tending to a singular point both when $t \to +\infty$, and when $t \to -\infty$ is called *a homoclinic trajectory* or *a loop of separatrix* of a singular point. Homoclinic trajectory belongs to intersection of stable W^s and unstable W^u invariant manifolds of a singular point.

The trajectory belonging to an intersection of stable invariant manifold of one singular point with unstable invariant manifold of other singular point is called *a heteroclinic trajectory*. It tends to the first singular point when $t \to +\infty$ and tends to the second singular point when $t \to -\infty$.

A closed (in phase space) curve (cycle) consisting of several separatrices connecting singular points is called *a separatrix contour*.

In two-dimensional systems with nondegenerate singular points there are possible only homoclinic trajectories of saddles and any heteroclinic trajectories and separatrix contours connecting saddles, nodes and focuses (Fig. 1.6).

Fig. 1.6   Homoclinic (a) and heteroclinic (b) contours.

Let us consider an example of system on a plane having at various values of parameter both homoclinic and heteroclinic trajectories

$$\dot{x} = y, \; \dot{y} = \gamma - \sin x.$$

The last system has saddles at points $(\pm\pi, 0)$ for $\gamma = 0$ and the center at a point $(0, 0)$. Two heteroclinic trajectories connecting saddles can be defined by direct integration of the system by the method of division of variables. They satisfy the equation $y^2 = 4\cos^2 \dfrac{x}{2}$. At $0 < \gamma < 1$ the system has a center at a point $(x_1, 0)$ and a saddle at a point $(x_2, 0)$, where $x_1 = \arcsin \gamma$, $x_2 = \pi - x_1$. Integrating the system, we shall find, that the separatrix loop of the saddle satisfies the equation

$$y^2 = 2\gamma x + 4\cos^2 \frac{x}{2} - C, \; \text{where} \; C = 2\gamma x_2 + 4\cos^2 \frac{x_2}{2}.$$

In three-dimensional systems with nondegenerate singular points separatrix loops both of saddle–nodes, and saddle–focuses, and also various heteroclinic trajectories and closed separatrix contours consisting of them are possible. Examples of such contours are presented in Sec. 2.4.2.

1.3 Periodic and Nonperiodic Solutions, Limit Cycles and Invariant Tori

1.3.1 *Periodic solutions*

The solution $x(t)$ of autonomous system of differential equations (1.1) is called *the periodic solution* if there exists a constant T, such that $x(t+T) = x(t)$ for all t. Minimal such value T is called *the period of the solution $x(t)$*, and the solution $x(t)$ is called *T-periodic solution.*

Phase curve (trajectory) of the periodic solution of the system (1.1) is closed and is called *a cycle*. Back, any cycle (*the closed phase curve*) of the systems (1.1) defines the periodic solution of the system with some period.

The theory of cycles developed, basically, in A. Poincare's works, enables to describe mathematically an evolution of a wide class of the natural phenomena and the social processes, consisting in an establishment in time of periodic regimes of their functioning or behaviour.

1.3.2 *Limit cycles*

Closed trajectories of systems of differential equations can be isolated and not isolated.

Definition 1.14 The isolated closed trajectory is called *a limit cycle* of an autonomous system of ordinary differential equations.

Not isolated closed trajectories existing, for example, in a neighbourhood of a singular point of the center type, do not represent interest for the theory of dissipative systems of differential equations as they are not limit trajectories in the sense that for each such trajectory there is no its neighbourhood from which all other trajectories tend to it when $t \to +\infty$ or when $t \to -\infty$.

Definition 1.15 A limit cycle is called as orbitally asymptotically stable (or simply *stable*) if for any its small neighbourhood U, all trajectories beginning in enough small neighbourhood of the cycle, do not leave in time U and tend to the cycle when $t \to +\infty$.

Investigation of a limit cycle on stability can be fulfilled with use of *the Floquet theory*. Let $x_0(t)$ be the T-periodic solution of the system (1.1), presented in phase space by its limit cycle. By linearizing the system (1.1) on its periodic solution similar to how it is made in Sec. 1.2.1, we shall obtain the linear non-autonomous system of ordinary differential equations

$$\dot{y} = A(t)y, \qquad \text{where} \qquad A(t) = \frac{\partial F}{\partial x}(x_0(t)) \tag{1.5}$$

with T-periodic matrix $A(t)$ and $y(t) = x(t) - x_0(t)$.

Theorem 1.9 (the Floquet theorem [Coddington and Levinson (1955); Haken (1983)]) *Each fundamental matrix solution of linear system* (1.5) *with periodic real coefficients can be presented in the form of* $Y(t) = P(t)\exp(Bt)$ *where* $P(t)$ *is some* T-*periodic complex matrix, and* B *is some constant complex matrix, and there exists an invertible real matrix* C *such, that* $C = \exp(BT)$.

The matrix C, named by *a matrix of monodromy*, is uniquely determined by a periodic matrix $A(t)$. Its eigenvalues λ_i are called multipliers of *the linearized system* (1.5) or *multipliers of a cycle* on which the system (1.5) is constructed. Eigenvalues α_i of the matrix B are called *Floquet exponents* of the linear system or original limit cycle. Their real parts also are determined uniquely. It is obvious, that $\lambda_i = \exp(\alpha_i T)$, and $C = Y(T)$, if $Y(0)$ is an unit matrix.

The real nonsingular matrix C can not have the real logarithm, i.e. not always there is a real matrix B such, that $C = \exp(BT)$. An example is the matrix C, having the simple negative multiplier. However the matrix C^2 already always has the real logarithm. Therefore each real fundamental matrix solution of the linear system (1.5) with T-periodic coefficients can be presented in the form of $Y(t) = \tilde{P}(t)\exp(\tilde{B}t)$, where $\tilde{P}(t)$ is some real $2T$-periodic matrix, and $\tilde{B}$ is some constant real matrix such, that $C^2 = \exp(2\tilde{B}T) = \exp(2BT)$.

The values of the multipliers or Floquet exponents obtained from a linear non-autonomous system (1.5) of the first approximation can be used for analysis of stability of the original periodic solution $x_0(t)$ of the nonlinear autonomous system (1.1).

Theorem 1.10 [Coddington and Levinson (1955)] *One simple multiplier of a cycle is always equal to* $+1$, *the corresponding Floquet exponent is equal to zero. If one of Floquet exponents is equal to zero, and all the others* $m - 1$ *Floquet exponents have negative real parts (or all multipliers of a cycle, except unit multiplier, have modules, smaller one, that is they lie inside of an unit circle of a plane of complex variable), then the periodic solution* $x_0(t)$ *of the system* (1.1) *is stable (asymptotically orbitally stable). If at least one Floquet exponent has a positive real part (or a multiplier of a* *cycle lies outside of an unit circle), the periodic solution $x_0(t)$ of the system* (1.1) *is unstable.*

It is possible to show, that in conditions of the formulated theorem the periodic solution $x_0(t)$ is not only stable, but each solution $x(t)$, lying close to its trajectory, possesses an asymptotic phase, i.e. there is a constant c such, that

$$\lim_{t \to \infty} \|x(t) - x_0(t + c)\| = 0.$$

It is easy to see, that the derivative $y(t) = \dot{x}_0(t)$ of the original periodic solution of the nonlinear autonomous system (1.1) is also one of solutions of the linear non-autonomous system (1.5), as

$$\dot{y} = \frac{dF(t)}{dt} = \frac{\partial F}{\partial x}(x_0(t))\dot{x}_0(t) = A(t)y(t).$$

Therefore the unit multiplier of the cycle corresponds to the eigenvector of the matrix of monodromy, touching the cycle. It is connected with movement along the cycle and does not influence on stability of the cycle. Stable and unstable limit cycles are presented in Fig. 1.7(a,b).

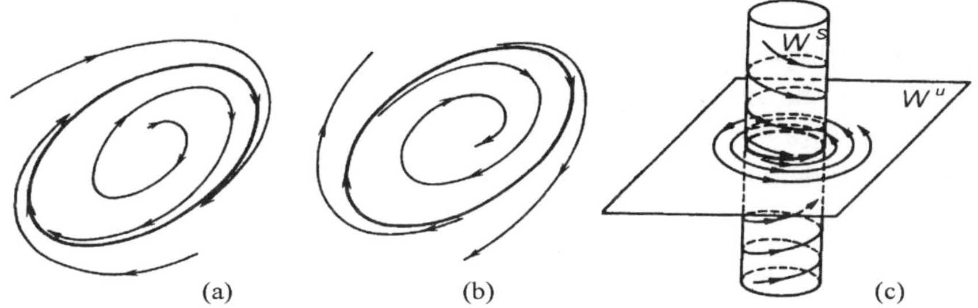

Fig. 1.7   Stable (a), unstable (b) and saddle (c) limit cycles.

A periodic solution cannot be asymptotically stable because solutions starting in different points of a cycle do not tend one to other when $t \to +\infty$.

Definition 1.16   The limit cycle is called *hyperbolic* if it has no multipliers lying on an unit circle, except one unit multiplier (there is exactly one Floquet exponent with the zero real part, which is equal to zero).

Definition 1.17   The limit cycle is called *non-degenerate* if it has no multipliers equal to +1, except one unit multiplier (there is exactly one simple zero Floquet exponent, but there can be nonzero Floquet exponents with zero real parts).

Definition 1.18   The limit cycle, having multipliers inside and on boundary of an unit circle, is called *semistable.*

The semistable limit cycle in a three-dimensional case has the following multipliers $\{|\lambda_1| < 1,\ \lambda_2 = +1,\ |\lambda_3| = 1\}$.

Definition 1.19   The hyperbolic limit cycle for which Floquet exponents have both negative, and positive real parts (or multipliers lie inside, and outside of an unit circle) is called *a saddle cycle.*

The concept of a saddle cycle is defined for the dimension of phase space $m > 2$. When $m = 3$ the saddle limit cycle has the following multipliers $\{|\lambda_1| < 1,\ \lambda_2 = +1,\ |\lambda_3| > 1\}$.

For a saddle limit cycle of autonomous system of ordinary differential equations some phase curves which are being close to the saddle trajectory, tend to it (are reeling on it) when $t \to +\infty$, forming its stable invariant manifold W^s. Other phase curves are reeling off it, forming its unstable invariant manifold W^u (Fig. 1.7c).

The given above definition of a hyperbolic limit cycle corresponds to the definition given in the book [Anosov (1985)], and naturally generalizes the concept of a hyperbolic singular point. From this definition in particular follows, that the stable cycle is also hyperbolic. We note, however, that in the literature devoted to the theory of hyperbolic dynamical systems, saddle in our definition limit cycles are named often hyperbolic cycles.

1.3.3   *Poincare map*

In addition to calculation of the Floquet exponents there exists also another method of studying qualitative behaviour of solutions of autonomous systems of ordinary differential equations in a neighbourhood of a limit cycle — construction of Poincare map and research of properties of this map in a neighbourhood of its fixed point. The basic sense of use of Poincare map for the analysis of dynamics of systems of differential equations consists in reducing of dimension of studied system by unit under transition to map.

Let γ be a limit cycle corresponding to T-periodic solution $x_0(t)$ of autonomous system (1.1). Take a $(m-1)$-dimensional secant hypersurface S passing through some point x^* of the closed curve γ transversally γ i.e. so that the vector tangent to the curve γ at the point x^* does not lie in S.

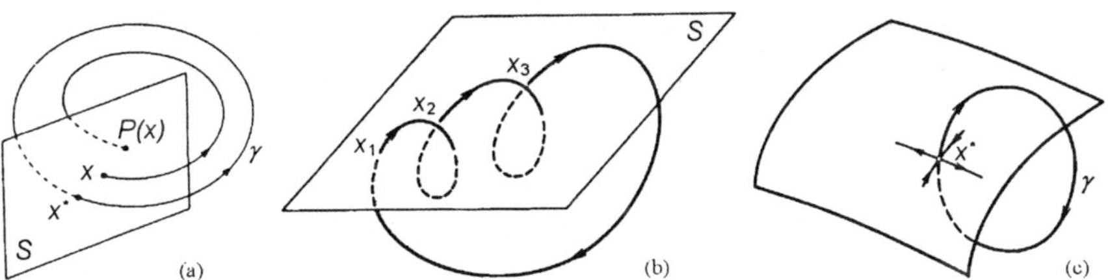

Fig. 1.8   Fixed point (a), cycle (b) and saddle point (c) of the Poincare map with saddle limit cycle corresponding to it.

As the curve γ is a phase curve of the T-periodic solution, the trajectory starting from the point x^* will return to the same point in time T. Any other trajectory, starting from any other point x, lying in a neighbourhood $U \subset S$ of the point x^*, will return and will cross S in the same direction at a point $P(x)$, lying in a neighbourhood $V \subset S$ of the point x^* (Fig. 1.8a). Local diffeomorphism $P : U \to V$, i.e. a one-to-one mapping differentiable together with its inverse mapping, is called *the Poincare map* or *the map of the first returning*.

As $P(x^*) = x^*$ then the point x^* is a fixed point of the Poincare map. To a cycle of the Poincare map of the order n (or n-multiple cycle) there corresponds sequence of points $x_1, x_2, \ldots, x_n$, satisfying the conditions (Fig. 1.8b)

$$x_2 = P(x_1), \ x_3 = P(x_2), \ \ldots, \ x_1 = P(x_n).$$

Definition 1.20   The fixed point x^* of a map $P(x)$ is *stable*, if there is its neighbourhood U such, that all iterations of the map P, starting in this neighbourhood, converge to the fixed point, i.e. if $x_0 \in U$ and $x_{k+1} = P(x_k)$, $k = 0, 1, \ldots$, then $x_k \to x^*$ when $k \to \infty$.

As the Poincare map is differentiable, then it is possible to linearize it at its fixed point

$$P(x) - x^* = A(x - x^*) + o(\|x - x^*\|), \quad \text{where} \quad A = D_x P(x^*) = \frac{\partial P}{\partial x}(x^*).$$

The linear operator A has a rank equal to $m - 1$. Under a choice of a corresponding hypersurface S, perpendicular to a vector, tangent to a cycle γ at a point x^*, and a system of the coordinates, for which one of orts coincides with a vector tangent to a cycle, the matrix of the operator A will have the dimension $(m - 1) \times (m - 1)$.

As well as in a case of systems of differential equations, map $P(x)$ can be conservative, i.e. save phase volume, and dissipative — to compress phase volume.

Definition 1.21   The map $P(x)$ is called *conservative* in the region $B \subset M$, if $|\det D_x P(x)| = 1$ for all $x \in B$. If for all $x \in B$ the condition $|\det D_x P(x)| < 1$ takes place, then the map $P(x)$ is called *dissipative* in the region $B \subset M$.

Stability of a fixed point of a map $P(x)$ is defined by eigenvalues of a matrix A.

Theorem 1.11   [Schuster (1984)] *If all $m-1$ eigenvalues α_k of the matrix A lie inside of an unit circle of a plane of complex variable, i.e. $|\alpha_k| < 1$, then the fixed point x^* of the map $P(x)$ is stable; if among eigenvalues α_k there exists at least one α_i with $|\alpha_i| > 1$, then the fixed point of the map $P(x)$ is unstable.*

Naturally all definitions and theorems given above are transferred without changes on any differentiable many-dimensional mapping $P(x) : \mathbb{R}^m \to \mathbb{R}^m$. To a stable (unstable) fixed point of Poincare map there corresponds a stable (unstable) cycle of corresponding system of differential equations, and to a saddle point having eigenvalues of the operator of a linear part, lying as inside, and outside of an unit circle of a plane of complex variable, there corresponds a saddle limit cycle (Fig. 1.8c).

Unfortunately, at the time of studying of concrete systems of differential equations, construction of Poincare map and its derivative is possible only numerically. On the other hand, the results obtained for abstract $(m-1)$-dimensional mappings, absolutely unessentially should be transferred to m-dimensional systems of ordinary differential equations. In other words, at present the problem of possibility of restoration of system of differential equations from its Poincare map is not solved, and it is very doubtful that to make it is easier, than to carry out full qualitative research of the system of differential equations. Besides, considered below in Chapters 3 and 5 numerous examples of analysis of scenarios of transition to chaos in concrete many-dimensional nonlinear systems of ordinary and partial differential equations show, that this transition looks like transition to chaos as a result of bifurcations of continuous one-dimensional mappings. Therefore the analysis of one-dimensional continuous mappings and their connection with nonlinear dissipative systems of ordinary and partial differential equations is represented as more interesting and important for us. This connection is realized not through Poincare map, but through transition to some $(m-1)$-dimensional (two-dimensional in a three-dimensional case) subspace, rotating transversally to an original singular cycle, and through construction in such subspace of a non-autonomous system of ordinary differential equations with rotor type singular point. Transition from such differential equation with rotor to a one-dimensional mapping possessing of chaotic dynamics, occurs already absolutely naturally. Thus, not Poincare map, but a new object of the theory of differential equations — a rotor type singular point of two-dimensional non-autonomous system of ordinary differential equations with periodic coefficients which theory is developed in [Magnitskii (2004); Magnitskii (2005)], is the bridge connecting the differential equations and one-dimensional mappings. Chapter 4 is devoted to consideration of these questions.

1.3.4 *Invariant tori*

In systems of differential equations T-periodic motion on a cycle is one of the most simple motions and is characterized by presence of one frequency $\omega = 2\pi/T$. Much more complex is *the multifrequency regime of motion*, described by presence of several independent frequencies $\omega_1,\ldots,\omega_n$. Motion in such regime can be presented as motion on a surface of n-dimensional invariant torus, given by angles $\alpha_i(t) = \alpha_{i0} + \omega_i t$, $i = 1,\ldots,n$ (Fig. 1.9a). Dimension m of phase space thus should be not less than $n + 1$.

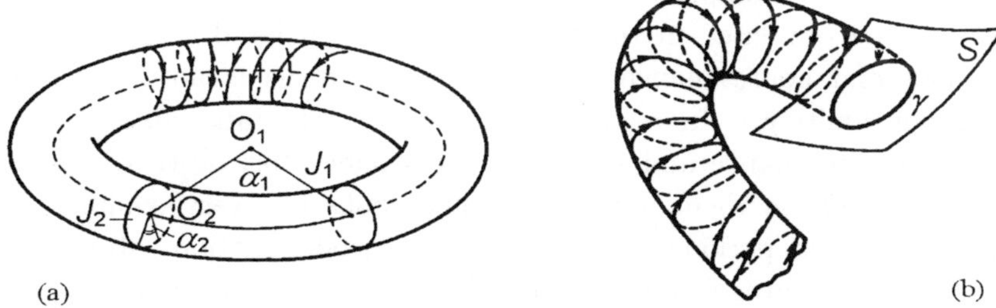

Fig. 1.9   Two-dimensional invariant torus (a) and quasiperiodic motion on it (b).

The behaviour of trajectories of a system on a surface of torus essentially depends on a ratio between frequencies ω_i, $i = 1,\ldots,n$. In the case of, for example, $n = 2$ motion on two-dimensional torus will be *periodic* in only case when the ratio of frequencies is rationally, i.e. $\omega_1/\omega_2 = k/m$, where

$k, m \in \mathbb{N}$. Thus $T = \pi(k/\omega_1 + m/\omega_2)$ and, hence, $\alpha_1(T) = \alpha_{10} + \omega_1 T = \alpha_{10} + 2\pi k$, and $\alpha_2(T) = \alpha_{20} + \omega_2 T = \alpha_{20} + 2\pi m$, i.e. angles α_1 and α_2 present the same point on a surface of torus and consequently to time T the trajectory comes back to an initial point.

In a case when frequencies ω_1 and ω_2 are incommensurable, i.e. their ratio is irrational, the phase trajectory will never become closed and will pass in time as much as close to any given point on a surface of torus. In this case the phase curve forms an everywhere dense winding of torus, and such motion is called *quasiperiodic*.

It is convenient to represent periodic and quasiperiodic motions on a surface of torus by Poincare map on a secant hypersurface S, given transversally to a surface of torus. Thus periodic motion will be presented in S by a finite number of points in a series passing each other under influence of Poincare map, and quasiperiodic — by infinite set of points densely filling some closed curve (Fig. 1.9b). Periodic and quasiperiodic motions on surfaces of many-dimensional invariant tori arise naturally in *Hamiltonian conservative systems* [Arnold *et al.* (1988)].

It is necessary to distinguish resonance and nonresonance tori.

Definition 1.22   Invariant torus of dimension $n \geq 2$ is called *resonance torus* if there exist not all equal to zero integers k_i such, that

$$\sum_{i=1}^{n} k_i \omega_i = 0.$$

In a case of completely integrable Hamiltonian systems all phase space can be presented in the form of a set of enclosed each into other resonance and nonresonance tori, so everyone torus is not neither isolated, nor limiting. In a case of not completely integrable Hamiltonian system resonance and some of nonresonance tori destroy, and motion with given on them initial conditions looks like very complex, different both from periodic, and from quasiperiodic motion. Regions of destroyed tori unite, forming a united network — *Arnold web*. Movement in this web, named by *Arnold diffusion*, is an example of chaotic behaviour of solutions in conservative systems. The explanation of this phenomenon is given by *the Kolmogorov–Arnold–Moser (KAM) theorem* fine represented in many monographies and textbooks [Arnold *et al.* (1988); Moser (2001)] to which we send a reader who is interested in these questions. Here we only note, that in any case tori in conservative Hamiltonian systems are not limiting sets and they can not be stable in the same sense, as limit cycles.

On the contrary, in dissipative systems to research of which the present work is devoted, stable two-dimensional tori play an essential role in scenarios of transition to dynamical chaos in many-dimensional systems of ordinary differential equations and in partial differential equations.

Definition 1.23 An invariant torus is called *stable torus* if for any its small neighbourhood U, all trajectories starting in its enough small neighbourhood, do not leave in time U and tend to torus when $t \to +\infty$.

Methods of analysis of stability of invariant tori in dissipative systems of nonlinear differential equations are now developed insufficiently. It is possible to try to analyze the stability of the closed curve filled by points of Poincare map on a secant hypersurface S. However, this curve is not a solution of any equation, and the analysis of its stability looks rather problematic. It is possible also to try to use the Lyapunov exponents considered in the following section, but that represents more likely theoretical, than practical value. As a whole the situation with dissipative systems looks much more complex, than in the case of Hamiltonian systems, and finding even two-dimensional stable tori in the concrete systems of differential equations is more likely art, than science. Some original methods of finding stable tori and analysis of their bifurcations are presented in Chapter 2.

1.3.5 *Nonperiodic solutions, Lyapunov exponents*

The concepts of stability formulated above for stationary and periodic solutions, are naturally transferred to any solutions of autonomous and non-autonomous systems of ordinary differential Eqs. (1.1) and (1.2).

Definition 1.24 The solution $x_0(t)$ of autonomous system of ordinary differential Eqs. (1.1) is called *Lyapunov stable* if all solutions of this system with initial conditions from enough small neighbourhood of the initial condition of the investigated solution are defined on all positive semi-axis of time and uniformly in time converge to the investigated solution when their initial conditions tend to the initial condition of the investigated solution.

In other words, the solution $x_0(t)$ of the system (1.1) is Lyapunov stable if for any $\varepsilon > 0$ there exists $\delta > 0$ such, that for all solutions $x(t)$ of the system (1.1) from a condition $\|x(0) - x_0(0)\| < \delta$ follows $\|x(t) - x_0(t)\| < \varepsilon$ for all $t > 0$.

Definition 1.25 The solution $x_0(t)$ of the autonomous system of differential Eqs. (1.1) is called *asymptotically stable* if it is Lyapunov stable and all solutions $x(t)$ of the system (1.1) with initial conditions close enough to the investigated solution $\|x(0) - x_0(0)\| < \delta$ tend to it when $t \to \infty$, i.e. $\|x(t) - x_0(t)\| \to 0$ when $t \to \infty$.

If under conditions of asymptotic stability any solution of a system of Eqs. (1.1) tends to the investigated solution $x_0(t)$ exponentially, i.e. $\|x(t) - x_0(t)\| \leq c\exp(-\gamma t)$ with some positive constants c and γ, then the solution $x_0(t)$ of the system (1.1) is called *exponentially asymptotically stable*.

Some kinds of asymptotic, but not exponential asymptotic stability (*power and fractionally-exponential stability*) are investigated by one of authors in the monograph [Magnitskii (1992)].

For investigation of stability of a solution $x_0(t)$ of the system (1.1) let us linearize the system at this solution similar to how it was made in Secs. 1.2.1 and 1.3.2. We shall obtain a linear non-autonomous system of ordinary differential equations

$$\dot{y} = A(t)y, \qquad \text{where} \quad A(t) = \frac{\partial F}{\partial x}(x_0(t)), \quad y(t) = x(t) - x_0(t). \quad (1.6)$$

In this case the linear operator $A(t)$ can have a matrix with any elements limited on semi-axis $0 \leq t < \infty$.

Let us consider a solution $y(t)$ of *the linearized system* (1.6) and define for it *the Lyapunov exponent* $\lambda(y)$ by the formula

$$\lambda(y) = \varlimsup_{t \to \infty} \frac{\ln \|y(t)\|}{t}.$$

In particular case, when $x_0(t) = x_0$ is a stationary solution, the operator of a linear part has a constant matrix $A(t) = A$. Then the fundamental matrix of solutions of linear system (1.6) looks like $Y(t) = \exp(At)$. Without restriction of generality we shall consider that the matrix A has the Jordan form. Hence, to everyone its Jordan cell of the order k with eigenvalue λ there corresponds a chain from k solutions of the linear system of a kind

$$y_i(t) = e^{\lambda t}\left(\xi_{i-1} + t\xi_{i-2} + \cdots + \frac{t^{i-1}}{(i-1)!}\xi_0\right), \quad i = 1, \ldots, k,$$

where ξ_0 is an eigenvector, and ξ_j is an attached vector for eigenvalue λ. Obviously, for each of solutions of a chain its Lyapunov exponent $\lambda(y_i)$ is equal to $\Re\{\lambda\}$. Thus, the linear system (1.6) with a constant matrix A has m Lyapunov exponents taking into account their multiplicity. Each exponent coincides with real part of some eigenvalue of matrix A. Multiplicity of each exponent is defined by the order of Jordan cell corresponding to it. By virtue of the Lyapunov theorem of stability on the first approximation, the stationary solution (a singular point) of the system (1.1) is asymptotically stable if all Lyapunov exponents of the linear system of the first approximation (1.6) are negative.

In other important special case when $x_0(t)$ is the periodic solution of the autonomous system (1.1), Lyapunov exponents coincide with real parts of Floquet exponents that follows from representation of a fundamental matrix of solutions of linear system (1.6) in the form of $Y(t) = P(t)\exp(Bt)$ with periodic and, hence, limited matrix $P(t)$. The system also has m exponents taking into account their multiplicities which are defined by orders of Jordan cells of matrix B. The following result takes place.

Theorem 1.12 [Haken (1983)] *If the trajectory of solution of autonomous system of ordinary differential Eqs. (1.1) remains in the limited region of the phase space and does not tend to a singular point when $t \to \infty$, then one of Lyapunov exponents of the system linearized at this solution is equal to zero.*

Thus, if one of Lyapunov exponents of the system, linearized at the periodic solution, is equal to zero, and all other exponents are negative, then the limit cycle is asymptotically orbitally stable. The zero exponent corresponds to a direction, tangent to the cycle (Fig. 1.10a).

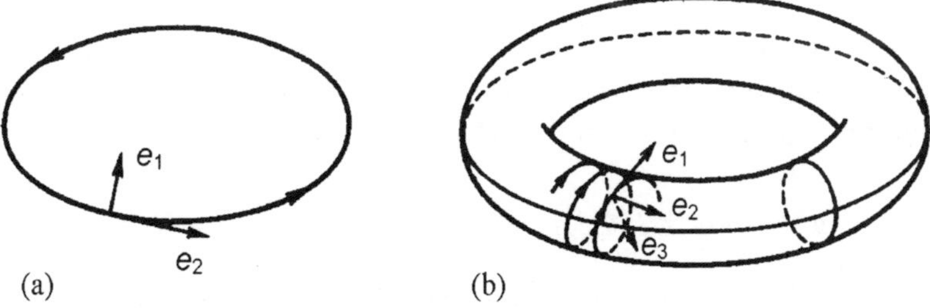

Fig. 1.10   In the cases of a limit cycle (a) and an invariant torus (b) the Lyapunov exponents corresponding accordingly to direction e_2 and to directions e_1, e_2 are equal to zero.

Generally the linear non-autonomous system (1.6) has m exponents

$$\lambda_m \leq \lambda_{m-1} \leq \ldots \leq \lambda_2 \leq \lambda_1,$$

among which there can be multiple exponents. In this case Lyapunov exponents are not eigenvalues of any constant matrix, including the matrix $A(\infty)$ even if the corresponding limit exists. The zero exponent corresponds to a direction, tangent to the limited nonperiodic solution which does not tend to a singular point. The exponent λ_1 is called *the senior characteristic exponent* of the system (1.6) and is designated through Λ. It is easy to show, that

$$\Lambda = \sup_{y \neq 0} \overline{\lim_{t \to \infty}} \frac{\ln \|y(t)\|}{t}.$$

Definition 1.26 The linear non-autonomous system (1.6) with real coefficients is called *correct* if the sum of its Lyapunov exponents coincides with the average value of a trace of matrix $A(t)$ which coincides with divergence of a vector field, i.e.

$$\sum_{k=1}^{m} \lambda_k = \lim_{t \to \infty} \frac{1}{t} \int_0^t \operatorname{tr} A(s)ds = \lim_{t \to \infty} \frac{1}{t} \int_0^t \sum_{k=1}^{m} a_{kk}(s)ds =$$

$$= \overline{\operatorname{tr} A(t)} = \lim_{t \to \infty} \frac{1}{t} \int_0^t \operatorname{div} F(x_0(s))ds.$$

It is considered to be, that all interesting systems from the practical point of view are correct. For such systems *the Lyapunov generalized theorem of stability on the first approximation* takes place.

Theorem 1.13 [Bilov *et al.* (1966)] *Let the solution $x_0(t)$ of non-autonomous system of ordinary differential equations (1.2) is such, that the difference $y(t) = x(t) - x_0(t)$ satisfies the equation*

$$\dot{y} = A(t)y + f(y,t), \quad where \quad A(t) = \frac{\partial F}{\partial x}(x_0(t),t),$$

$\|f(y,t)\| \leq K\|y\|^q$, $\quad q > 1$. *Then, if the linear system of the first approximation $\dot{y} = A(t)y$ is correct and has a negative senior characteristic exponent Λ, then the solution $x_0(t)$ of the system (1.2) is asymptotically stable. If the exponent $\Lambda > 0$, then the solution $x_0(t)$ of the system (1.2) is unstable.*

There are the numerous generalizations of the formulated theorem connected, basically, with generalization of a concept of senior characteristic exponent Λ. So in [Bilov *et al.* (1966); Daletskii and Krein (1974)] the concepts of the top central exponent and the general exponent are considered. In [Magnitskii (1992)] the concept of *characteristic function* $\Lambda(t)$ of the linear system (1.6), generalizing both concept of the senior characteristic exponent, and concept of the top central exponent is proposed. Negativity of characteristic function provides an asymptotic stability of solution $x_0(t)$ of the system (1.2) without the requirement of correctness of linear system of the first approximation and without the requirement of negativity of the senior characteristic exponent (it can be equal to zero).

Let us notice, that negativity of the senior characteristic exponent is not characteristic property of stable solutions of nonlinear systems of ordinary differential equations. Simple enough stable periodic solution already has a zero senior characteristic exponent. More complex stable and semistable nonperiodic solutions possess also the same property. Therefore application of the theory of Lyapunov exponents for the analysis of complex irregular dynamics of nonlinear systems of ordinary differential equations is rather limited.

Definition 1.27 The trajectory of a limited solution of the autonomous system (1.1) is called *hyperbolic* if the system (1.6) linearized at this solution has exactly one simple zero Lyapunov exponent.

Last definition, obviously, generalizes the concepts of a hyperbolic singular point and a hyperbolic limit cycle given above.

Definition 1.28 The hyperbolic trajectory having both positive, and negative Lyapunov exponents, is called *a saddle trajectory*.

For saddle trajectory of non-autonomous system of ordinary differential equations also as well as for saddle limit cycle it is possible to define its stable invariant manifold W^s and its unstable invariant manifold W^u.

Definitions of *asymptotically orbital stability* (simple stability) and *semistability* of a limited nonperiodic solution can be given similarly to how it has been made for limit cycle in Sec. 1.3.2. In the first case the system linearized at the solution has one zero and other negative Lyapunov exponents, and in the second case it has multiple zero and other negative exponents.

If the trajectory of periodic or nonperiodic solution $x_0(t)$ of the system (1.1) lies on a surface of n-dimensional invariant torus, then such solution cannot be asymptotically stable since two solutions starting in different points of trajectory cannot approach when $t \to +\infty$. Thus n Lyapunov exponents of the system linearized at such solution will be equal to zero. These exponents correspond to the independent directions in a hyperplane which is tangent to a torus surface (Fig. 1.10). Torus itself will be stable if all other Lyapunov exponents are negative. Thus, any trajectory on a surface of a stable torus is semistable.

Note that for a nonperiodic solution of the autonomous system (1.1) which trajectory does not lie on a surface of an invariant torus, i.e. the solution is not also quasiperiodic, nothing forbids to be realized to a case when one (senior) or several Lyapunov exponents are equal to zero, and other exponents are negative.

In conclusion of this section we shall note one important property of dissipative systems of autonomous differential equations, following from a condition of correctness of the system (1.6), linearized at the solution of the system (1.1). As along such solution the divergence of a vector field is negative, then the sum of Lyapunov exponents of the system (1.6) also is negative, i.e. $\sum_{k=1}^{m} \lambda_k < 0$.

1.4   Attractors of Dissipative Systems of Ordinary Differential Equations

1.4.1   *Basic definitions*

As it was already mentioned above, the basic distinctive property of dissipative system of ordinary differential equations is compression of its phase volume in time. As a result when $t \to \infty$ all solutions of such system or a part of solutions tend to some compact (closed and limited) subset B of phase space M, named an attractor. Thus, attractor contains "the set of established regimes" of the system.

Now there is no generally accepted strict definition of attractor. It is connected first of all with the reason that till now it is not clear what is an irregular (chaotic or any else) attractor and how it is arranged.

Definition 1.29   The point y is called ω-*limiting point* for $x \in M$ if there is a sequence $t_n \to \infty$, such, that $\varphi^{t_n}(x) \to y$. The set of all ω-limiting points for a trajectory starting at a point x, is called ω-*limiting set* for x and is designated $\omega(x)$.

If $y \in \omega(x)$ then $\varphi^t(y) \in \omega(x)$, i.e. $\omega(x)$ is invariant set. The union of all sets $\omega(x)$ for all $x \in G \subset M$ is designated as $\omega(G)$. Similarly for negative values of t *α-limiting points* and *sets* are determined.

Definition 1.30 Compact invariant in relation to a flow φ^t set $B \subset M$ is called *an attractive set* if there is its neighbourhood U (the open set containing B) such, that $B \subset \omega(U)$ and for almost all $x \in U$, $\varphi^t(x) \to B$ when $t \to \infty$ (i.e. $\mathrm{dist}(\varphi^t(x), B) = \inf_{y \in B} \|\varphi^t(x) - y\| \to 0$ when $t \to \infty$).

The greatest set U, satisfying to this definition, is called *a domain of attraction* for B.

In traditional definition of attractive set [Malinetskii and Potapov (2000)] we have replaced words "for all" with words "for almost all". Otherwise even the elementary Feigenbaum attractor (see Chapters 2–4) would not satisfy the definition of attractive set.

Not all attractive sets are attractors, but only those from them which possess *a property of indecomposability* into two separate compact invariant subsets. We shall give the most popular on today definition of attractor.

Definition 1.31 The indecomposable attractive set is called *attractor*.

Often an attractor is named an attractive set B, containing an everywhere dense trajectory [Malinetskii and Potapov (2000)], i.e. containing a point x, for which $\omega(x) = B$. It is really valid for many attractors, born as a result of cascades of bifurcations of stable cycles. However, even simple resonance two-dimensional stable torus has no everywhere dense trajectory, and contains infinite number of periodic semistable trajectories.

Dissipative system of differential equations can have both finite, and infinite number of various attractors. All points in the phase space, except for a set of zero measure, lie in the domain of attraction of one of them.

In many cases it is not possible to find an attractor, but it is important to know, whether it exists. For this purpose the concept of an absorbing set is used.

Definition 1.32 The compact invariant set $G \subset M$ is called *absorbing* if there exists its neighbourhood U such, that all trajectories starting in U, for finite time enter G and remain there forever.

It is possible to show, that the system possessing an absorbing set G, has also an attractor $B \subset \omega(G)$.

Attractors of nonlinear systems of ordinary differential equations, satisfying the definition given above, happen to be simple (regular) and complex (irregular).

1.4.2   *Classical regular attractors of dissipative systems of ordinary differential equations*

In the modern literature there is no strict definition of a regular attractor. Intuitively this concept is connected with sufficient simplicity of behaviour in solutions of systems of differential equations on such attractors and with sufficient smoothness of attractor. It is considered, that ergodic motion can be the most complex motion on a regular attractor.

Definition 1.33   A motion of a system of differential equations on its invariant set $B \subset M$ is called *an ergodic motion* if the relative time spent by a phase trajectory inside any region $\Omega \subset B$ is equal to relative volume of this region in B and does not depend on a choice of initial conditions. In other words, for almost all $x \in B$

$$\lim_{T \to \infty} \frac{1}{T} \int_0^T \chi_\Omega(\varphi^t(x)) dt = V_\Omega / V_B,$$

where χ_Ω is characteristic function and V_Ω is Euclidean volume of the Ω.

Quasiperiodic motion with incommensurable frequencies on invariant torus is an example of ergodic motion. In the course of time the phase trajectory uniformly and densely covers the surface of the torus. But influence of a phase flow φ^t on a small region Ω_0, belonging to a surface of two-dimensional invariant torus, is reduced simply to moving of Ω_0 along a torus surface (Fig. 1.11a). Hence, as well as in case of periodic motion, any two trajectories which were close in initial moment of time remain close and at all subsequent moments of time.

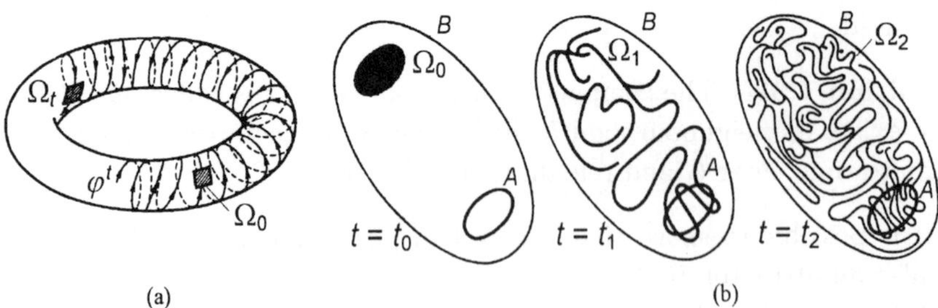

Fig. 1.11   Influence of a phase flow φ^t on region Ω in a case of quasiperiodic motion (a) and in a case of mixing (b).

Thus, only stable singular points, stable (asymptotically orbitally stable) limit cycles and stable invariant tori are considered to be as simple (regular) attractors. All these attractors are submanifolds of the phase space (for example, the limit cycle and two-dimensional invariant torus are, accordingly, one-dimensional and two-dimensional submanifolds). Dynamics of systems with such attractors is not chaotic, but has asymptotically converging, periodic or, the most complex, ergodic character. The main thing is that trajectories of the systems with regular attractors are globally stable in relation to small perturbations meaning their global predictability.

Lyapunov exponents of the system of the first approximation linearized at this trajectory allow to determine a type of regular attractor to which the investigated trajectory of the system belongs. If all exponents are negative, then a trajectory is an attractor, being a stable singular point (the stable node or stable focus). If one (senior) exponent is equal to zero, and all other exponents are negative, then the trajectory most likely is a stable limit cycle. The word-combination "most likely" is used here in the sense, that in this case nothing forbids the trajectory to be as well the stable nonperiodic trajectory which does not tend to a singular point. If n Lyapunov exponents are equal to zero, and all others $m - n$ exponents are negative, then the trajectory most likely lies on the surface of n-dimensional invariant stable torus. In the latter case anything also does not forbid a trajectory to be a semistable nonperiodic trajectory lying on some n-dimensional invariant stable surface.

Here it is necessary to note, that as follows from the results of the paper [Ruelle and Takens (1971)], stable invariant tori with dimensions $n > 2$ as a rule are destroyed under influence of small perturbations which always are present in the system. Besides, occurrence of a regime of motion with a great number of incommensurable frequencies is resisted by the phenomenon of *synchronization of oscillations* [Anishchenko *et al.* (1999)]. Synchronization lies in the fact that in many-dimensional systems oscillations with independent frequencies feel a complex influence each against other. That leads to disappearance of a quasiperiodic motion and to establishment of a periodic regime of motion with commensurable frequencies, that is limit cycle on a torus. Therefore occurrence of many-dimensional invariant stable torus in the phase space of dissipative system of autonomous differential equations is more likely exception, than a rule.

1.4.3   *Classical irregular attractors of dissipative dynamical systems*

Modern scientific literature uses various definitions of complex (irregular) attractors, reflecting the different sides of an irregularity of behaviour of trajectories belonging to them.

The basic sense of all definitions lies in the fact that on the most irregular attractor the motion should be unstable: trajectories of a system should diverge quickly, remaining on the attractor. Thus the behaviour of solutions of dissipative systems with irregular attractors will be characterized by the combination of global compression of phase volume with local instability of separate phase trajectories.

However, existence of the majority classical irregular attractors is proved only for discrete dissipative dynamical systems (mappings or cascades). The last circumstance defines the contents of the present section, forcing to consider some important interesting examples not from the area of dissipative systems of ordinary differential equations.

Often the attractor, the dynamics of which is characterized by Lyapunov positive exponent is called *the chaotic attractor* [Malinetskii and Potapov (2000)]. Thus it is considered, that dissipative system is correct, and, hence, should negative exponents also exist, on the sum of absolute values exceeding a positive exponent. If dimension of phase space is equal to three, then Lyapunov exponents should be the following: $\lambda_3 < 0$, $\lambda_2 = 0$, $\lambda_1 > 0$. However, as it was already noticed above, complex attractor can have a zero senior characteristic exponent. Such is, for example, Feigenbaum attractor (see Chapter 2). Hence, exponential scattering of trajectories is not necessary for a chaotic motion. Besides, as Lyapunov exponents can be found only numerically, and the approached calculation of solutions of system of differential equations on its complex attractor is a strongly ill-posed problem, then it is not possible to trust positivity of a Lyapunov exponent. For example, in case of a well-known Lorenz system from positivity of the Lyapunov exponent found numerically does not follow, as a rule, the presence of chaotic dynamics in the system. Actually, in this case the system can have quite a stable limit cycle [Magnitskii and Sidorov (2001c)]. Thus, we come to conclusion, that positive value of a Lyapunov exponent numerically found on any trajectory is not, first, correct, and is not, secondly, a characteristic feature of chaotic motion, and, in the third, nothing speaks neither about the nature, nor about structure of irregular attractor.

The majority of researchers fairly connect a concept of chaotic motion with presence of more complex, than ergodic, regimes of behaviour on an attractor. In this case the initial area Ω_0 is thus distributing in time on all invariant set B, that its separate parts can be found in any as much as small open subset $A \subset B$ independently of its size, form or location in the initial area Ω_0. It is claimed, that such systems or motions possess property of mixing (Fig. 1.11b). The condition of mixing is formulated strictly as follows. We shall designate $\Omega_t = \varphi^t(\Omega_0)$. Let $\Omega_t \cap A$ represent the intersection of the sets A and Ω_t.

Definition 1.34 The autonomous system of ordinary differential Eqs. (1.1) is called *mixing* (and accordingly the flow is called *mixing*) on invariant compact set B if for any regions $\Omega_0 \subset B$ and $A \subset B$ there is a limit

$$\lim_{t \to \infty} \frac{V(\Omega_t \cap A)}{V(A)} = \frac{V(\Omega_0)}{V(B)},$$

where $V(G)$ is the Euclidean volume of a region G.

The attractor, motion on which possesses the property of mixing, is called *a stochastic attractor* [Malinetskii and Potapov (2000)]. For mixing systems trajectories being close in the initial moment of time, they do not remain close during the subsequent moments of time. Scattering of phase trajectories means *unpredictability of behaviour of solutions* of a system. Any error in calculation of a trajectory in the initial moment of time can lead in time to absolutely unexpected results. On the other hand, movement along on stochastic attractor except for unpredictability possesses as well *the property of irreversibility*. Knowing the position of a phase point in the final moment of time, it is impossible to say, where the point was in the initial moment.

An example of a classical attractor of dissipative system, possessing the mixing property, is *the Henon attractor* of the two-dimensional cascade [Henon (1976)]

$$x_{n+1} = 1 - ax_n^2 + y_n, \ y_{n+1} = bx_n, \ a = 1.4, \ b = 0.3. \tag{1.7}$$

Henon mapping P is dissipative in all range of definition, as $|\det D_x P| = |-b| = 0.3 < 1$. All trajectories of the cascade (1.7) tend to attractor, represented in Fig. 1.12a (from [Malinetskii and Potapov (2000)]). Moreover, any small volume in phase space under action of Henon map will be found uniformly spread along the whole attractor.

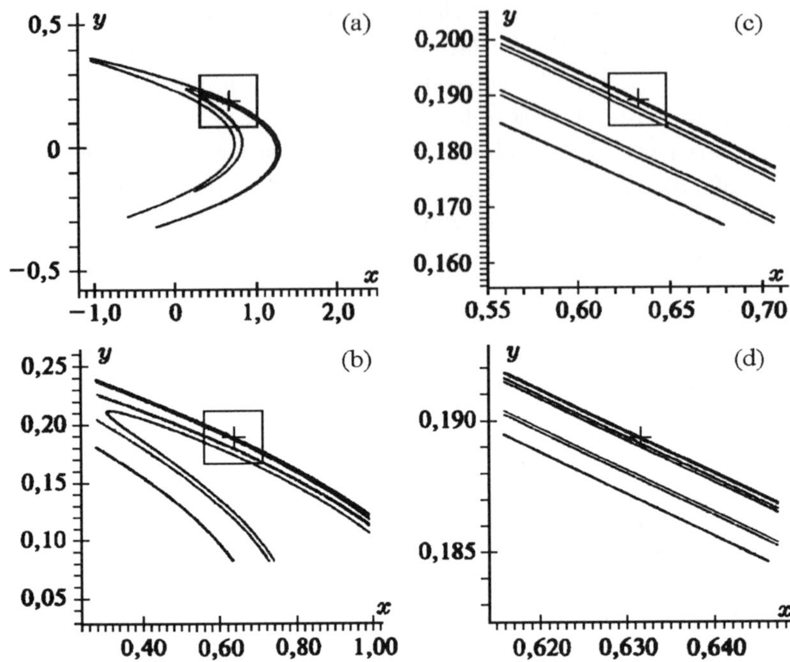

Fig. 1.12 Illustration of fractal structure of Henon attractor. The square gets out in a neighbourhood of the fixed point designated by a dagger.

From Fig. 1.12b,c,d also it is seen, that Henon attractor possesses scale invariance, i.e. the increased part of attractor appears similar to all attractor. It is claimed, that in this case the set possesses *a fractal structure*. Attractors, having fractal structure, are called *strange attractors* [Ruelle and Takens (1971)]. Such attractors are not finite unions of submanifolds of phase space (as a cycle or torus) and have fractional dimensions (see below).

However, if for ergodicity there are the theorems showing, that the majority of real systems possess this property, then mixing demands the proof in each separate case. As far as we know, now there are no examples of attractors of dissipative systems of ordinary differential equations for which property of mixing is strictly proved. Existence of classical strange attractors is also proved only for discrete dissipative dynamical systems, but not for systems of ordinary differential equations. Except for that there are examples of strange attractors with zero Lyapunov senior exponents, i.e. attractors, having a fractal structure, but not possessing property of exponential divergency of trajectories. The elementary example of such attractor is the invariant set (*Feigenbaum attractor*) of one-dimensional *logistic map*

$$x_{n+1} = \mu x_n(1 - x_n), \qquad x \in [0; 1] \tag{1.8}$$

at the some $\mu = \mu_\infty$ (in detail the map (1.8) will be considered in the following chapters where it will be shown, that its properties are typical not only for mappings, but also for continuous systems of differential equations). The same logistic map at $\mu = 4$ sets an example of attractor, coincident with all interval $[0; 1]$ and, hence, not having a fractal structure and not being strange [Jakobson (1981)]. Therefore neither strangeness of attractor, nor its stochasticity also are not characteristic features of a chaotic motion.

In some works, in particular under the theory of discrete dynamical systems (cascades) the concept of *hyperbolic attractor*, being simultaneously by attractor and hyperbolic set, is actively used [Anosov (1985); Eckmann and Ruelle (1985)]. This attractor entirely consists of only saddle trajectories. Hyperbolicity of an attractor provides splitting of the tangent space on stretching and compressing subspaces. In a neighbourhood of hyperbolic attractor the dynamical system alongside with instability of trajectories finds out as well strong stochastic properties. Therefore all hyperbolic attractors are stochastic attractors.

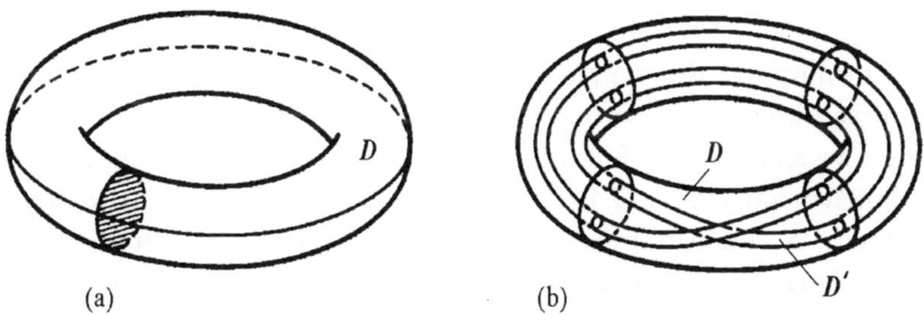

Fig. 1.13   Construction of Smale-Williams solenoid.

The property of hyperbolicity imposes even more strong requirements on an attractor, than its stochasticity. Therefore it is no wonder, that by present time only some exclusive modelling examples of existence of hyperbolic attractors of only discrete dynamical systems were constructed. In dissipative autonomous systems of nonlinear differential equations hyperbolic attractors were not found. An example of hyperbolic attractor is *Smale–*

Williams solenoid. The general design of construction of the solenoid can be found in [Nitecki (1971)]. Here we present only the scheme of construction. We shall consider a toroidal region D, i.e. an interior of two-dimensional torus in space of dimension not less than three (Fig. 1.13a).

Let us stretch it, then compress it along a meridian, then overwind and bend it so that it will pass in area D', lying in D (Fig. 1.13b). We shall apply analogical transformation to the region D' and we shall obtain a region D'', *etc.* As a result of infinite sequence of such transformations in a section of toroidal region by a vertical plane we shall obtain a hierarchy of structures, represented in Fig. 1.14. Thus initially closed points will diverge exponentially quickly, and the volume of initial region D will tend to zero. In a limit we obtain the invariant attractive set as attractor. It represents a line, infinite number of times crossing a secant plane and having in section a fractal structure of Cantor set.

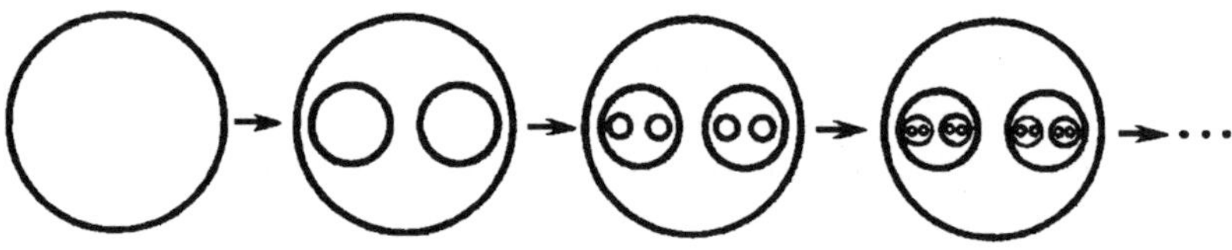

Fig. 1.14   Hierarchy of the structures arising at construction of Smale-Williams solenoid (in section).

Another important example of hyperbolic set though not being attractor, is a well-known *Smale horseshoe*. A statement of construction of a horseshoe we shall spend, following [Kuznetzov (2001)]. We shall consider a two-dimensional region G in the form of a stadium, consisting of three parts, square S and two halves of a circle D_1 and D_0. We shall compress this region horizontally more than twice, then we shall even more strongly stretch it on a vertical, then we shall bend it in the form of a horseshoe and we shall impose it on the initial region how it is shown in Fig. 1.15. The obtained mapping of two-dimensional region G in itself is the mapping F of Smale horseshoe.

We shall be interested in the set

$$\Lambda = \bigcap_{-\infty < n < \infty} F^n(S).$$

It is easy to see, that the set $F(S) \cap S$ consists of two vertical zones V_0 and

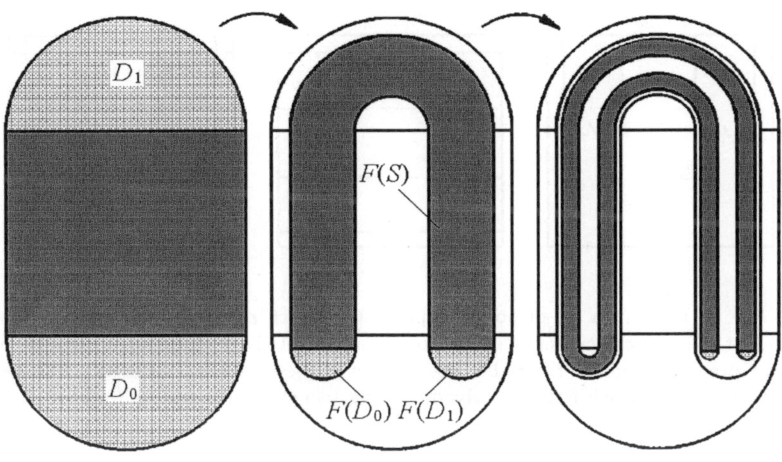

Fig. 1.15   Construction of Smale horseshoe.

V_1, corresponding of two halves of horseshoe and turning out as a result of horizontal compression of the region S. The set $F^{-1}(S) \cap S$ consists of two horizontal zones H_0 and H_1 which at a vertical stretching should be made equal to width of the region S. Hence, the set $(F^{-1}(S) \cap F(S)) \cap S$ consists of four squares which can be coded according to indexes of their horizontal and vertical zones. Up to a dividing point we put an index of a horizontal zone, and an index of a vertical zone after a point (Fig. 1.16).

For two steps in iterations of mappings F and F^{-1} we shall obtain a set $(F^{-2}(S) \cap F^{-1}(S) \cap F(S) \cap F^2(S)) \cap S$, consisting of sixteen squares, lying in intersection of four horizontal zones: H_{00}, H_{01}, located inside a zone H_0, and H_{10}, H_{11} located inside a zone H_1; and four vertical zones: V_{00}, V_{01}, located inside a zone V_0, and V_{10}, V_{11}, located inside a zone V_1. We code all squares, accordingly, in two figures up to a dividing point and two figures after it.

Continuing this process further, after an infinite number of iterations we shall obtain in a limit a Cantor lattice, the two-dimensional not anywhere dense closed set Λ, invariant concerning to the map F. Elements of set Λ are coded by infinite in both sides from a dividing point binary sequences.

Application of map F to them leads to a shift of a dividing point on one position to the right, and F^{-1} — to the left. Hence, the set Λ contains a countable set of cycles of a horseshoe map to which there correspond periodic binary sequences, and a continuum of different nonperiodic trajectories of the cascade F.

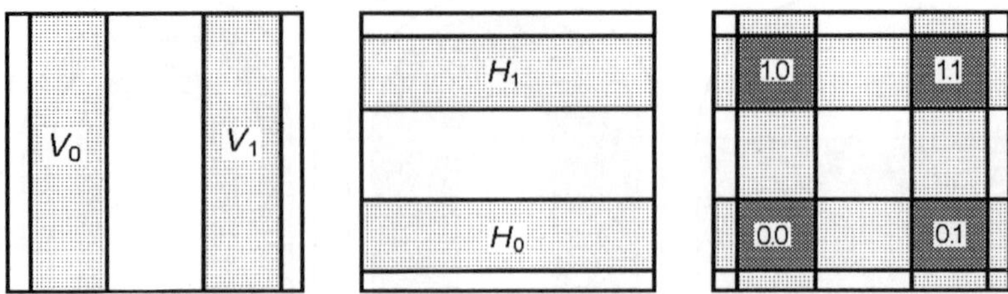

Fig. 1.16   At the left, a set of points is presented which remain in the region S on the subsequent step; in the center, a set of points is presented which belonged to the region S on the previous step; on the right, intersection of these sets is presented which defines a set of the points which have "survived" during two steps of iterative process.

It is proved in [Smale (1967); Nitecki (1971)] that the set Λ is a hyperbolic and, hence, a stochastic set. It is possible to show, that the Smale horseshoe is presented at Henon map, and also in Poincare map of a neighbourhood of a saddle–focus separatrix loop of system of three ordinary differential equations (see Sec. 2.4.1). But, as already it was noted above, the set Λ is not an attractor. Therefore, though the presence of a horseshoe allows to draw a conclusion on the complex nature of dynamics of considered system, but, unfortunately, does not solve a problem of a substantiation of presence in the system of irregular attractor.

Sometimes an attractor containing infinite everywhere dense set of unstable periodic trajectories is understood as chaotic attractor. In the literature the concept of *quasiattractor* [Neimark and Landa (1992)], containing simultaneously with unstable as well as stable periodic trajectories, having very small domains of attraction also is used. As we shall see below, these properties also are not typical for dissipative systems of ordinary differential equations possessing chaotic dynamics. Attractors of such systems, on the one hand, can not contain periodic trajectories, and on the other hand, can contain everywhere dense sets of more complex nature, than infinite sets of unstable periodic trajectories.

Thus, classification of irregular attractors of complex dynamical systems existing now only in exceptional cases describes separate features of irregular attractors of dissipative systems of ordinary differential equations, not giving the general representation about their nature, structure and principles of formation. In Chapter 4, the new theory and, in our opinion, more natural classification of the wide family of irregular attractors of dissipative systems of ordinary differential equations, based on various scenarios (cascades of bifurcations) of their appearance from stable limit sets (regular attractors) will be offered to attention of the reader. As the numerous examples considered in works from [Magnitskii and Sidorov (2001c)] to [Magnitskii and Sidorov (2005c)], and also the content of Chapter 3 show, that a wide class of irregular attractors of autonomous dissipative systems of ordinary differential equations, containing all classical chaotic attractors, are born as a result of the same cascades of soft bifurcations of regular attractors (cycles, tori). The beginning always is the cascade of the period doubling bifurcations of some original cycle, passing in complete or incomplete subharmonic cascade of bifurcations which then proceeds by complete or incomplete homoclinic cascade of bifurcations. As the name chaotic attractor is already occupied by the attractors with Lyapunov positive exponents, we shall name irregular attractors, born at all stages of all mentioned above cascades of bifurcations at points of accumulation of values of bifurcation parameter as *singular attractors*. Foundations and main results of the new theory of singular attractors will be presented in the following chapters.

1.4.4 *Dimension of attractors, fractals*

Regular attractors of dissipative dynamical systems as it was noted above, are smooth submanifolds of phase space and have the whole dimension. At the same time there exist attractors, such, for example, as attractor of Henon map or invariant set of the map of Smale horseshoe which do not possess even a simple continuity, but possess geometrical (scale) invariance, i.e. they are *fractals* [Mandelbrot (1982)]. The elementary example of a fractal is *the Cantor perfect set* which scheme of construction is presented in Fig. 1.17. We shall divide an interval $[0; 1]$ on three equal parts and we shall cut out average of them — an interval $(1/3; 2/3)$. We shall act with each of the remained parts in the same way, and this procedure we shall repeat an infinite number of times. At the first stage of construction we shall have two intervals in length $1/3$ everyone, at the second stage — four intervals in length $1/9$ everyone, *etc.*

On k-th stage we shall have 2^k intervals in length $(1/3)^k$ everyone not connected with each other. In a limit when $k \to \infty$, there will remain the set of points in the original interval named by Cantor set. This set is not dense anywhere in the original interval, i.e. does not contain any interval of as much as small length. But it is closed and is dense in itself, i.e. does not contain the isolated points, and, hence, is the perfect set. Moreover, the Cantor set has the continuum power, but the zero Borel measure. It is easy to verify the last statement, having counted up the sum of lengths of all cut out intervals:

$$L = 1/3 + 2/9 + 4/27 + \cdots = \frac{1}{3} \sum_{k=0}^{\infty} \left(\frac{2}{3}\right)^k = \frac{1/3}{1 - 2/3} = 1.$$

Direct generalization of the Cantor set on a plane and spatial cases are *a carpet* and *a cube of Sierpinski* [Schuster (1984)].

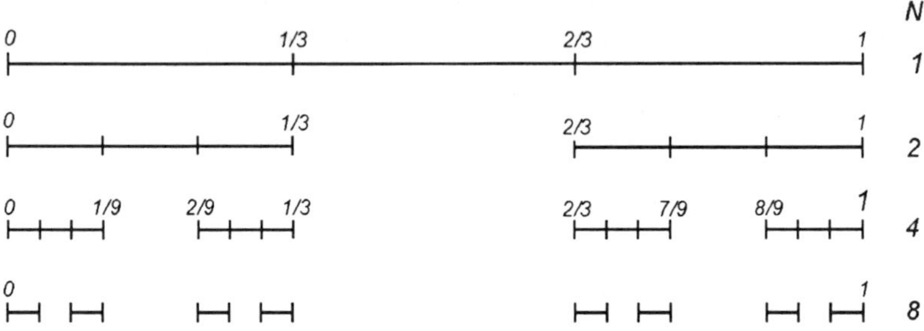

Fig. 1.17   Construction of the Cantor set.

It is easy to see, that if on each step we shall cut out not a third, but q-th part from the remained intervals, the set obtained as a result also will have a zero measure for any $0 < q < 1$. However, at various values q these sets, obviously, will essentially differ. Hence, for their measurement other quantitative characteristics are necessary which are different from a usual measure.

There are some definitions of dimension of the sets possessing fractal structure: Hausdorff dimension, capacity of a set, probabilistic dimension, entropic dimension, *etc.* [Anishchenko *et al.* (1999)]. We shall define as the most widespread dimension — capacity or fractal dimension of a set which most often is used as the quantitative characteristic of strange attractors of dissipative dynamical systems.

Let us consider in m-dimensional phase space M of a dynamical system some set B. Cover this set by m-dimensional cubes with side ε so that these cubes must contain all points of set B. Let $N(\varepsilon)$ be the minimal number of such cubes necessary for a covering B. If there is a limit

$$d_F(B) = \lim_{\varepsilon \to 0} \frac{\ln N(\varepsilon)}{\ln(1/\varepsilon)}, \tag{1.9}$$

then this limit $d_F(B)$ is called *capacity* or *fractal dimension* of the set B.

Let us show, that fractal dimension of regular attractors of dissipative dynamical systems is equal to an integer and it coincides with their usual dimension. Really, we shall designate through $d = 1$, 2, 3 usual dimensions of one-dimensional, two-dimensional and three-dimensional compact submanifolds of three-dimensional Euclidean space. As the quantity of cubes with the side ε, necessary for covering of a unit segment, is proportional to $1/\varepsilon$, for covering of a unit square to $1/\varepsilon^2$, and for covering of a unit volume to $1/\varepsilon^3$, then for everyone d the number of cubes necessary for a covering of compact submanifold of dimension d, is equal $N(\varepsilon) = C\varepsilon^{-d}$ where C is some constant. Substituting $N(\varepsilon)$ in the formula (1.9), we shall obtain, that $d_F = d$, i.e. fractal dimension of a compact submanifolds of Euclidean space coincides with their usual (*topological*) dimension. Hence, the fractal dimension of a stable limit cycle is equal to unit, and of a stable invariant torus — to two.

However, for irregular sets and strange attractors, possessing scale-invariant structures, fractal dimension has a fractional value. We shall show it on a considered above example of Cantor set B. We shall consider sequence $\varepsilon_k = (1/3)^k \to 0$ when $k \to \infty$. It follows from construction of Cantor set, that for any k (number of a stage of construction) the Cantor set becomes completely covered by $N(k) = 2^k$ segments in length $\varepsilon_k = (1/3)^k$ everyone. Then it follows from (1.9), that

$$d_F(B) = \lim_{k \to \infty} \frac{\ln N(k)}{\ln(1/\varepsilon_k)} = \lim_{k \to \infty} \frac{\ln 2^k}{\ln 3^k} = \frac{\ln 2}{\ln 3} \approx 0.631.$$

Clearly, that considered above strange attractors of two-dimensional discrete dissipative dynamical systems (cascades) have a fractional fractal dimension the value of which lies in an interval between the unit and the two, i.e. these sets already are not lines, but yet are not as well surfaces. We shall note, that the finding of value of fractal dimension is a difficult computing problem and can not be always realized successfully. As if to irregular attractors of dissipative systems of ordinary differential equations, they, as it was already noted above, not need be strange and to have not a whole fractal dimension (see Chapter 4).

Attractors of some chaotic (mixing) mappings can be not strange and can have no fractal structure. The classical example is *the modified Arnold "cat map"*

$$x_{n+1} = (x_n + y_n + \delta \cos 2\pi y_n) \bmod 1,$$
$$y_{n+1} = (x_n + 2y_n) \bmod 1. \tag{1.10}$$

The map (1.10) at $\delta < 1/2\pi$ one-to-one translates a unit square of a plane (x, y) in itself and is dissipative in the region of $1/2 < y < 1$, i.e. at each iteration the element of the area of this region is compressed. But, despite of compression of the area, the map (1.10) is ergodic and mixing. Therefore all unit square is an attractor of the map (1.10), and its fractal dimension is equal to two.

The points, being by consecutive iterations of the map (1.10), practically completely cover the unit square, but the density of their distribution is essentially non-uniform. A quantitative measure of this irregularity is the value of *information dimension d_I*, defined as follows

$$d_I(B) = \lim_{\varepsilon \to 0} \frac{I(\varepsilon)}{\ln(1/\varepsilon)}, \qquad I(\varepsilon) = -\sum_{i=1}^{N(\varepsilon)} P_i \ln P_i,$$

where $N(\varepsilon)$ is a quantity of m-dimensional cubes with the side ε, covering the set $B \subset M$, P_i is a probability of visiting of i-th cube by a trajectory of a system, and $I(\varepsilon)$ is the Shannon's entropy.

As a result of heterogeneity of density of distribution of probabilities of points in the unit square, information dimensions of attractors of the map (1.10) will be various at different δ and all of them will lie in an interval of $1 < d_I < 2$.

As it will follow from the results of Chapter 4, it is more natural to distinguish the entered above in consideration singular attractors of autonomous dissipative systems of ordinary differential equations by their information, instead of fractal dimensions.

Chapter 2

Bifurcations in Nonlinear Systems of Ordinary Differential Equations

The questions considered in Chapter 1 have been connected with behaviour of trajectories of a system of ordinary differential equations in a phase space. Thus the vector field F was fixed, and we simply studied its properties and the property of a system of differential equations set by it. We shall pass now to studying properties of a family of systems of differential equations in relation to perturbations of the vector field F.

2.1   Structural Stability and Bifurcations

2.1.1   *Structural stability*

The concept of structural stability of a vector field or a system of differential equations was offered in [Andronov and Pontryagin (1937)]. It demands definitions of perturbation of a vector field and topological equivalence of vector fields.

Definition 2.1   ([Guckenheimer and Holmes (1983)]) As *perturbation* of amplitude ε of a vector field $F(x) \in C^1$ we shall name any vector field $F_1(x) \in C^1$ for which there is a compact set K outside of which $F_1(x) \equiv F(x)$, and on K

$$\|F(x) - F_1(x)\| < \varepsilon \qquad \text{and} \qquad \|\partial(F(x) - F_1(x))/\partial x\| < \varepsilon.$$

Definition 2.2   Two systems of differential equations (or, that the same — two vector fields) are *topologically equivalent*, if there exists homeomorphism of phase space of one system on phase space of the second system, transforming the oriented trajectories of the one system in the oriented trajectories of the other system.

45

That is definition of global topological equivalence, fair in all phase space. It naturally generalizes the definition of local topological equivalence of vector fields (systems of the differential equations) in neighbourhoods of their singular points (see Sec. 1.2.1). We shall notice, that in case of topological equivalence of vector fields the direction of time along equivalent trajectories should be saved, but the time scale can change. So motions, for example, on equivalent cycles can occur, generally speaking, with different periods.

Definition 2.3 The smooth vector field $F(x)$ (or the differential Eq. (1.1)), given on a smooth compact manifold M, is called *structurally stable* if there exists $\varepsilon > 0$ such, that all perturbations $F_1(x)$ with amplitudes, smaller than ε, are topologically equivalent to $F(x)$.

Theorem 2.1 [Arnold (1978a); Andronov and Pontryagin (1937)] *Structurally stable vector fields form the opened and everywhere dense set in space of all continuously differentiable vector fields on two-dimensional compact manifold.*

Hence structurally stable vector fields fill the whole areas in space of all fields, and in any neighbourhood of any structurally unstable field it is possible to find the field, being structurally stable. Thus, structural stability is *typical property* or a case of the general position for two-dimensional vector fields. In this case it is possible to expect, that errors in definition of vector fields will not lead to change of a qualitative picture of solutions of systems of differential equations given by these fields.

Necessary and sufficient conditions of structural stability of two-dimensional autonomous systems of differential equations are given by the following theorem.

Theorem 2.2 [Bautin and Leontovich (1990)] *The system of differential Eq. (1.1) is structurally stable (global structurally stable) on two-dimensional compact manifold M in only case when:*

(1) the number of singular points is finite and all of them are hyperbolic;
(2) the number of limit cycles is finite and all of them are nondegenerate;
(3) in M there are no saddle connections, i.e. there are no separatrices, going from a saddle into saddle.

The performance of first two conditions of the theorem provides by virtue of the Grobman–Hartman theorem local structural stability of system of differential equations in neighbourhoods of hyperbolic singular points and nondegenerate limit cycles, and the additional condition of absence of saddle homoclinic and heteroclinic trajectories is already a sufficient condition in a two-dimensional case for global structural stability of system as a whole.

For systems of higher, than two, dimensions the conditions listed in the theorem with replacement of nondegenerate cycles on hyperbolic cycles also are necessary conditions of global structural stability as at default of first two conditions the system will not be not only global, but even local structurally stable, and any saddle connection collapses under small perturbation of a vector field. But, unfortunately, these conditions are not sufficient for global structural stability of systems of high (more than two) dimensions.

Moreover, for such systems the set of structurally stable vector fields is not everywhere dense in space of all continuously differentiable vector fields, i.e. in this space there are areas, free from structurally stable vector fields, and, hence, typical property is structural instability. The example of such area for fields on three-dimensional torus for the first time was constructed in [Smale (1966)]. Below it will be shown, that in space of three-dimensional autonomous systems of ordinary differential equations in any neighbourhoods of values of parameters at which systems possess homoclinic or heteroclinic trajectories, there can be both structurally stable, and structurally unstable systems. Therefore the typical nature of property in this case should be determined by a measure or dimension of a set of systems possessing this property.

2.1.2 *Bifurcations*

Structural stability of systems of differential equations is the stability in relation to any small smooth perturbations of vector fields. However the systems of differential equations which are derived from various applications, always contain some number of system parameters. Therefore from the point of view of applications the analysis of stability of systems of differential equations in relation to narrower class of perturbations, that is for any small perturbations of parameters of such systems, is more natural and interesting. In this case we shall interpret the very space of parameters as a finite-dimensional space of systems of differential equations of some special kind, and the perturbation of concrete system as some perturbation of its parameters.

The bifurcation theory of systems of differential equations, originating in A. Poincare works, describes qualitative, spasmodic changes of phase portraits of systems of differential equations at continuous, smooth changes of their parameters. Values of parameters at which there are these qualitative changes of phase portraits, are called *bifurcation values* or *points of bifurcations.*

Everywhere in Chapters 2 and 3 we shall consider a smooth family of many-dimensional autonomous systems of ordinary differential equations

$$\dot{x} = F(x, \mu), \quad x \in M \subset \mathbb{R}^m, \quad \mu \in L \subset \mathbb{R}^k, \quad F \in C^\infty, \qquad (2.1)$$

given in phase space M by smooth vector fields F, depending on coordinates of vectors of system parameters μ, lying in the region L of the space $\mathbb{R}^k$.

Definition 2.4   A vector field $F(x, \mu_0)$ of system (2.1) (or the differential Eq. (2.1)), is called as *rough* if there is a neighbourhood $U \subset L$ of a vector μ_0 such, that for all $\mu \in U$ vector fields $F(x, \mu)$ are topologically equivalent to a vector field $F(x, \mu_0)$.

According to the given definition by points of bifurcations are those and only those sets of values of parameters at which the system is *not rough* i.e. at which there is no continuous dependence of a phase portrait of a system on its parameters. We shall notice, that in some papers the concept of roughness is identified with concept of structural stability of a vector field and the differential equation corresponding to it [Palis and Melo (1982)]. We shall use the term of *roughness of a system* only for designation of its stability in relation to the narrower, but at the same time more important, family of perturbations of vector fields — to parametrical perturbations.

Let us assume, that in a space of parameters there are no areas filled by exclusively not rough systems of a kind (2.1). Then the full qualitative research of family of systems (2.1) is reduced to an establishment of splitting the space of parameters on areas with identical (rough) qualitative structure and to establishment of this qualitative structure. Splitting the space of parameters into rough areas and separating them $(k-1)$-dimensional *bifurcation films*, corresponding to not rough systems, is called *the bifurcation diagram.*

Bifurcation films are $(k-1)$-dimensional smooth surfaces in the space of parameters. They are defined by one condition $G_1(\mu) = 0$ with a gradient not equal to zero. Therefore it is claimed, that bifurcation, connected with picking of a film by a vector of bifurcation parameters, has *codimension* 1. A transversal intersection of two films is the smooth surface of dimension $k-2$, given by two conditions $G_1(\mu) = 0$ and $G_2(\mu) = 0$. Bifurcation, connected with transition of a vector of bifurcation parameters through such surface filled by not rough systems, has codimension 2. In general it is claimed, that bifurcation has codimension n if it is connected with transition of a vector of bifurcation parameters through a smooth surface of dimension $k - n$, being a transversal intersection of n smooth hypersurfaces given by n conditions and filled by not rough systems.

Thus, codimension of bifurcation shows, from how many parameters the system of differential equations should depend that the bifurcation was typical for it. If codimension is more, than corresponding bifurcation is more atypical for the system. In three-dimensional space of parameters which will represent for us the greatest interest, bifurcation of codimension 1 will occur on some smooth two-dimensional surface, codimension 2 — on a line, and codimension 3 — at a point.

Often, except for bifurcation diagrams of family of systems in a space of parameters, so-called *phase-parametrical diagrams* are used for clearness of representation. In this case some coordinate axes correspond to values of parameters, others correspond to the dynamical variables or some values connected with them. As a result we have the surface which points correspond to the certain dynamical regimes of family of the systems, varying at change of parameters.

As structurally stable vector fields fill areas in space of all fields, they fill areas and in finite-dimensional space of parameters. Hence, structurally stable fields are rough. From this follows, that candidates for points of bifurcations are first of all those values of parameters μ_0, at which the vector field $F(x, \mu_0)$ has nonhyperbolic singular points, nonhyperbolic cycles or separatrix contours.

As roughness as a special case of structural stability can be both local, and nonlocal (global), than bifurcations also can be local and nonlocal. Bifurcations of nonhyperbolic singular points, cycles and tori which lead to local qualitative change of a phase portrait of a system are called as *local bifurcations*. Bifurcations of separatrix contours, nonlocal changing a phase portrait of a system, are usually called as *nonlocal bifurcations*. Besides, in systems of dimension above two there are possible nonlocal bifurcations of various irregular attractors.

From the beginning we shall consider the most simple local bifurcations, having codimension 1. For the analysis of such bifurcations it will be enough to consider a family of systems of differential Eqs. (2.1), having one-dimensional space of parameters in which these bifurcations are dotty.

Investigation of nonlocal (global) bifurcations is more difficult. First, such bifurcations can have codimension, greater than 1. Secondly, that the most important, in the modern qualitative theory of differential equations and modern theory of dynamical systems methods of finding *bifurcation surfaces* (films) for nonlocal bifurcations in a space of parameters, and methods of definition of their codimension (number and a kind of conditions) are completely absent. Some original approaches developed by authors in this direction, will be presented in Sec. 2.4.3.

2.2 One-Parametrical Local Bifurcations

Let value $\mu = 0$ be bifurcation value of parameter μ, that is at this value the phase portrait of family of systems of differential Eqs. (2.1) qualitatively varies. The most interesting from the point of view of various applications are bifurcations of stable limit sets (attractors) as they lead to changes of the established regimes observabled in real experiments. It is accepted to share bifurcations of attractors into *soft* (*internal*) and *rigid* (*crises of attractors*). Soft bifurcations lead to topological changes of attractors, but do not lead to their disappearance. Rigid bifurcations lead to disappearance of attractors. In the present section one-parametrical local bifurcations of regular attractors will be considered.

So, let systems of the differential equations from the family (2.1) at all values of parameter μ, lying in some neighbourhood U of bifurcation value $\mu = 0$, have as solutions or singular points (equilibrium states, fixed or stationary points) $x_0(\mu)$, or limit cycles $x_0(t, \mu)$, or invariant two-dimensional tori, which are stable for all $\mu < 0$. Let us describe the main bifurcations which can occur in these cases in family (2.1) at transition of parameter through the value $\mu = 0$.

2.2.1 *Bifurcations of stable singular points*

A singular (stationary) point or equilibrium state of one-parametrical family (2.1) satisfies, obviously, a condition $F(x_0(\mu), \mu) = 0$. Therefore, linearizing family (2.1) in a neighbourhood of a singular point, we shall obtain a system of ordinary differential equations depending on parameter

$$\dot{y} = A(\mu)y + O(|y|^2), \quad A(\mu) = D_x F(x_0(\mu), \mu) = \frac{\partial F}{\partial x}(x_0(\mu), \mu), \qquad (2.2)$$

where $y(t) = x(t) - x_0(\mu)$. Vector $y = 0$ is the solution of system (2.2) for all $\mu \in U$.

Bifurcations, connected with loss of stability of a singular point of the family (2.1), can occur in transition of parameter through value at which the point is not hyperbolic. We shall consider two main cases, most often meeting in applications: one eigenvalue of a matrix $A(0)$ is equal to zero or two complex conjugated eigenvalues of a matrix $A(0)$ lie on an imaginary axis, and all other eigenvalues have negative real parts. It is possible to show, using the theorem on the central manifold, that bifurcations, leading to loss of stability of a singular point of family (2.1), are defined exclusively by those coordinates of system (2.2) which correspond to eigenvalues of a matrix $A(0)$, lying on an imaginary axis. The system of equations which have been written down in these coordinates, usually is named *a normal form of family* (2.1) in a neighbourhood of a singular point. The following four types of bifurcations of singular points are most widespread.

2.2.1.1 *Transcritical (exchange of stability) bifurcation*

Bifurcation has a normal form of $\dot{y} = \mu y + y^2$ (or $\dot{y} = \mu y - y^2$). It is easy to see, that two stationary solutions $y = 0$ and $y = -\mu$ ($y = \mu$) coexist together and exchange stability at transition of parameter through bifurcation value $\mu = 0$ (Fig. 2.1a). Bifurcation appears soft.

To obtain analogue of *transcritical bifurcation* for systems with phase space of greater dimension, it is necessary to add to the considered real eigenvalue of a matrix of linearization, responsible for bifurcation, corresponding number of eigenvalues with negative real parts and to interpret the transition between the obtained singular points. So analogue to transcritical bifurcation for two-dimensional systems of differential equations is the bifurcation at which the stable node becomes a saddle, and the saddle becomes a stable node. In a case of the space of dimension more than two, as a result of considered bifurcation the stable node or stable focus becomes a saddle–node or a saddle–focus, and a saddle–node or a saddle–focus becomes a stable node or focus.

2.2.1.2 *Saddle–node bifurcation*

This bifurcation has not so successful name which is connected exclusively with its interpretation in two-dimensional phase space. That is a birth of degenerate plane saddle–node. It is not necessary to confuse such saddle–node with a nondegenerate saddle–node in three-dimensional phase space about which there will be a talk in the description of a pitchfork type bifurcation.

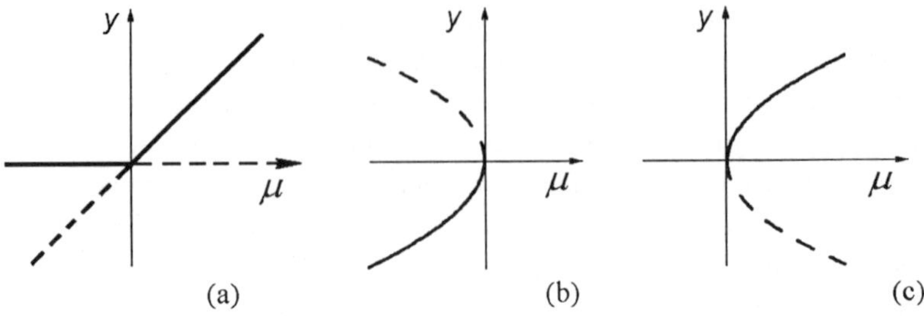

Fig. 2.1   Examples of bifurcations: transcritical (a), saddle–node type with simultaneous disappearance (b) and with birth (c) of stable and unstable singular points.

The normal form of *a saddle–node bifurcation* looks like $\dot{y} = \mu + y^2$. Last equation at $\mu < 0$ has two stationary solutions $y_{12} = \pm\sqrt{-\mu}$, one of which is asymptotically stable, and another is not stable. At $\mu = 0$ both solutions merge in one stationary solution $y = 0$, being asymptotically stable (unstable) for trajectories starting at the left (to the right) from zero. At $\mu > 0$ the equation has no singular points, and, hence, attractor disappears, i.e. considered bifurcation is crisis (Fig. 2.1b).

As it has been noted above, analogue of this bifurcation for two-dimensional systems of differential equations is a bifurcation at which available in system at $\mu < 0$ stable node (Fig. 2.2a) and a saddle merge at $\mu = 0$ into a degenerate singular point, *the degenerate saddle–node* (Fig. 2.2b), collapsing at $\mu > 0$ (Fig. 2.2c).

In case of dimension of space more than two as a result of considered bifurcation a stable node and a saddle–node (nondegenerate) merge at $\mu = 0$ into a degenerate singular point disappearing at $\mu > 0$.

Inverse saddle–node bifurcation has also great value. It has a normal form $\dot{y} = \mu - y^2$ at which there is a simultaneous birth of stable and unstable singular points or a stable node and a saddle (Fig. 2.1c and Fig. 2.2).

2.2.1.3 *Pitchfork type bifurcation*

There are two kinds of this bifurcation: *supercritical*, having a normal form $\dot{y} = \mu y - y^3$, and *subcritical*, having a normal form $\dot{y} = \mu y + y^3$. In case of supercritical bifurcation stationary solutions look like: $y = 0$ and $y_{2,3} = \pm\sqrt{\mu}$ (the last two solutions are defined only for $\mu > 0$). The stable stationary solution, becoming unstable, generates two other stable stationary solutions (Fig. 2.3a). Such bifurcation is also soft.

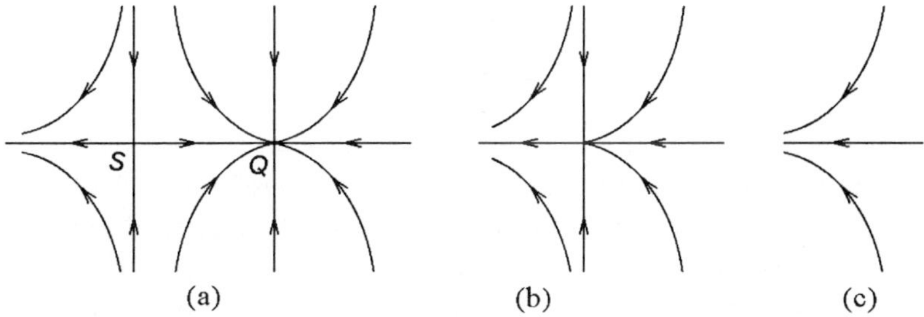

Fig. 2.2   The saddle–node bifurcation on a plane.

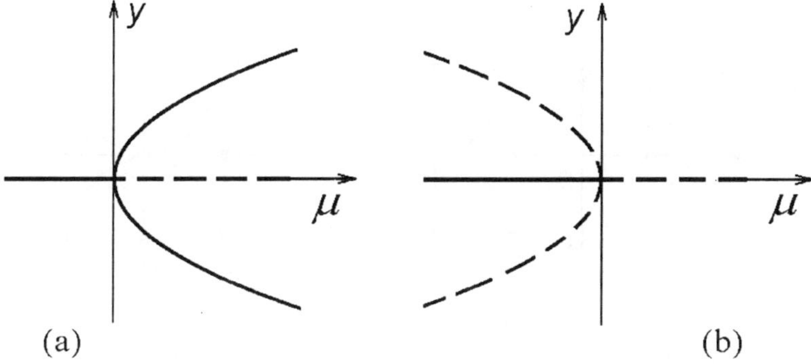

Fig. 2.3   Examples of pitchfork type bifurcations: (a) supercritical and (b) subcritical.

An analogue to this bifurcation for two-dimensional systems of differential equations is the bifurcation at which the stable node becomes a saddle, in which neighbourhood symmetrically two new stable nodes are born (Fig. 2.4).

In case of phase space of dimension more than two as a result of considered bifurcation the stable node becomes a saddle–node, and the stable singular points which have born in its neighbourhood can be not only nodes, but also focuses, i.e. they can have two complex conjugate eigenvalues with negative real parts alongside with one negative real eigenvalue of a matrix of linearization. For example, *pitchfork type bifurcation* takes place in well-known *Lorenz system* of three ordinary differential equations [Lorenz

(1963)]

$$\dot{x} = \sigma(y - x),$$
$$\dot{y} = x(r - z) - y,$$
$$\dot{z} = xy - bz,$$

(2.3)

at transition of parameter r through the value $r = 1$. The system (2.3) has an infinite spectrum of various bifurcations, therefore we shall repeatedly come back to this system in future. In case of subcritical bifurcation stationary solutions look like: $y = 0$ and $y_{1,2} = \pm\sqrt{-\mu}$ (the last two solutions are defined only for $\mu < 0$). Thus, the stable stationary solution (node) becomes an unstable saddle (a saddle–node, a saddle–focus), and together with it other two unstable stationary solutions (saddles, saddle–nodes, saddle–focuses) disappear (Fig. 2.3b). Such bifurcation is a crisis.

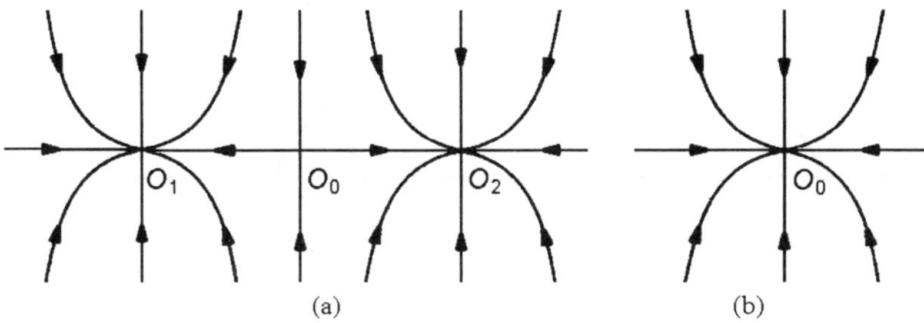

Fig. 2.4   Pitchfork type bifurcation on a plane: (a) — two stable nodes and a saddle after bifurcation; (b) — one stable node up to bifurcation.

2.2.1.4   *Andronov–Hopf bifurcation (cycle birth bifurcation)*

Three bifurcations considered above correspond to a case when at increasing the values of parameter μ exactly one real eigenvalue of a matrix $A(\mu)$ passes at $\mu = 0$ from left to right through an imaginary axis of a plane of complex variable along a real axis, and all other eigenvalues have negative real parts, i.e. remain in left half-plane. *Andronov–Hopf bifurcation* corresponds to a case when at increasing the values of parameter μ two complex conjugate eigenvalues of a matrix $A(\mu)$ pass at $\mu = 0$ from left to right through an imaginary axis of a plane of complex variable, and all other eigenvalues have negative real parts [Marsden and McCracken (1976)].

There are two kinds of Andronov–Hopf bifurcation: soft or *supercritical bifurcation* and rigid or *subcritical bifurcation*. Supercritical bifurcation has a normal form

$$\dot{y}_1 = -\nu y_2 + [\mu - (y_1^2 + y_2^2)]y_1,$$
$$\dot{y}_2 = \nu y_1 + [\mu - (y_1^2 + y_2^2)]y_2, \qquad (2.4)$$

that is equivalent to more compact record $\dot{y} = (\mu + i\nu)y - y|y|^2$ in terms of a complex variable $y = y_1 + iy_2$.

The solution of system (2.4) can be written down in the form of:

$$y_1(t) = u(t)\cos \nu t, \quad y_2(t) = u(t)\sin \nu t,$$

where function $u(t)$ satisfies the equation $\dot{u} = \mu u - u^3$. Therefore, as it is easy to see, at $\mu < 0$ the system has the unique stable stationary solution, focus $y_1 = y_2 = 0$ (or $u = 0$). At $\mu = 0$ the zero solution also is stable focus, as thus $u(t) = (c + 2t)^{-1/2} \to 0$, $t \to \infty$. At $\mu > 0$ besides an unstable focus, the stationary solution $y_1 = y_2 = 0$, there is also another solution $y_1(t) = \sqrt{\mu}\cos \nu t$, $y_2(t) = \sqrt{\mu}\sin \nu t$ in the system. For this solution $y_1^2 + y_2^2 = \mu$, whence follows, that its trajectory on a phase plane (y_1, y_2) is a limit cycle, a circle with radius $\sqrt{\mu}$ (Fig. 2.5a). Moreover, the limit cycle is stable, that follows from stability of a singular point $u = \sqrt{\mu}$ of the equation $\dot{u} = \mu u - u^3$.

Thus, as a result of soft or supercritical Andronov–Hopf bifurcation there is a change of stability of the stationary point, accompanied by a birth from it of a stable limit cycle which amplitude is proportional to $\sqrt{\mu}$, and the period is $T \approx 2\pi/\nu$ at $\mu \to 0$. This bifurcation plays an important role in the theory of nonlinear dynamical systems and as an independent mathematical object which is present practically in all classical systems with irregular dynamics, such as the Lorenz system (2.3), *the Rossler system* [Rossler (1976)]

$$\dot{x} = -(y + z),$$
$$\dot{y} = x + ay, \qquad (2.5)$$
$$\dot{z} = b + z(x - c),$$

the Chua system [Chua *et al.* (1986)]

$$\dot{x} = \alpha[y - h(x)],$$
$$\dot{y} = x - y + z, \qquad (2.6)$$
$$\dot{z} = -\beta y,$$

and many other systems [Akhromeeva *et al.* (1992); Anishchenko *et al.* (1999); Zhou and Chen (2005); Magnitskii and Sidorov (2003a)] and as a starting point of various cascades of bifurcations of transition to chaos (see Chapter 3).

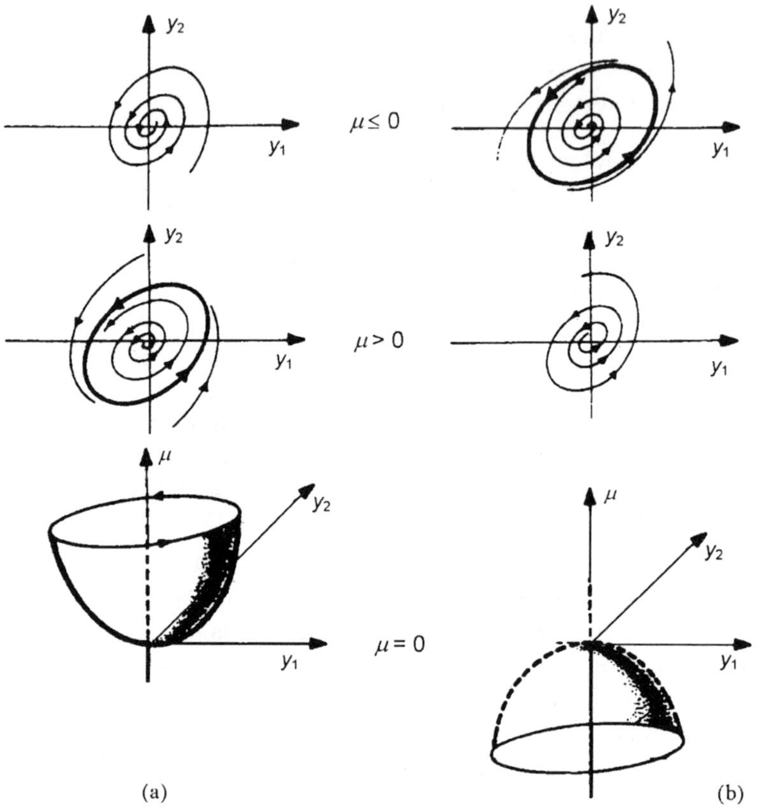

Fig. 2.5   Two kinds of Andronov–Hopf bifurcation on a plane: (a) — supercritical and (b) — subcritical.

Let us notice, that only conditions on eigenvalues of a matrix $A(0)$ do not provide a presence in system of soft Andronov–Hopf bifurcation. An additional sufficient condition is the presence in a system of a stable focus not only for $\mu < 0$, but also for $\mu = 0$, that is executed for the normal form of bifurcation considered above. If the singular point is the center at $\mu = 0$ under the same conditions on eigenvalues of matrix $A(0)$ there cannot be born a limit cycle at all, that takes place in the system

$$\dot{x} = (\mu + 1)x - 2xy + x^3,$$
$$\dot{y} = -2y + 2x^2,$$

or can occur a simultaneous change of stability both at a stationary point, and in a limit cycle surrounding it. Last bifurcation, for example, together with the Andronov–Hopf bifurcation, is a starting point of the scenario of transition to chaos in *the Magnitskii system* of equations

$$\dot{x} = bx[(1 - \sigma)z - \delta y],$$
$$\dot{y} = x[1 - (1 - \delta)y + \sigma z], \qquad (2.7)$$
$$\dot{z} = a(y - dx),$$

describing behaviour of macroindices of a market economy [Magnitskii (1991); Magnitskii and Sidorov (2005b)].

Subcritical (rigid) Andronov–Hopf bifurcation has a normal form

$$\dot{y}_1 = -\nu y_2 + [\mu + (y_1^2 + y_2^2)]y_1,$$
$$\dot{y}_2 = \nu y_1 + [\mu + (y_1^2 + y_2^2)]y_2.$$

The solution of system can be written down in the form of:

$$y_1(t) = u(t)\cos\nu t, \quad y_2(t) = u(t)\sin\nu t,$$

where function $u(t)$ satisfies the equation $\dot{u} = \mu u + u^3$. Therefore at $\mu < 0$ the system has the stable stationary solution, focus $y_1 = y_2 = 0$ or $(u = 0)$ and an unstable limit cycle: $y_1(t) = \sqrt{-\mu}\cos\nu t$, $y_2(t) = \sqrt{-\mu}\sin\nu t$. At $\mu \geq 0$ the limit cycle compresses into a zero solution which thus becomes an unstable focus. Thus there is a disappearance of attractor, and, hence, such bifurcation is a crisis (Fig. 2.5b).

Soft Andronov–Hopf bifurcation takes place also in general case of any dimension of phase space $m > 2$ and even in infinitely-dimension case [Hassard *et al.* (1981)]. In a case of rigid bifurcation it is necessary to speak about disappearance of a saddle limit cycle.

2.2.2 *Bifurcations of stable limit cycles*

In case of, when a limit cycle $x_0(t, \mu)$, having period T, is the solution of a family of Eqs. (2.1) at all $\mu \in U$, then, linearizing family (2.1) on a cycle, we shall obtain a system of non-autonomous ordinary differential equations depending on the parameter μ and with the periodic matrix of its linear part

$$\dot{y} = A(t,\mu)y + O(|y|^2), \quad A(t,\mu) = D_x F(x_0(t,\mu),\mu), \qquad (2.8)$$

where $y(t) = x(t) - x_0(t,\mu)$, and $A(t+T,\mu) = A(t,\mu)$. Thus the vector $y = 0$ is the solution of system (2.8) for all $\mu \in U$.

As it follows from the Floquet theory, each fundamental matrix solution of linear system with periodic coefficients

$$\dot{y} = A(t,\mu)y, \quad A(t+T,\mu) = A(t,\mu), \qquad (2.9)$$

can be represented in the form of $Y(t,\mu) = P(t,\mu)U(t,\mu)$ where $P(t,\mu)$ is some, generally speaking, complex T-periodic matrix, and a matrix $U(t,\mu) = \exp(B(\mu)t)$ is a fundamental matrix of the linear system of equations with constant, generally speaking, complex coefficients

$$\dot{u} = B(\mu)u(t). \qquad (2.10)$$

Transformation $P(t,\mu)$, thus, carries out a reduction of linear system (2.9) with periodic coefficients to linear system (2.10) with constant coefficients. As it has been noted in Sec. 1.3.2 stability (instability) of periodic solution is defined by eigenvalues of a matrix B, Floquet exponents of an original cycle or, that is equivalent, eigenvalues of the real matrix $C = \exp(BT)$, multipliers of the cycle. As the cycle is stable for all $\mu < 0$, one Floquet exponent is equal to zero, and all the others $m - 1$ Floquet exponents have negative real parts (one simple multiplier is equal to $+1$, and all other multipliers have modules, smaller than 1, that is they lie inside a unit circle of a plane of complex variable).

Hence, bifurcations, connected with the loss of stability of periodic solution of the family (2.1), can occur only when at $\mu = 0$ one or several Floquet exponents from $m - 1$ cross an imaginary axis from left to right or, that is equivalent, one or several multipliers lying inside a unit circle at $\mu < 0$, cross this circle at $\mu = 0$. Bifurcations of limit cycles can take place, obviously, only in systems of equations of dimension $m > 1$.

It is possible to show, using the theorem on central manifold, that, as well as in a case of a singular point, bifurcations leading to the loss of stability of periodic solution of family (2.1), are defined exclusively by those coordinates of system (2.10) which correspond to eigenvalues of a matrix $B(0)$, lying on an imaginary axis. The system of equations which was written down in these coordinates, will be named *a normal form of family* (2.1) in the neighbourhood of a periodic solution.

Definition of a normal form and, accordingly, a kind of occurring bifurcation of the periodic solution, is an extremely difficult task. By replacement of variables $y(t) = Q(t,\mu)z(t)$ with T-periodic matrix $Q(t,\mu)$ a solution of this task can be simplified by transition to the system of coordinates connected with a cycle. In such system one of coordinate vectors is the vector $\dot{x}_0(t)$, tangent to a cycle, another is a vector of a cycle $x_0(t)$. As the multiplier of a cycle corresponding to a vector $\dot{x}_0(t)$, is always equal to a unit, and the Floquet exponent is equal to zero and is not bifurcational, coordinates of a normal form obviously lie in a hyperplane S, transversal to a vector $\dot{x}_0(t)$, and given by the last $m-1$ components of a vector $z(t)$. We shall designate this vector as $v(t)$. It has one unit smaller dimension, than an initial vector $y(t)$ of system (2.9), linearized on a cycle. The vector $v(t)$ satisfies a system of $m-1$ linear differential equations

$$\dot{v}(t) = D(t,\mu)v(t), \tag{2.11}$$

where real matrix $D(t,\mu)$ is obtained from the real matrix

$$G(t,\mu) = Q^{-1}A(t,\mu)Q(t,\mu) - Q^{-1}(t,\mu)\dot{Q}(t,\mu)$$

by deleting of its first line and first column. The system (2.11) has the same Floquet exponents, except the zero exponent, as the original linearized system (2.9).

Two essentially different cases are possible: a case of a constant matrix $D(t,\mu) = D(\mu)$ and a case of a variable T-periodic matrix $D(t,\mu)$. The first case means, that matrices $Q(t,\mu) = P(t,\mu)$ and $G(t,\mu) = B(\mu)$ are real, and the Floquet exponents of the cycle, distinct from zero, are eigenvalues of the matrix $D(\mu)$. In other words, the first case means, that transition by transformation $Q(t,\mu)$ to the system of coordinates connected with a cycle, already carries out the reduction of system (2.9) with periodic coefficients to a system with constant real coefficients. Vector $z(t)$ in this case coincides with vector $u(t)$ in (2.10), and components of vector $v(t)$ are the last $m-1$ components of vector $u(t)$. In the second case a more complex situation takes place. We shall consider separately two cases described above.

In the case of a constant matrix $D(t,\mu) = D(\mu)$ a typical bifurcation of a cycle corresponds to transition through an imaginary axis of one real eigenvalue, or two complex conjugate eigenvalues of a matrix $D(\mu)$. Thus all other eigenvalues should have negative real parts. We shall consider four main, most often meeting in applications, bifurcations of cycles corresponding to this case.

2.2.2.1 *Bifurcation of birth of a pair of stable closed trajectories*

In this case at $\mu < 0$ in a system there is a stable cycle $x_0(t, \mu)$ for which one multiplier is equal to $+1$, and all other multipliers lie inside a unit circle. At transition of values of parameter μ through a point $\mu = 0$ one simple multiplier of the cycle passes through the point $+1$ of the unit circle that corresponds to crossing of an imaginary axis by one simple real Floquet exponent of linearized systems (2.9) or one real eigenvalue of a matrix $D(\mu)$. Therefore the normal form of such bifurcation of a cycle coincides with an one-dimensional normal form of pitchfork type bifurcation of a singular point: $\dot{u} = \mu u - u^3$. To a cycle, obviously, there corresponds the zero solution of this equation.

As a result of this bifurcation the cycle $x_0(t, \mu)$ loses stability (but does not disappear), and simultaneously near to it in the distance of $u_{1,2} = \pm\sqrt{\mu}$ in a direction of eigenvector corresponding to an exponent, passing through an imaginary axis, a pair of stable limit cycles is born. In Fig. 2.6 the born cycles lie in planes parallel to an original cycle as the vector u is directed vertically. Another case where the vector u lies in a plane of an original cycle is possible also. Then the born cycles will lie in the same plane, one inside an original cycle, and another outside it.

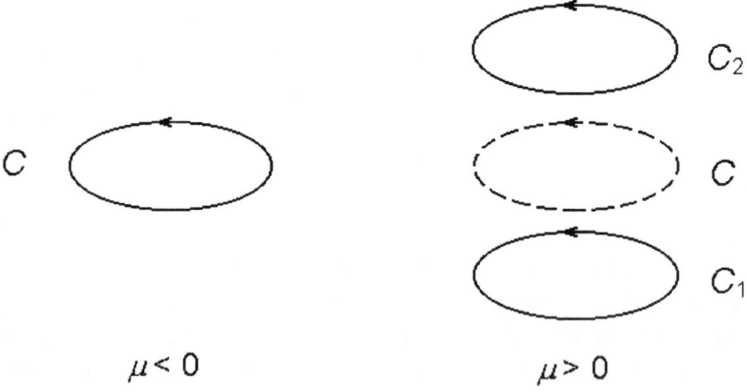

Fig. 2.6   The birth from a stable cycle (left) of two other stable cycles (right).

In a case of phase space of dimension $m > 2$, the cycle which has lost stability, becomes a saddle cycle. Bifurcation is soft.

Example 2.1   As an example we shall consider a system of three differential equations

$$
\begin{aligned}
\dot{x}_1 &= -\nu x_2 - x_2 x_3 + (\mu + 1)(1 - x_1^2 - x_2^2)x_1, \\
\dot{x}_2 &= \nu x_1 + x_1 x_3 + (\mu + 1)(1 - x_1^2 - x_2^2)x_2, \\
\dot{x}_3 &= \mu x_3 - x_3^3.
\end{aligned}
\tag{2.12}
$$

The system (2.12) has a singular point $O = (0,0,0)$. It is a stable solution for $\mu < -1$. For $\mu > -1$ from it as a result of Andronov–Hopf bifurcation the stable limit cycle $x_0(t) = (\cos \nu t, \sin \nu t, 0)^{\mathrm{T}}$, lying in a plane of variables (x_1, x_2), is born. We shall conduct the further research of system (2.12) by the method described above. The system linearized on a cycle will become:

$$
\begin{aligned}
\dot{y}_1 &= -2y_1(\mu + 1)\cos^2 \nu t - y_2(\nu + (\mu + 1)\sin 2\nu t) - y_3 \sin \nu t + f_1, \\
\dot{y}_2 &= y_1(\nu - (\mu + 1)\sin 2\nu t) - 2y_2(\mu + 1)\sin^2 \nu t + y_3 \cos \nu t + f_2, \\
\dot{y}_3 &= \mu y_3 - y_3^3,
\end{aligned}
$$

where expansion of functions $f_1(y_1, y_2)$ and $f_2(y_1, y_2)$ in a series at point $(0,0)$ begins with members of the second order. By replacement of variables $y(t) = Q(t)z(t)$ with $2\pi/\nu$-periodic matrix

$$
Q(t) = \big(\dot{x}_0(t),\ x_0(t),\ (0,\ 0,\ 1)^{\mathrm{T}}\big) = \begin{pmatrix} -\nu \sin \nu t & \cos \nu t & 0 \\ \nu \cos \nu t & \sin \nu t & 0 \\ 0 & 0 & 1 \end{pmatrix}
$$

we shall pass to the system which is written down in coordinates connected to a cycle:

$$
\begin{aligned}
\dot{z}_1 &= z_3/\nu + \widetilde{f}_1(z_1, z_2), \\
\dot{z}_2 &= -2(\mu + 1)z_2 + \widetilde{f}_2(z_1, z_2), \\
\dot{z}_3 &= \mu z_3 - z_3^3.
\end{aligned}
$$

Linear part of the last system has the real constant matrix $G(t, \mu) = G(\mu) = B(\mu)$ with a zero first column, i.e. the coordinate z_1 is uniquely defined by coordinates z_2 and z_3 of the system, lying in a plane S perpendicular to the plane of a cycle. Therefore stability of a cycle and its possible bifurcations are defined exclusively by a two-dimensional system of equations

$$
\begin{aligned}
\dot{u}_2 &= -2(\mu + 1)u_2 + \widetilde{f}_2(u_2, u_3), \\
\dot{u}_3 &= \mu u_3 - u_3^3,
\end{aligned}
\tag{2.13}
$$

written down in coordinates $u_2 = z_2$ and $u_3 = z_3$ with a diagonal matrix $D(\mu) = \mathrm{diag}(-2(\mu + 1), \mu)$. We shall notice, that diagonal elements of the matrix $D(\mu)$ are Floquet exponents of a limit cycle of original system (2.12).

As zero singular point of system (2.13) corresponds to a cycle of system (2.12), the cycle is stable in an interval of change of parameter $-1 < \mu < 0$. At $\mu > 0$ the zero singular point of system (2.13) and, accordingly, the cycle of system (2.12) become unstable. Two new stable singular points $u_3 = z_3 = y_3 = x_3 = \pm\sqrt{\mu}$ of system (2.13) are born, lying on a segment, perpendicular to the plane of a cycle. Simultaneously two stable cycles appear lying in planes parallel to the original cycle plane in a distance of $\pm\sqrt{\mu}$ from it.

Bifurcation of birth of two other stable cycles from one stable cycle forestalls *a homoclinic cascade of bifurcations* of birth of stable cycles in the Lorenz system (2.3) [Magnitskii and Sidorov (2001c)].

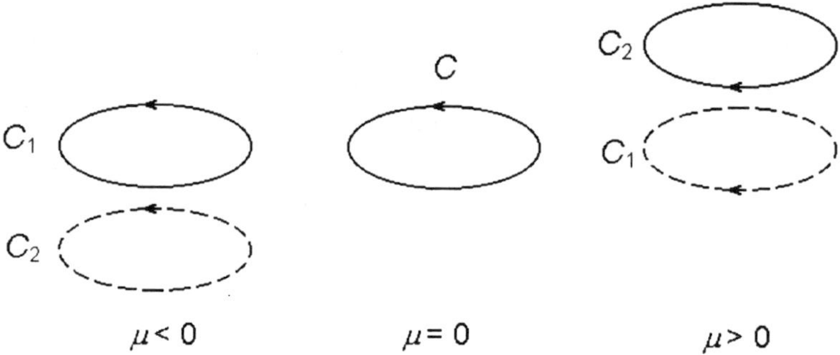

Fig. 2.7   Bifurcation of an exchange of stability between cycles.

2.2.2.2   *Transcritical (exchange of stability between cycles) bifurcation*

In this case at $\mu < 0$ together with a stable cycle $x_0(t, \mu)$ in the system (2.1) there exists as well an unstable (saddle) cycle $x_1(t, \mu)$, lying in a distance of $u = \mu$ from a stable cycle in a direction of eigenvector corresponding to the Floquet exponent of a stable cycle (to simple real eigenvalue of a matrix $D(\mu)$), passing through an imaginary axis. At transition of values of parameter μ through a point $\mu = 0$ one simple multiplier of the cycle corresponding to the indicated Floquet exponent, passes through the point

+1 of the unit circle. As cycles exchange their stability, the stable cycle $x_0(t,\mu)$ becomes unstable (saddle), and the unstable (saddle) cycle $x_1(t,\mu)$ becomes stable (Fig. 2.7).

The normal form of such bifurcation of cycle coincides with an one-dimensional normal form of transcritical bifurcation of a singular point: $\dot{u} = \mu u - u^2$ (in Fig. 2.7 direction of a vector u coincides with a vertical). Bifurcation is soft. If the direction of vector u coincides with horizontal, then both cycles should lie in one plane (one inside another). As a result of such bifurcation an internal, for example, stable cycle becomes an unstable cycle, and an external unstable cycle becomes a stable cycle.

Example 2.2   As an example we shall consider a system of three differential equations

$$
\begin{aligned}
\dot{x}_1 &= -\nu x_2 - x_2 x_3 + (\mu + 1)(1 - x_1^2 - x_2^2)x_1, \\
\dot{x}_2 &= \nu x_1 + x_1 x_3 + (\mu + 1)(1 - x_1^2 - x_2^2)x_2, \\
\dot{x}_3 &= \mu x_3 - x_3^2,
\end{aligned}
\tag{2.14}
$$

differing from system (2.12) only by the third equation. At $-1 < \mu < 0$ the system (2.14) except for a stable limit cycle $x_0(t) = (\cos \nu t,\ \sin \nu t,\ 0)^{\mathrm{T}}$, lying in a plane of variables (x_1, x_2), has also an unstable limit cycle $x_1(t) = (\cos \nu t, \sin \nu t, \mu)^{\mathrm{T}}$, lying in a distance of μ from it along an axis x_3. At $\mu > 0$ there is *an exchange of stability between cycles*. We shall notice, that the analysis of stability of the cycle of system (2.14) is reduced by the method described in the previous section to the analysis of stability of a zero singular point of two-dimensional system

$$
\begin{aligned}
\dot{u}_2 &= -2(\mu + 1)u_2 + \widetilde{f}_2(u_2, u_3), \\
\dot{u}_3 &= \mu u_3 - u_3^2
\end{aligned}
$$

with a diagonal matrix of its linear part (Floquet exponents of an original cycle $x_0(t)$) $D(\mu) = \mathrm{diag}(-2(\mu + 1), \mu)$.

2.2.2.3   *Bifurcation of disappearance (appearance) of a pair of closed trajectories*

In this case at $\mu < 0$ together with a stable cycle $x_0(t,\mu)$ in system (2.1) there exists as well unstable saddle cycle $x_1(t,\mu)$. At increasing of values μ these cycles approach each other and at $\mu = 0$ merge in one degenerate cycle. At $\mu > 0$ both of cycles disappear (Fig. 2.8). Therefore the normal form of such bifurcation of a cycle coincides with a one-dimensional normal form of a saddle–node bifurcation of a singular point: $\dot{u} = \mu + u^2$ (in Fig. 2.8 direction of a vector u coincides with vertical). Bifurcation is rigid (crisis). As well as in cases of previous bifurcations cycles can lie in one plane one inside another.

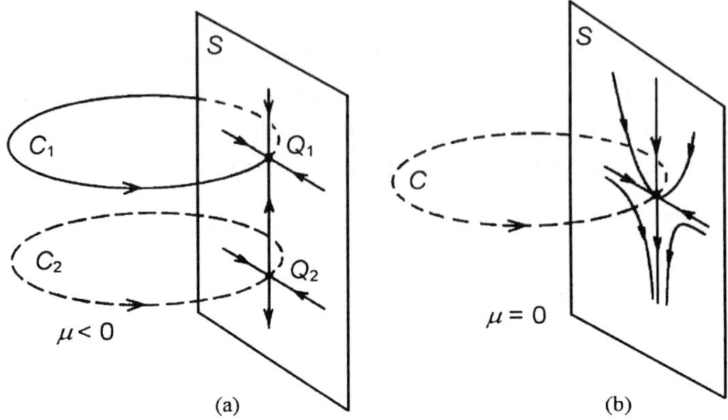

Fig. 2.8   The saddle–node bifurcation of a limit cycle.

Example 2.3   As an example we shall consider a system of three differential equations

$$\dot{x}_1 = -\nu x_2 - x_2 x_3 + (\mu + 1)(1 - x_1^2 - x_2^2)x_1,$$
$$\dot{x}_2 = \nu x_1 + x_1 x_3 + (\mu + 1)(1 - x_1^2 - x_2^2)x_2, \qquad (2.15)$$
$$\dot{x}_3 = \mu + x_3^2,$$

differing only by the third equation from systems (2.12) and (2.14). At $-1 < \mu < 0$ the system (2.15) except for a stable limit cycle $x_0(t) = (\cos(\nu - \sqrt{-\mu})t, \sin(\nu - \sqrt{-\mu})t, -\sqrt{-\mu})^{\mathrm{T}}$, lying in a plane $x_3 = -\sqrt{-\mu}$, has as well an unstable limit cycle $x_1(t) = (\cos(\nu + \sqrt{-\mu})t, \sin(\nu + \sqrt{-\mu})t, \sqrt{-\mu})^{\mathrm{T}}$, lying at distance $2\sqrt{-\mu}$ from it in plane $x_3 = \sqrt{-\mu}$. We shall notice, that movement on cycles occurs with different frequencies. At $\mu = 0$ both of cycles merge, forming an unstable cycle lying in plane $x_3 = 0$ and disappearing at $\mu > 0$. The analysis of stability of a cycle of system (2.15) also is reduced to the analysis of stability of a zero singular point of two-dimensional system

$$\dot{u}_2 = -2(\mu + 1)u_2 + \widetilde{f}_2(u_2, u_3),$$
$$\dot{u}_3 = \mu + u_3^2. \qquad (2.16)$$

If we consider the process of change of parameter μ from greater values to smaller values, then described bifurcation means, that at $\mu = 0$ suddenly there arises closed trajectory (cycle) which at the further reduction of parameter is split at two closed trajectories, one of which is stable, and another is unstable. In other words, there is a sudden simultaneous appearance of stable and saddle limit cycles. The cascade of bifurcations of such type, generating infinite sequence of various pairs of stable and saddle *homoclinic cycles*, is observed in the Lorenz system (2.3) [Magnitskii and Sidorov (2001c); Magnitskii and Sidorov (2005a)].

2.2.2.4   *Bifurcation of birth (destruction) of two-dimensional torus*

This bifurcation can occur only in phase space of dimension $m > 2$. It is connected with transition at $\mu = 0$ of two complex conjugate multipliers of an original limit cycle $x_0(t, \mu)$, having period T, through points $\exp(\pm i\varphi)$, $0 < \varphi < \pi$, of a unit circle or, that is equivalent, with transition of two complex conjugate Floquet exponents of a cycle (eigenvalues of a matrix $D(\mu)$ of system (2.11)) through points $\pm i\varphi/T$ of an imaginary axis of a plane of complex variable. This bifurcation is a special case of Andronov–Hopf bifurcation in a hyperplane S, transversal to an original cycle, and has a normal form

$$\dot{u}_1 = (\mu - u_1^2 - u_2^2)u_1 - \frac{\varphi}{T}u_2,$$
$$\dot{u}_2 = \frac{\varphi}{T}u_1 + (\mu - u_1^2 - u_2^2)u_2.$$

Thus the cycle $x_0(t, \mu)$ loses stability (but does not disappear), and simultaneously there appears a new stable motion on a two-dimensional torus T^2, given by the basic frequency of original cycle $\omega_0 = 2\pi/T$ and frequency $\omega_1 = \varphi/T = \omega_0\varphi/2\pi$ of a cycle $u(t) = \sqrt{\mu}\exp(i\varphi/T)$ born as a result of this bifurcation (Fig. 2.9).

The number equal to the relation between frequencies $\theta = \omega_1/\omega_0 = \varphi/2\pi$, is called *the number of rotation on torus T^2*. If the number of rotation is rational, i.e. $\theta = p/q$, where q and $p < q/2$ are the integer positive numbers, then the born stable motion on torus will be periodic. If the number of rotation is irrational, the phase curve of the born stable motion forms an everywhere dense winding of torus (see Sec. 1.3.4).

At further increasing of values of parameter μ the phase portrait of system will qualitatively vary, showing transitions from ergodic motions on a torus to regimes of resonances. Born torus is stable in the sense that any trajectory of system tend to some trajectory located on torus with some asymptotic phase.

Example 2.4   As an example we shall consider a system of three differential equations

$$\dot{x}_1 = -\nu x_2 - x_2 x_3 - 2x_1 x_3 - \mu(1 - x_1^2 - x_2^2)x_1,$$
$$\dot{x}_2 = \nu x_1 + x_1 x_3 - 2x_2 x_3 - \mu(1 - x_1^2 - x_2^2)x_2, \qquad (2.17)$$
$$\dot{x}_3 = 2\mu x_3 - x_3^3 + x_1^2 + x_2^2 - 1.$$

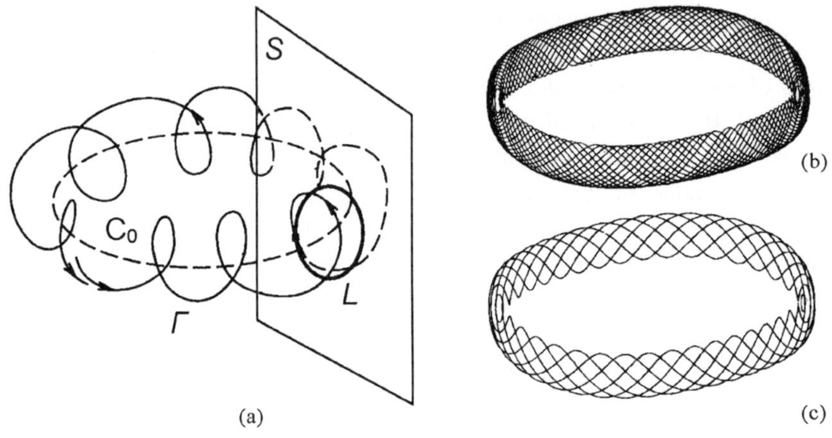

Fig. 2.9   Bifurcation of birth of two-dimensional torus from a limit cycle C_0: (a) trajectory on a torus in a neighbourhood of the cycle C_0 which has lost its stability ; (b) an ergodic torus; (c) resonance on a torus.

The system (2.17) has a limit cycle $x_0(t) = (\cos \nu t, \sin \nu t, 0)^{\mathrm{T}}$, lying in plane of variables (x_1, x_2). Linearized on a cycle system looks like:

$$\dot{y}_1 = 2y_1 \mu \cos^2 \nu t - y_2(\nu - \mu \sin 2\nu t) - y_3(\sin \nu t + 2\cos \nu t) + f_1,$$
$$\dot{y}_2 = y_1(\nu + \mu \sin 2\nu t) + 2y_2 \mu \sin^2 \nu t + y_3(\cos \nu t - 2\sin \nu t) + f_2, \qquad (2.18)$$
$$\dot{y}_3 = 2y_1 \cos \nu t + 2y_2 \sin \nu t + 2\mu y_3 - y_3^3,$$

where expansions of functions $f_1(y_1, y_2)$ and $f_2(y_1, y_2)$ in a series at point $(0,0)$ begin with members of the second order. By replacement $y(t) = Q(t)z(t)$ with $2\pi/\nu$-periodic matrix $Q(t)$, described in Sec. 2.2.2.1, we shall reduce the system (2.18) to the coordinates which are connected with cycle

$$\dot{z}_1 = z_3/\nu + \widetilde{f}_1(z_1, z_2),$$
$$\dot{z}_2 = 2\mu z_2 - 2z_3 + \widetilde{f}_2(z_1, z_2),$$
$$\dot{z}_3 = 2z_2 + 2\mu z_3 - z_3^3.$$

The analysis of stability of cycle is reduced to the analysis of stability of a zero singular point of two-dimensional system

$$\dot{u}_2 = 2\mu u_2 - 2u_3 + g(u_2, u_3),$$
$$\dot{u}_3 = 2u_2 + 2\mu u_3 - u_3^3, \tag{2.19}$$

Matrix $D(\mu)$ of the linear part of last system has complex conjugate eigenvalues $2\mu \pm 2i$. Hence, at $\mu < 0$ the zero singular point of system (2.19) and the cycle $x_0(t)$ of system (2.17) are stable. As a result of Andronov–Hopf bifurcation the stable limit cycle of the period $T \approx \pi$ is born at $\mu > 0$ in system (2.19), that corresponds to birth of a stable two-dimensional torus in the system (2.17) with frequencies approximately equal to ν and 2. Clear transitions from resonance to nonresonance tori can be observed in the system (2.17) at small $\mu > 0$ and at variation of the parameter ν.

Bifurcation of birth of stable torus from a cycle can be observed also at some values of parameters in a complex system of the Lorenz equations [Magnitskii and Sidorov (2002)]

$$\dot{X} = \sigma(Y - X),$$
$$\dot{Y} = X(r - Z) - Y, \tag{2.20}$$
$$\dot{Z} = (X^*Y + XY^*)/2 - bZ.$$

In system (2.20) variables X, Y and parameter $r = r_1 + ir_2$ are complex. Hence, the system (2.20) is a five-dimensional system of ordinary differential equations which at $r_2 = 0$ is reduced to the classical three-dimensional system of the Lorenz Eqs. (2.3).

The considered bifurcation of birth of stable torus T^2 is a supercritical (soft) bifurcation. Together with it, as well as in the case of Andronov–Hopf bifurcation, a subcritical (rigid) bifurcation can also occur. In this case at $\mu < 0$ the system has a stable cycle lying inside an unstable torus. At $\mu \geq 0$ the torus compresses to a stable cycle which after that becomes unstable. That is a disappearance of attractor, and, hence, such bifurcation is a crisis.

The case of variable T-periodic matrix $D(t, \mu)$ means, that transition by transformation $Q(t, \mu)$ to the system of coordinates connected with a cycle, carries out a reduction of the system (2.8) with periodic coefficients to a system (2.11) of smaller dimension, but having also T-periodic real coefficients. In this case, as it follows from the Floquet theorem, the fundamental matrix solution of linear system (2.11) can be represented in the form of $V(t, \mu) = R(t, \mu) \exp(E(\mu)t)$, where $R(t, \mu)$ is some T- periodic complex matrix, and $E(\mu)$ is some constant complex matrix, which eigenvalues are the Floquet exponents of an original cycle. We shall consider the most interesting and important from the point of view of various applications case of period doubling bifurcation of the cycle to which there corresponds the transition of one complex eigenvalue of matrix $E(\mu)$ through an imaginary axis at $\mu = 0$. All other eigenvalues should have negative real parts at that.

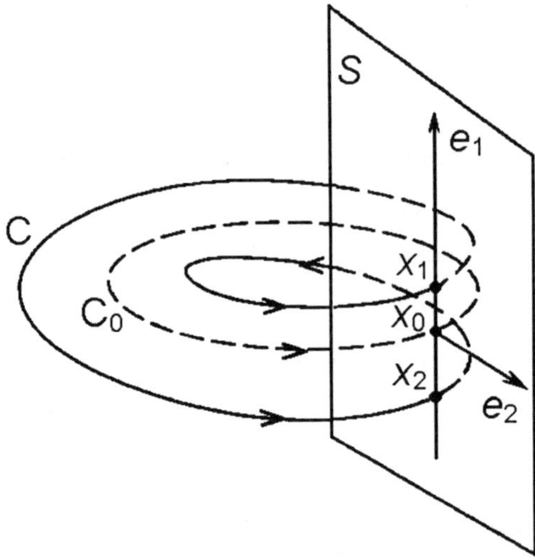

Fig. 2.10 Supercritical period doubling bifurcation of a cycle: C_0 is an original cycle; C is a cycle of the double period after bifurcation.

2.2.2.5 *Period doubling bifurcation of a cycle*

This bifurcation can occur only in a phase space of dimension $m > 2$. It is connected with transition through an imaginary axis from left to right of one complex Floquet exponent $\alpha(\mu)$ of the original cycle $x_0(t, \mu)$ at $\mu = 0$.

At that the corresponding real multiplier of the cycle at $\mu = 0$ looks like

$$\lambda = \exp(\alpha(0)T) = \exp(i\pi) = -1.$$

But on the other hand, that too follows from the Floquet theorem, the fundamental matrix solution of the linear system (2.9) is represented in the form of product of some $2T$-periodic real matrix on a real matrix $\exp(H(\mu)t)$, one eigenvalue of which leaves a unit circle through the point $+1$ at $\mu = 0$. Hence, transition of one complex Floquet exponent of a cycle $x_0(t, \mu)$, having the period T, through an imaginary axis that is equivalent to transition of the multiplier of a cycle through the point -1 of a unit circle, means, that a cycle $x_0(t, \mu)$ loses stability (but does not disappear), and simultaneously there arises another stable cycle having the same amplitude and the double period $2T$ (Fig. 2.10).

Example 2.5 As an example we shall consider a system of three differential equations

$$\begin{aligned}
\dot{x}_1 &= -\nu x_2 - \nu x_1 x_3/2 - ((\mu - 1)x_1 + 1)(1 - x_1^2 - x_2^2), \\
\dot{x}_2 &= \nu x_1 + 2x_3(1 - \nu x_2/4) - (\mu - 1)x_2(1 - x_1^2 - x_2^2), \qquad (2.21) \\
\dot{x}_3 &= 2(\mu - 1 - x_1)x_3 + (x_2 + \nu/4)(x_1^2 + x_2^2 - 1).
\end{aligned}$$

The system (2.21) has a limit cycle $x_0(t) = (\cos\nu t, \sin\nu t, 0)^{\mathrm{T}}$, lying in a plane of variables (x_1, x_2). The system linearized on a cycle looks like:

$$\begin{aligned}
\dot{y}_1 &= 2y_1((\mu - 1)\cos\nu t + 1)\cos\nu t \\
&\quad + y_2(((\mu - 1)\cos\nu t + 1)2\sin\nu t - \nu) - y_3\frac{\nu}{2}\cos\nu t + f_1(y_1, y_2, y_3), \\
\dot{y}_2 &= y_1(\nu + (\mu - 1)\sin 2\nu t) \\
&\quad + 2y_2(\mu - 1)\sin^2\nu t + 2y_3(1 - \frac{\nu}{4}\sin\nu t) + f_2(y_1, y_2, y_3), \\
\dot{y}_3 &= 2y_1\left(\frac{\nu}{4} + \sin\nu t\right)\cos\nu t \\
&\quad + 2y_2\left(\frac{\nu}{4} + \sin\nu t\right)\sin\nu t + 2y_3(\mu - 1 - \cos\nu t) + f_3(y_1, y_2, y_3),
\end{aligned}$$

where expansions of functions $f_1(y_1, y_2, y_3)$, $f_2(y_1, y_2, y_3)$ and $f_3(y_1, y_2, y_3)$ in series at point $(0, 0, 0)$ begin with members of the second order. By replacement $y(t) = Q(t)z(t)$ with $2\pi/\nu$-periodic matrix $Q(t)$, described in Sec. 2.2.2.1, we shall present the last system in coordinates connected with the cycle

$$
\dot{z}_1 = -\frac{2}{\nu}z_2 \sin \nu t + \frac{2}{\nu}z_3 \cos \nu t + \tilde{f}_1(z_1, z_2, z_3),
$$

$$
\dot{z}_2 = 2(\mu - 1 + \cos \nu t)z_2 + (2\sin \nu t - \frac{\nu}{2})z_3 + \tilde{f}_2(z_1, z_2, z_3),
$$

$$
\dot{z}_3 = (2\sin \nu t + \frac{\nu}{2})z_2 + 2(\mu - 1 - \cos \nu t))z_3 + f_3(z_1, z_2, z_3).
$$

The analysis of stability of the cycle is reduced to the analysis of stability of the zero solution of the linear two-dimensional system

$$
\dot{u}_2 = 2(\mu - 1 + \cos \nu t)u_2 + (2\sin \nu t - \frac{\nu}{2})u_3,
$$
$$
\dot{u}_3 = (2\sin \nu t + \frac{\nu}{2})u_2 + 2(\mu - 1 - \cos \nu t)u_3.
$$

$$(2.22)$$

Matrix $D(t, \mu)$ of coefficients of the system (2.22) is $2\pi/\nu$-periodic matrix. It is possible to be convinced by direct substitution that the system (2.22) has a fundamental matrix of solutions of a kind

$$
\begin{pmatrix} \cos \dfrac{\nu t}{2} & -\sin \dfrac{\nu t}{2} \\ \sin \dfrac{\nu t}{2} & \cos \dfrac{\nu t}{2} \end{pmatrix} \cdot \begin{pmatrix} \exp(2\mu t) & 0 \\ 0 & \exp(2(\mu - 2)t) \end{pmatrix}.
$$

$$(2.23)$$

Hence, at $\mu < 0$ the zero solution of the system (2.22) and, accordingly, a cycle $x_0(t)$ of the system (2.21) are stable. At $\mu > 0$ in system (2.21) there appears a stable solution with frequency $\nu/2$ or with double period $4\pi/\nu$. Expression (2.23) is representing of the solution of linear system with $2\pi/\nu$-periodic matrix $D(t, \mu)$ in the form of a product of $4\pi/\nu$-periodic real matrix on a real matrix $\exp(H(\mu)t)$, one eigenvalue of which leaves a unit circle at $\mu = 0$ through the point $+1$. To turn to the Floquet representation, we shall write down (2.23) in the form of a product of $2\pi/\nu$-periodic complex matrix on a complex matrix $\exp(E(\mu)t)$

$$
\begin{pmatrix} \dfrac{1 + e^{-i\nu t}}{2} & -\dfrac{1 - e^{-i\nu t}}{2i} \\ \dfrac{1 - e^{-i\nu t}}{2i} & \dfrac{1 + e^{-i\nu t}}{2} \end{pmatrix} \cdot \begin{pmatrix} e^{(2\mu + \frac{i\nu}{2})t} & 0 \\ 0 & e^{(2\mu - 4 + \frac{i\nu}{2})t} \end{pmatrix}.
$$

Diagonal elements of the matrix $E(\mu)$: $i\nu/2 + 2\mu$ and $i\nu/2 + 2(\mu - 2)$ are Floquet exponents of the cycle $x_0(t)$ of system (2.21). At $\mu = 0$ the first of them passes from left to right through an imaginary axis, and the second remains in the left half-plane. Multiplier corresponding to the first Floquet exponent is equal to

$$\lambda = \exp\!\left(\left(2\mu + \frac{i\nu}{2}\right)\frac{2\pi}{\nu}\right) = \exp\!\left(\frac{4\pi\mu}{\nu} + i\pi\right).$$

At $\mu = 0$ it, obviously, crosses boundary of a unit circle at point -1. The second multiplier is equal to $-\exp(-8\pi/\nu)$ and remains lying on a real axis inside a unit circle.

The period doubling bifurcation of a cycle plays a basic role during formation of chaotic attractors of nonlinear dissipative systems of ordinary differential equations. It begins the infinite *cascade of period doubling bifurcations*, discovered for the first time by M. Feigenbaum for one-dimensional mappings. This cascade leads to occurrence of the simplest irregular attractor, a Feigenbaum attractor. The period doubling bifurcation can be observed as well in all other more complex cascades of bifurcations, leading to occurrence of more complex irregular attractors, such, for example, as Lorenz, Rossler, Chua and Magnitskii attractors. It is found out as well in a large number of nonlinear dynamical systems described not only by ordinary differential equations, but also partial differential equations and differential equations with delay arguments (see in detail in Chapters 3–5).

2.2.3 *Bifurcations of stable two-dimensional tori*

Definition of a normal form and, accordingly, a kind of bifurcations occurring on torus, is even more a difficult task, than a similar problem for stable limit cycles. While for the analysis of bifurcations of limit cycles there is let not quite constructive, but a fundamental basis in the form of the theory of Floquet exponents, any theory of bifurcations of stable two-dimensional tori is practically absent now. We shall consider one of possible approaches to construction of such theory.

Let us assume, that stable two-dimensional torus T^2 of the family of systems (2.1) were born as a result of Andronov–Hopf bifurcation of stable T-periodic limit cycle $x_0(t)$ considered in the previous sections. It means, that linearized on a cycle, the system (2.8) by replacement of variables $y(t) = Q(t,\mu)z(t)$ with T-periodic matrix $Q(t,\mu)$ can be reduced to a system of $m-1$ order, written down in the system of coordinates connected with a cycle. In such system, we shall remind, one of coordinate vectors is the vector $\dot{x}_0(t)$, tangent to a cycle, another vector is a vector of the cycle $x_0(t)$. As the multiplier of the cycle, corresponding to a vector $\dot{x}_0(t)$, is always equal to unit, and the Floquet exponent is equal to zero and is not bifurcational, then coordinates of a normal form of formed two-dimensional torus obviously lie in a hyperplane S, transversal to a vector $\dot{x}_0(t)$ and given by the last $m-1$ components of a vector $z(t)$. This vector which we shall designate through $v(t)$, satisfies the system of $m-1$ differential equations

$$\dot{v}(t) = D(t,\mu)v(t) + P(v,\mu), \tag{2.24}$$

where expansion of a vector-function $P(v,\mu)$ in a series on degrees of v in zero begins with members of the second order, and T periodic matrix $D(t,\mu)$ is obtained from a matrix

$$G(t,\mu) = Q^{-1}(t,\mu)A(t,\mu)Q(t,\mu) - Q^{-1}(t,\mu)\dot{Q}(t,\mu)$$

by deletion of its first row and first column.

The birth of stable two-dimensional torus T^2 in system (2.1) follows from the birth of a stable cycle $v_0(t)$ in the system (2.24) as a result of Andronov–Hopf bifurcation (see Sec. 2.2.2 and an example in it). Considering now the system (2.24) as a new original system of type (2.1), having one unit smaller dimension and possessing a stable limit cycle, we can apply to it the theory of Floquet exponents, stated in Sec. 2.2.2. Clearly, that bifurcations of a cycle $v_0(t)$ of system (2.24) lead to corresponding bifurcations of two-dimensional torus T^2 of system (2.1). Thus, the following five types of bifurcations of stable two-dimensional tori in nonlinear systems of differential equations are possible.

2.2.3.1 *Bifurcation of birth of pair of stable two-dimensional tori*

In this case at $\mu < 0$ in system (2.24) there is a stable cycle $v_0(t,\mu)$ one multiplier of which is equal to $+1$, and all other multipliers lie inside of unit circle. At transition of values of parameter μ through a point $\mu = 0$ one simple multiplier of a cycle passes through the point $+1$ of a unit circle that corresponds to crossing of an imaginary axis by one simple real Floquet exponent of the system linearized on a cycle. Therefore the normal form of such bifurcation of two-dimensional torus coincides with a one-dimensional normal form of a pitchfork type bifurcation of a singular point: $\dot{u} = \mu u - u^3$. A torus, obviously, corresponds to the zero solution of this equation.

As a result of this bifurcation two-dimensional torus T^2 loses stability (but does not disappear), and simultaneously a pair of stable two-dimensional tori is born near to it at a distance $u_{1,2} = \pm\sqrt{\mu}$ from it (inside and outside of it in a three-dimensional case) in a direction of eigenvector corresponding to an exponent of the cycle $v_0(t,\mu)$, passing through an imaginary axis. Bifurcation is soft.

2.2.3.2 *Transcritical (exchange of stability between tori) bifurcation*

In this case at $\mu < 0$ together with a stable two-dimensional torus T_0^2 of system (2.1) and corresponding to it a stable cycle $v_0(t, \mu)$ of system (2.24) there exists as well as unstable (saddle) two-dimensional torus T_1^2 of system (2.1) and corresponding to it a saddle cycle $v_1(t, \mu)$ of system (2.24), lying in a distance of $u = \mu$ inside of it. At transition of parameter values μ through the point $\mu = 0$ one simple multiplier of a cycle $v_0(t, \mu)$ passes through the point $+1$ of a unit circle, and tori exchange their stability. Stable external torus T_0^2 and, accordingly, a stable cycle $v_0(t, \mu)$ become unstable (saddle), and unstable internal (saddle) torus T_1^2 and, accordingly, an unstable (saddle) cycle $v_1(t, \mu)$ become stable. The normal form of such bifurcation of two-dimensional torus coincides with a one-dimensional normal form of transcritical bifurcation of a singular point: $\dot{u} = \mu u - u^2$. Bifurcation is soft.

2.2.3.3 *Bifurcation of disappearance (appearance) of a pair of two-dimensional tori*

In this case at $\mu < 0$ together with stable two-dimensional torus T_0^2 of the system (2.1) and corresponding to it a stable cycle $v_0(t, \mu)$ of the system (2.24) there exists as well an unstable (saddle) two-dimensional torus T_1^2 of the system (2.1) and corresponding to it a saddle cycle $v_1(t, \mu)$ of the system (2.24). At increasing the parameter values μ these tori approach each other and at $\mu = 0$ merge into one degenerate torus. At $\mu > 0$ both tori disappear. Therefore the normal form of such bifurcation of a stable two-dimensional torus coincides with a one-dimensional normal form of saddle–node bifurcation of a singular point: $\dot{u} = \mu + u^2$. Bifurcation is rigid (crisis).

If we consider a process of change of parameter values μ from great values to smaller values, then described bifurcation means, that at $\mu = 0$ two-dimensional torus is suddenly appeared, and at further reduction of parameter it is divided to two tori, one of which is stable, and another is unstable. In other words, there is a sudden simultaneous appearance of stable and saddle two-dimensional tori, located one inside of another.

2.2.3.4   *Bifurcation of birth (destruction) of three-dimensional torus*

This bifurcation can occur only in phase space of dimension $m > 3$. It is connected with simultaneous transition at $\mu = 0$ of two complex conjugate multipliers of a cycle $v_0(t, \mu)$ of the systems (2.24) through a unit circle, that is equivalent to transition through an imaginary axis from left to right by two complex conjugate Floquet exponents of a cycle $v_0(t, \mu)$. Thus the cycle $v_0(t, \mu)$ of system (2.24) loses its stability (but does not disappear), and simultaneously around it in a hyperplane S transversally to the cycle there arises a two-dimensional torus T^2 as a result of Andronov–Hopf bifurcation. Simultaneously obviously, a stable three-dimensional torus T^3 is born in the system (2.1). The two-dimensional torus T^2 existing before in the system becomes unstable, but does not disappear. Depending on the ratio of frequencies of rotation on cycles of original two-dimensional torus and on a cycle born in a hyperplane S, the motion on arisen three-dimensional stable torus can be either periodic, or quasiperiodic. In the first case any trajectory on torus is closed, in the second, an ergodic case, any motion on a torus has not a closed trajectory everywhere densely filling the surface of torus.

At the further increasing of parameter values μ the phase portrait of system will qualitatively vary, showing transition from ergodic motion on a torus to regimes of resonances.

Considered bifurcation of birth of stable torus T^3 is a supercritical (soft) bifurcation. Along with it, as well as in the case of bifurcation of two-dimensional torus, subcritical (rigid) bifurcation can occur as well there. In this case at $\mu < 0$ the system has a stable two-dimensional torus, lying inside of unstable three-dimensional torus. At $\mu \geq 0$ unstable three-dimensional torus merges into a stable two-dimensional torus which after that becomes unstable. There is a disappearance of attractor, and, hence, such bifurcation is a crisis.

2.2.3.5   *Period doubling bifurcation of two-dimensional torus*

This bifurcation also can occur only in the phase space of dimension $m > 3$. It is connected with transition through an imaginary axis from left to right of one complex Floquet exponent of a cycle $v_0(t, \mu)$ of the system (2.24) at $\mu = 0$. Thus the corresponding real multiplier of a cycle at $\mu = 0$ passes through a point -1 of a unit circle. The cycle $v_0(t, \mu)$ of the system (2.24) loses stability (but does not disappear), and simultaneously there arises in system another stable cycle having the same amplitude and double period. Obviously at the same time two-dimensional torus T^2 of original system (2.1) loses its stability (but does not disappear), and simultaneously with this there arises in system (2.1) a stable two-dimensional torus having double period on one of its frequencies. As a result of such bifurcation a doubling of the torus surface takes place inside the limited volume of phase space. At that torus remains two-dimensional.

Bifurcation of birth of a double period stable two-dimensional torus can be observed at some values of parameters in a complex system of the Lorenz Eqs. (2.20) [Magnitskii and Sidorov (2002)].

Except for the above considered main local one-parametrical bifurcations some more complex insufficiently known bifurcations of stable cycles and tori can occur in nonlinear systems of differential equations such, for example, as recently discovered in the Lorenz system bifurcations of self-organization of cycles [Magnitskii and Sidorov (2001c)]; or bifurcations, connected with various cases of degeneration of multipliers of linear systems of the first approximation on cycles, or with resonances of frequencies of born tori [Arnold *et al.* (1999)].

2.3   The Simplest Two-Parametrical Local Bifurcations

The analysis of bifurcations of systems of differential equations in many-dimensional space of parameters is considerably more difficult in comparison to the analysis of bifurcations in one-parametrical systems spent above. It is connected, first, by that a problem of finding of boundaries of bifurcational surfaces (films) in a space of parameters is very difficult even for bifurcation of codimension 1. Secondly, at families of such systems there are more complex bifurcations of codimension 2 and more. The problem of finding the conditions determining such bifurcations, and their bifurcational surfaces is solved now only for the most elementary modelling equations.

Let us consider two most often meeting cases of analysis of two-parametrical local bifurcations in families of equations, generalizing the equations considered in the previous section. We shall notice, that such many-parametrical local bifurcations are a subject of study of *the catastrophe theory* [Arnold (1990); Gilmor (1981); Poston and Stewart (1972)], and their normal forms and phase-parametrical diagrams have already generally accepted names which will be used by us too.

2.3.1 *The normal form of a fold*

We shall consider a two-parametrical normal form of the following kind

$$\dot{y} = \mu_1 + \mu_2 y + y^2. \tag{2.25}$$

It is easy to see, that at $\mu_1 = 0$ this normal form coincides with a normal form of transcritical bifurcation, and at $\mu_2 = 0$ with a normal form of saddle–node bifurcation. We shall find out, what is the codimension of each of these bifurcations in two-dimensional space of parameters (μ_1, μ_2) of the family of differential Eqs. (2.25).

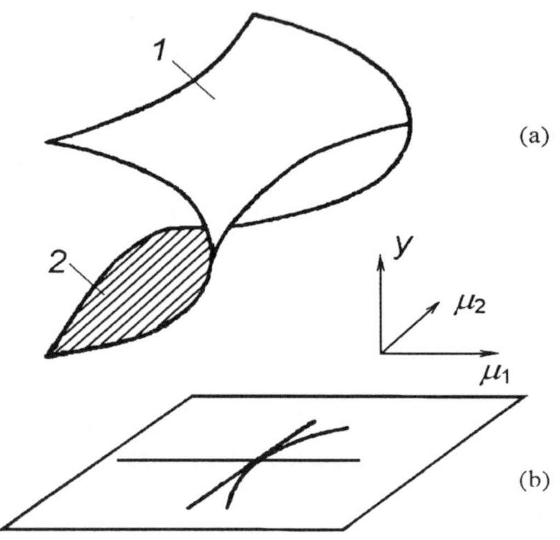

Fig. 2.11   The phase-parametrical diagram of a fold (a) and its projection to a plane of parameters (μ_1, μ_2) (b): 1 — unstable points, 2 — stable points.

Eq. (2.25) has two singular points

$$y_{12} = \frac{-\mu_2 \pm \sqrt{\mu_2^2 - 4\mu_1}}{2}.$$

As a curve in space of parameters, at transition through which singular points disappear, is defined by condition $\mu_2^2 = 4\mu_1$, then this condition is the condition defining a one-dimensional bifurcational curve for a saddle–node bifurcation in two-dimensional space of parameters (μ_1, μ_2). Hence, the saddle–node bifurcation has the codimension 1. In Fig. 2.11 there is presented the phase-parametrical diagram of a surface of singular points of two-parametrical family (2.25), having fold type features along a line, projection of which to a plane of parameters is a bifurcational parabola $\mu_2^2 = 4\mu_1$.

Let us consider now the conditions defining the transcritical bifurcation in the space of parameters (μ_1, μ_2). At this bifurcation there is an exchange of stability between singular points of a family of Eqs. (2.25). Projections of a phase-parametrical surface of singular points of a family on the plane (y, μ_2) for cases $\mu_1 < 0$, $\mu_1 = 0$ and $\mu_1 > 0$ are presented in Fig. 2.12. It is obvious, that exchange of stability can occur only in the case of $\mu_1 = 0$. Thus as we saw above at the analysis of transcritical bifurcation in one-parametrical families, it is necessary to put also $\mu_2 = 0$. Hence, transcritical bifurcation in two-parametrical family of Eqs. (2.25) has codimension 2.

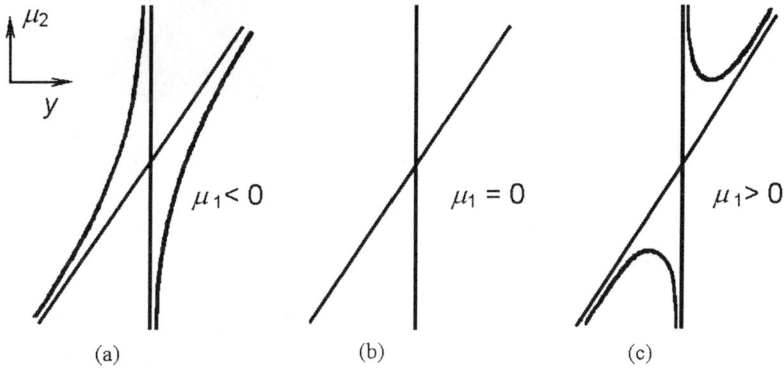

Fig. 2.12 Projections of a fold to a plane (y, μ_2) at $\mu_1 < 0$ (a), $\mu_1 = 0$ (b), $\mu_1 > 0$ (c).

2.3.2 *The normal form of an assembly*

We shall consider a two-parametrical normal form

$$\dot{y} = \mu_1 + \mu_2 y - y^3, \tag{2.26}$$

coinciding at $\mu_1 = 0$ with a normal form of considered earlier supercritical bifurcation of pitchfork type. Eq. (2.26) depending on parameters μ_1 and μ_2 can have one or three singular points. The phase-parametrical diagram of a surface of singular points of two-parametrical family (2.26) is presented in Fig. 2.13a. This surface has an assembly type feature in the field, projection of which to a plane of parameters (μ_1, μ_2) is represented in Fig. 2.13b.

At $\mu_2 < 0$ and any μ_1 the Eq. (2.26) has an unique asymptotically stable stationary solution. At $\mu_2 > 0$ there is an area of values μ_1 (the shaded area G on the bifurcational diagram, Fig. 2.13b) where the system has three singular points lying on three sheets of a surface of assembly. At that, the top and bottom sheets correspond to asymptotically stable singular points, and an average sheet corresponds to unstable singular points.

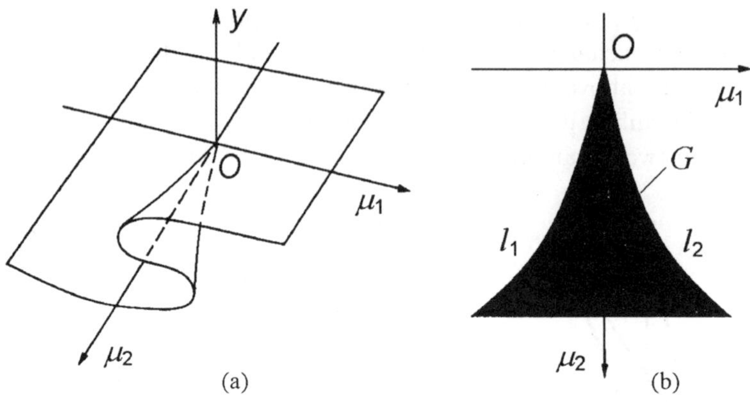

Fig. 2.13   The phase-parametrical diagram of assembly (a) and its projection to a plane of parameters (μ_1, μ_2) (b).

If we change parameter μ_1 from left to right in a direction of increasing its values, then at transition through the left boundary l_1 of an area G there appear one more stable and one unstable singular points on the bifurcational diagram in addition to the already existing one stable singular point. In the field G the family (2.26) has two attractors (Fig. 2.13a). At transition through the right boundary l_2 of the area there occurs an exchange of attractor. Two of three singular points (stable and unstable) merge and disappear. At that there disappears not that stable singular point which has appeared at crossing the left boundary of the area, but that which existed up to the approach to the left boundary. If we change parameter μ_2 in a direction of increasing its values at $\mu_1 \neq 0$, also at transition through a boundary of area G there appear one more stable and one unstable singular points on the bifurcational diagram in addition to the already existing one stable singular point (Fig. 2.13a).

Thus, on boundaries of area G at $\mu_1 \neq 0$ there occurs a saddle–node bifurcation described above in Sec. 2.2.1 and having codimension 1. Hence, a unique point in a plane of parameters at which a pitchfork type bifurcation can occur is the point $\mu_2 = \mu_1 = 0$. Pitchfork type bifurcation in this case has codimension 2.

2.4 Nonlocal Bifurcations

Except the most widespread local bifurcations of singular points, cycles and tori, considered above in nonlinear systems of differential equations there are more complex and insufficiently known nonlocal bifurcations of *homoclinic and heteroclinic contours*, being separatrices of saddle limit sets, the same singular points, cycles and tori. Such bifurcations do not lead to local topological changes of saddle limit sets, but have a key influence on nonlocal change of dynamics of system in the areas of phase space covering limit sets, connected by separatrix contours. Bifurcations of such type, named the homoclinic butterfly, the point–cycle bifurcation and the homoclinic separatrix loop of a saddle–focus are presented, for example, at system of the Lorenz Eqs. (2.3). In nonlinear systems of differential equations nonlocal bifurcations of irregular attractors can occur also born as a result of cascades of soft bifurcations of regular attractors. We shall consider now the main nonlocal bifurcations known today.

2.4.1 *Bifurcations of homoclinic separatrix contours*

At present only bifurcations of homoclinic contours of singular points are investigated to some extent. Bifurcations of more complex contours practically are not investigated today neither theoretically, nor experimentally. Exception is discovered by authors in [Magnitskii and Sidorov (2004c)] a separatrix loop of a singular saddle cycle to which there corresponds a homoclinic separatrix loop of a rotor type singular point of a two-dimensional non-autonomous system (see Chapter 4). We shall consider some bifurcations of homoclinic contours most important from our point of view.

2.4.1.1 *Separatrix loop of a saddle type singular point*

This bifurcation is possible in a phase plane at $m = 2$. Let there is a saddle singular point O of system (2.1) with real eigenvalues of a matrix of linearization in it $\lambda_1(\mu) < 0$ and $\lambda_2(\mu) > 0$. Let stable W^s and unstable W^u one-dimensional manifolds of a singular point O at increasing of parameter $\mu < 0$ approach, and at $\mu = 0$ concern each other. During the moment of a contact there occurs a bifurcation and the special asymptotic in two ways trajectory Γ is formed, named *a separatrix loop of a saddle* (Fig. 2.14a).

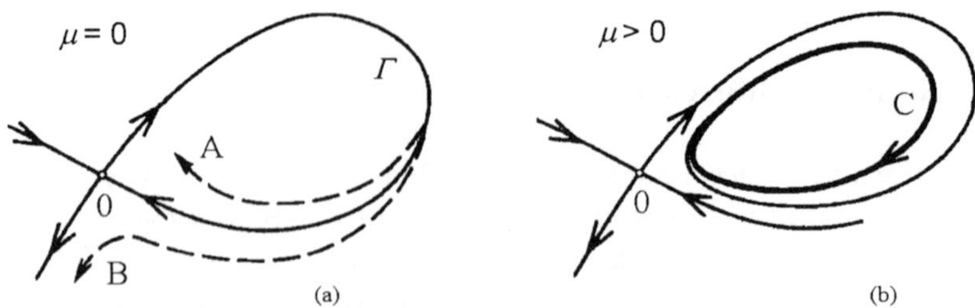

Fig. 2.14   Bifurcation of birth of the separatrix loop of a saddle: (a) the moment of bifurcation; (b) after bifurcation.

The separatrix loop of a saddle is not a rough formation and at $\mu > 0$ it collapses. If *the saddle value* $\sigma(\mu) = \lambda_1(\mu) + \lambda_2(\mu)$ is negative at $\mu = 0$, i.e. $\sigma(0) < 0$, then the loop is stable, and at its destruction in A direction (Fig. 2.14a), a stable limit cycle is born from it (Fig. 2.14b). At destruction of a loop in B direction a cycle is not born. If $\sigma(0) > 0$ the loop is unstable, and at its destruction only unstable cycle can be born from it.

Bifurcation of the separatrix loop of a saddle considered in reverse order can mean a crisis of attractor, a stable limit cycle as a result of its contact with the saddle O. The length of a periodic orbit remains limited at approach to the point of bifurcation while its period tends to infinity.

2.4.1.2   *Separatrix loop of a degenerate two-dimensional saddle–node*

This bifurcation is also possible in a phase plane $m = 2$. Let there be two singular points at $\mu < 0$ in system (2.1): a saddle O_1 and a stable node O_2, and unstable separatrices of a saddle, closing on the node, forming a separatrix contour (Fig. 2.15a).

In the point of bifurcation at $\mu = 0$ there exists a not rough equilibrium state, *a degenerate saddle–node*, having a homoclinic separatrix loop Γ (Fig. 2.15b). At $\mu > 0$ the saddle–node collapses, and a stable limit cycle is born from a separatrix loop (Fig. 2.15c).

Considered in reverse order the given bifurcation can mean a crisis of attractor, a stable limit cycle as a result of birth of a degenerate saddle–node on it. The length of a periodic orbit remains limited at approach to the point of bifurcation while its period tends to infinity.

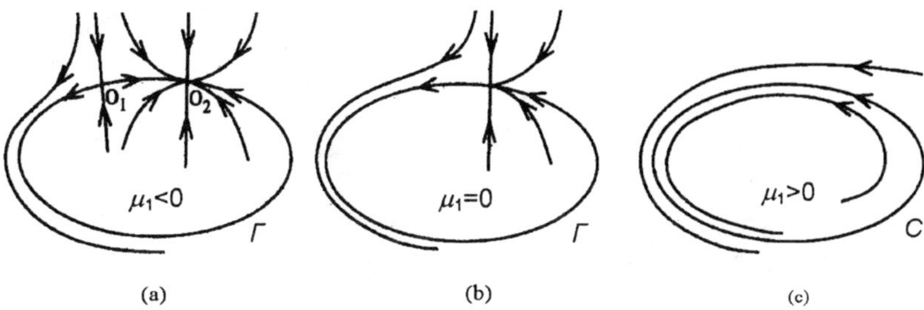

Fig. 2.15   Bifurcation of the birth of a saddle–node separatrix loop.

2.4.1.3   *Separatrix loop of a saddle–node*

This bifurcation is possible only in phase space of dimension $m > 2$. Let there be a singular point O of the system (2.1) of a saddle–node type with real eigenvalues of a matrix of linearization at it, such that $\lambda_i(\mu) < 0$, $i = 1, \ldots, m - 1$, and $\lambda_m(\mu) > 0$. Let one of the separatrix of unstable one-dimensional manifold W^u of a singular point approach with a stable $(m - 1)$-dimensional manifold W^s at increasing of a parameter $\mu < 0$ and touch it at $\mu = 0$. During the moment of contact there occurs a bifurcation and a special trajectory Γ asymptotic in two ways is formed. It is named by *a separatrix loop of a saddle–node* (Fig. 2.16a). The separatrix loop of a saddle–node also is not a rough formation and at $\mu > 0$ it collapses. If *the saddle value*

$$\sigma(\mu) = \lambda_m(\mu) + \max_{i=1,\ldots,m-1} \lambda_i$$

is negative at $\mu = 0$, i.e. $\sigma(0) < 0$, then the loop is stable, and a stable limit cycle can be born from it at its destruction similarly how it takes place in case of the separatrix loop of a saddle. If $\sigma(0) > 0$ the loop is unstable, and at its destruction only an unstable limit cycle can be born from it.

Considered in reverse order the bifurcation of birth of separatrix loop of a saddle–node can mean a crisis of attractor, a stable limit cycle as a result of its contact with saddle–node O. The length of a periodic orbit remains limited at approach to the point of bifurcation while its period tends to infinity.

2.4.1.4   *Separatrix loop of a degenerate saddle–node*

This bifurcation also is possible only in phase space of dimension $m > 2$. Let at $\mu < 0$ there be two singular points in system (2.1) : a saddle–node

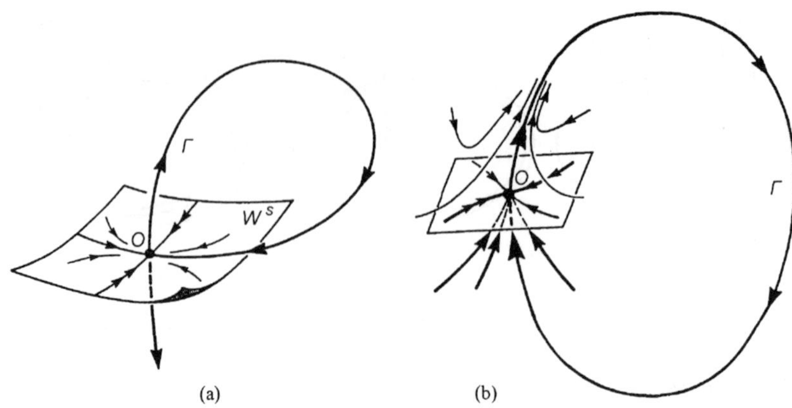

Fig. 2.16   The separatrix loop of a saddle–node (a) and a degenerate saddle–node (b).

O_1, having one-dimensional unstable manifold, and a stable node O_2 , and let the unstable separatrices of a saddle–node, closing on the stable node, form a separatrix contour. At a point of bifurcation at $\mu = 0$ there is a birth of a not rough equilibrium state, *a degenerate saddle–node*, having a homoclinic separatrix loop Γ (Fig. 2.16b). At $\mu > 0$ the degenerate saddle–node collapses, and a stable limit cycle can be born from a separatrix loop.

Considered in reverse order the given bifurcation can mean a crisis of attractor, a stable limit cycle as a result of birth of a degenerate saddle–node on it. The length of a periodic orbit remains limited at approach to the point of bifurcation while its period tends to infinity.

2.4.1.5   *Homoclinic butterfly*

This bifurcation also is possible only in phase space of dimension $m > 2$. Let there be a singular point O of the system (2.1) of a saddle–node type with real eigenvalues of a matrix of linearization at it, such that $\lambda_i(\mu) < 0$, $i = 1, \ldots, m - 1$, and $\lambda_m(\mu) > 0$. Let both separatrices of unstable one-dimensional manifold W^u of a singular point approach a stable $(m-1)$-dimensional manifold W^s at increasing of parameter $\mu < 0$ and touch it at $\mu = 0$. During the moment of contact there occurs a bifurcation of a birth of a special separatrix contour consisting of two asymptotic in two ways trajectories, named as *a homoclinic butterfly* (Fig. 2.17).

Homoclinic butterfly also is not a rough formation and at $\mu > 0$ it can collapse to unstable limit cycle in the form of the eight. Bifurcation of such type is characteristic for systems with symmetry. It is one of the main bifur-

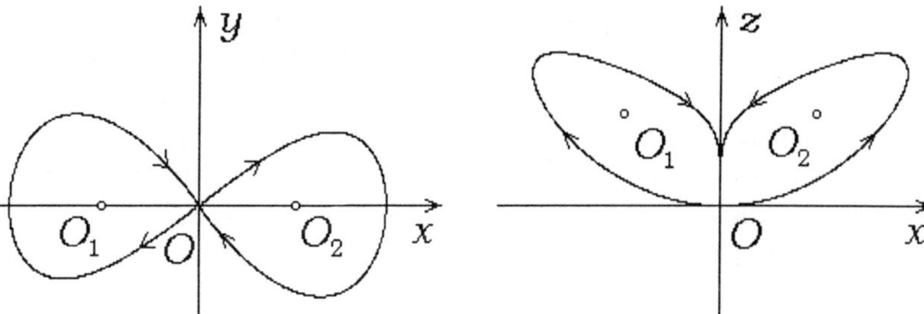

Fig. 2.17   The homoclinic butterfly of a three-dimensional saddle–node in projections to planes (x, y) and (x, z).

cations in the already repeatedly mentioned well-known system of Lorenz Eqs. (2.3). It is considered to be, that exactly this bifurcation is an original cause of rigid birth of chaotic Lorenz attractor. A great number of works is devoted to research of this problem (see [Guckenheimer (1976); Guckenheimer and Williams (1979); Sparrou (1982); Shil'nikov (1980); Tucker (2002); Williams (1979)]). The main sense of all these works comes to attempts of construction in a neighbourhood of a separatrix contour of invariant set of mapping of Smale horseshoe type with fractal structure. These attempts till now have not brought success in understanding of the nature of Lorenz attractor for reasons which will be in detail considered in Chapters 3 and 4.

2.4.1.6   *Separatrix loop of a saddle–focus*

This bifurcation is also possible only in phase space of dimension $m > 2$. Let there be a saddle–focus type singular point O of the system (2.1) with one real eigenvalue $\lambda(\mu) > 0$ and two complex conjugate eigenvalues $\rho(\mu) \pm i\omega(\mu)$ with $\rho < 0$. Let at $\mu = 0$ *a separatrix loop of a saddle–focus* be formed in system (2.1). That will occur at $\mu > 0$ depends on in what direction will the separatrices of a saddle–focus after destruction of a loop go and what is *the first saddle value* $\sigma_1(\mu) = \rho(\mu) + \lambda(\mu)$. If $\sigma_1(0) < 0$ then a stable limit cycle can be born from a loop similarly to the birth of a cycle from a separatrix loop of a saddle–node. If $\sigma_1(0) > 0$ then in a neighbourhood of a loop during the moment of its existence, and also at its destruction there is a complex structure of phase trajectories presumably consisting of a countable set of periodic trajectories and a continuum set of nonperiodic trajectories, that indicates on a presence of complex irregular dynamics in the system. This result is connected with the presence in system at $\sigma_1(0) > 0$ of the Smale horseshoes (see Chapter 1), that in case of dimension of space $m = 3$ has been proved analytically in the paper [Shil'nikov (1970)]. *Shil'nikov theorem* though does not explain the nature of chaotic attractors of nonlinear dissipative systems of ordinary differential equations, nevertheless it has great value as in this theorem for the first time attention was paid that in systems with saddle–focus type singular points the existence of chaotic dynamics is possible.

Following the works [Shil'nikov (1970); Kuznetzov (2001)], we shall consider the autonomous system of three differential equations having a saddle–focus type singular point with two-dimensional stable and one-dimensional unstable manifolds. For convenience we shall place the origin of coordinates into this singular point. The matrix of linearization of a system at a saddle–focus has one real eigenvalue $\lambda > 0$ and two complex conjugate eigenvalues $\rho \pm i\omega$ with $\rho < 0$. The system of coordinates can be chosen so that the equations of a system have a following kind:

$$\begin{aligned}
\dot{x} &= \rho x - \omega y + P(x, y, z), \\
\dot{y} &= \omega x + \rho y + Q(x, y, z), \\
\dot{z} &= \lambda z + R(x, y, z),
\end{aligned} \tag{2.27}$$

where P, Q, R are some functions, expansion of which in a series on degrees of x, y, z contains members, starting with the second degree. Let the system (2.27) have a homoclinic separatrix loop of a saddle–focus, schematically represented in Fig. 2.18a. Let us show, that if $|\rho| < \lambda$, i.e. speed of leaving from a singular point along its unstable manifold prevails the speed of approach to it along its stable manifold, then the presence of a separatrix loop of a saddle–focus implies an existence of a Smale horseshoe and, hence, complex irregular dynamics.

Let us surround a singular point with the cylinder of height $2h$ and radius r with a generatrix, parallel axis z. The size of the cylinder is considered so small that it is possible to use a linear approximation by virtue of the Grobman–Hartman theorem inside of it for the analysis of a flow of trajectories. A point in which the separatrix pierces through the top basis of the cylinder, we shall designate as p, and a point in which it pierces through the lateral surface of the cylinder at returning , as q. We shall choose on a lateral surface of the cylinder, a rectangular area D narrow in a vertical direction. We shall characterize the position of a point in this area by two variables ξ and θ, $0 < \xi \leq \varepsilon$, $|\theta| < \theta_{\max}$, where $\xi = z$, and θ is an angle counted from a point q (Fig. 2.18b).

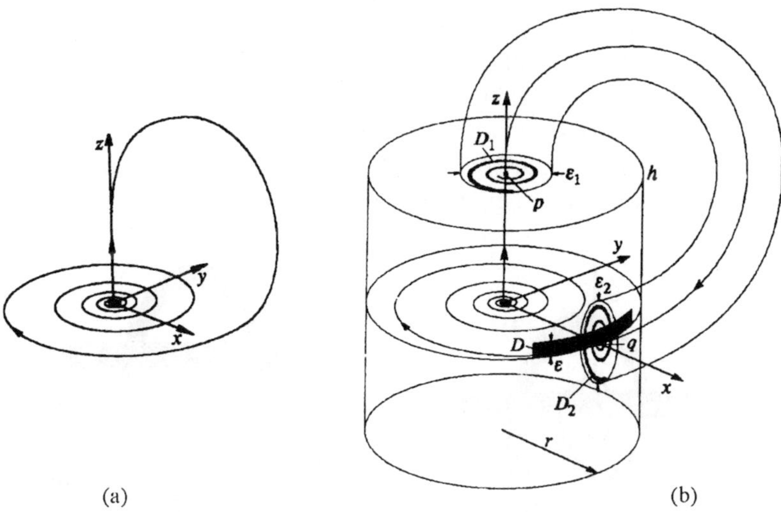

(a)   (b)

Fig. 2.18   The separatrix loop of a saddle–focus (a) and the construction illustrating
the proof of the Shil'nikov theorem (b).

Movement along the trajectory which started from the point (z, θ), will
be defined inside the cylinder by virtue of validity of linear approximation
by the following expressions:

$$x = r \exp(\rho t) \cos(\omega t + \theta),$$
$$y = r \exp(\rho t) \sin(\omega t + \theta),$$
$$z = \xi \exp(\lambda t).$$

Supposing $z = h$, we shall find from the third equation the moment of an
exit of a trajectory from the cylinder $t = \lambda^{-1} \ln(h/\xi)$. We shall obtain then
coordinates of an exit from first two equations:

$$x = r \left(\frac{\xi}{h}\right)^{-\rho/\lambda} \cos\left(\theta - \frac{\omega}{\lambda} \ln \frac{\xi}{h}\right),$$
$$y = r \left(\frac{\xi}{h}\right)^{-\rho/\lambda} \sin\left(\theta - \frac{\omega}{\lambda} \ln \frac{\xi}{h}\right). \tag{2.28}$$

Let us vary ξ and θ within the limits of area D. Then obtained according
to the formula (2.28) points x and y will be placed in the area D_1 on
top surface of the cylinder, having the form of a spiral twisted to point
p (Fig. 2.18). Area D_1, obviously, is situated inside a circle of diameter

$\varepsilon_1 = 2r(\varepsilon/h)^{|\rho|/\lambda}$. The area D_1 will be displayed by a flow of trajectories along the separatrix in some area D_2 on a lateral surface of the cylinder. Thus its central point p will pass in point q so the new area will be imposed the original area D. We shall make an assumption that the mapping of area D_1 in area D_2 can be approximated by linear equations. Then it is possible to consider, that area D_2 is also a spiral lying inside a circle diameter of which will change in comparison with diameter of the circle containing area D_1, in k times and will be equal to $\varepsilon_2 = 2kr(\varepsilon/h)^{|\rho|/\lambda}$.

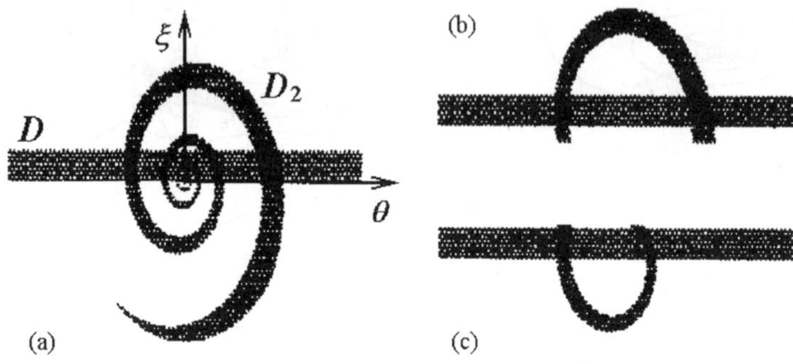

Fig. 2.19　The explanatory of presence of horseshoes in a neighbourhood of a separatrix loop of a saddle–focus.

If the exponent $|\rho|/\lambda$ is more than the unit, then at small ε the value $\varepsilon_2 < 2\varepsilon$, so the top part of spiral D_2 does not leave the limits of original area D. On the contrary, at $|\rho|/\lambda < 1$ at small ε it will always be $\varepsilon_2 > 2\varepsilon$, so there is obvious a presence of horseshoe in mapping of original area D in area D_2 (Fig. 2.19a). Moreover, at reduction ε in the constructed mapping there appear all new and new horseshoes (see Fig. 2.19b,c). Passing to a limit at $\varepsilon \to 0$, it is possible to find out the presence of countable set of horseshoes at this mapping.

If we enter a small perturbation leading to destruction of a separatrix loop of the system, the finite number of horseshoes will be saved at presence of perturbation too. Thus, it is possible to confirm, that in the neighbourhood of a separatrix loop of a saddle–focus with one-dimensional unstable manifold a complex irregular dynamics takes place for $|\rho| < \lambda$.

The alternative version of the Shil'nikov theorem takes place if we consider the system with homoclinic separatrix loop of a saddle–focus, having one-dimensional stable and two-dimensional unstable manifolds. At that $\lambda < 0$, $\rho > 0$, and the problem is reduced to the previous one by the time reversal $t \to -t$. In this case a condition of presence of a horseshoe and complex irregular dynamics in the system is the inequality $\rho < |\lambda|$.

Remark 2.1 *Generally speaking in Shil'nikov theorem, existence of a horseshoe is proved, but the existence of invariant set of a horseshoe mapping which should possess the certain properties is not proved. Therefore from the theorem, generally speaking, it is impossible to draw a conclusion that the complex irregular dynamics which takes place in the neighbourhood of a separatrix loop of a saddle–focus is the same dynamics which is the characteristic feature of an invariant set of the Smale horseshoe (see Chapter 1). Moreover, as numerous examples of systems of differential equations considered in Chapters 3 and 4 show, complex irregular and even chaotic dynamics is presented at these systems not only in a neighbourhood of a separatrix loop of a saddle–focus, but also at any distance from it in a space of parameters. A necessary condition of chaotic dynamics is not neither presence of a separatrix loop, nor presence of a saddle–focus itself. It is already can not be explained in any way by neither the proved theorem, nor mapping of a horseshoe type. Hence, there should be some other mechanism providing presence of complex irregular dynamics in systems of ordinary differential equations as with singular points of a saddle–focus or a saddle–node type, and without them (see Chapter 4).*

2.4.1.7 *Separatrix loop of a singular saddle cycle*

This bifurcation was discovered by authors at studying cascades of bifurcations of *a singular saddle cycle* of a three-dimensional autonomous system of ordinary differential equations (see in detail in Chapter 4).

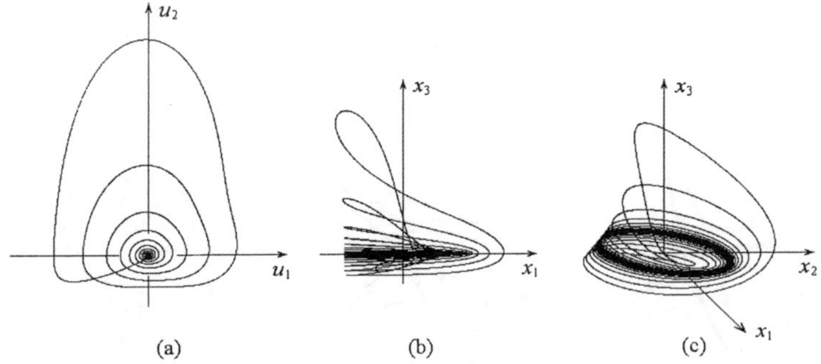

Fig. 2.20 A separatrix loop of a rotor type singular point (a), and a singular cycle (b,c) corresponding to it.

Separatrix of a cycle, being untwisted on Mobius band, comes back to a cycle along a two-dimensional surface containing it (Fig. 2.20b,c). This bifurcation is inseparably connected with bifurcation of *a homoclinic separatrix loop of a rotor* type singular point of a two-dimensional non-autonomous system of ordinary differential equations discovered by authors and investigated in detail in Chapter 4 (Fig. 2.20a). In any neighbourhood of a loop during the moment of its existence, and also at its destruction there is a complex structure of phase trajectories in system with infinite number of unstable periodic orbits.

2.4.2　*Bifurcations of heteroclinic separatrix contours*

Unlike bifurcations of homoclinic separatrix contours, bifurcations of heteroclinic contours in phase space of dimension $m > 2$ now are not investigated practically even in the case when they connect different singular points of a system of ordinary differential equations. Therefore the facts stated in this section concerning bifurcations of heteroclinic contours in systems of a large dimension, are based mainly on results of numerical experiments. We shall consider some bifurcations of heteroclinic contours most important from our point of view.

2.4.2.1　*Separatrix going from a saddle into a saddle*

This bifurcation is possible in a phase plane at $m = 2$. Let there be two saddle singular points O and O_1 of the system (2.1) with real eigenvalues of matrices of linearization at them, having different signs. Let unstable one-dimensional manifold W^u of a singular point O and stable one-dimensional manifold W_1^s of a singular point O_1 at increasing of the parameter $\mu < 0$ approach, and at $\mu = 0$ touch each other. During the moment of contact there occurs a bifurcation of formation of a special heteroclinic trajectory Γ named by *separatrix, going from a saddle into a saddle* (Fig. 2.21a).

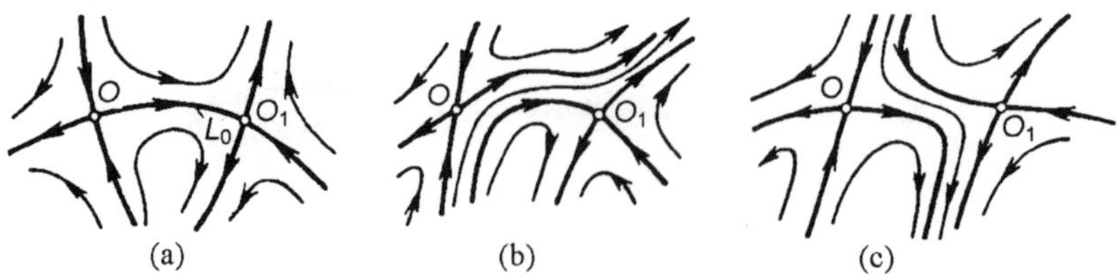

Fig. 2.21　Bifurcation of a separatrix, going from a saddle into a saddle.

Such separatrix contour is not a rough formation and at $\mu > 0$ it simply collapses in two possible ways represented in Fig. 2.21b,c. We shall notice, that in a two-dimensional case there can exist also separatrices of the saddles, tending in one side to the node, focus or a limit cycle (Fig. 2.22a,b), but such separatrix contours are rough and their kind does not vary at small changes of values of system parameters [Bautin and Leontovich (1990)].

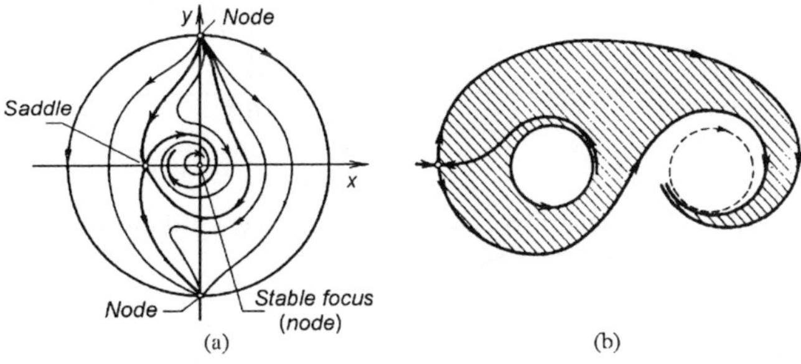

Fig. 2.22   Examples of rough separatrix contours in a two-dimensional case.

2.4.2.2   *Separatrix contour connecting a saddle–node with a saddle–focus*

This bifurcation is possible only in phase space of dimension $m > 2$. Let in system (2.1) there be two singular points of a saddle–node type and a saddle–focus type, one of which has one-dimensional stable W^s and $(m-1)$-dimensional unstable W^u manifolds, and another, on the contrary, has one-dimensional unstable W^u and $(m-1)$-dimensional stable W^s manifolds. Let one of two separatrices of one-dimensional stable manifold of the first singular point at increasing of parameter $\mu < 0$ approach the separatrix of one-dimensional unstable manifold of the second singular point and touch it at $\mu = 0$. Simultaneously, one of separatrices of unstable manifold of the first singular point touches the stable manifold of the second singular point. During the moment of contact there occurs a bifurcation of formation of a special heteroclinic closed *contour connecting a saddle–node with a saddle–focus* (Fig. 2.23a). Numerical experiments spent with the Lorenz system show that such separatrix contour is not a rough formation and that in its neighbourhood a complex irregular dynamics of trajectories can be observed with presence of infinite number of unstable and stable limit cycles. More in detail about it see in Chapter 3.

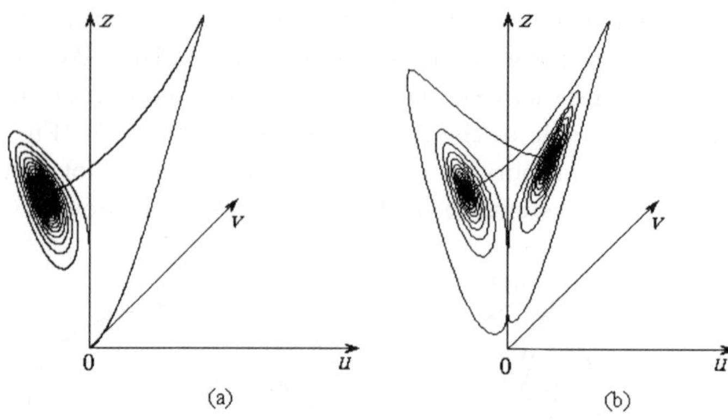

Fig. 2.23   The heteroclinic separatrix contours connecting a saddle–node with a saddle–focus (a) and a saddle–focus with a saddle–focus (b).

2.4.2.3   *Separatrix contour connecting a saddle–focus with a saddle–focus*

This bifurcation also is possible only in the phase space of dimension $m > 2$. Let there be two singular points of a saddle–focus type in system (2.1), one of which has one-dimensional stable W^s and $(m-1)$-dimensional unstable W^u manifolds, and another also has one-dimensional stable W^s and $(m-1)$-dimensional unstable W^u manifolds. Let one of two separatrices of one-dimensional stable manifold of the first singular point at increasing of the parameter $\mu < 0$ approach with unstable manifold of the second singular point and touch it at $\mu = 0$. Simultaneously, one of two separatrices of one-dimensional stable manifold of the second singular point approaches with unstable manifold of the first singular point and at $\mu = 0$ touches it. During the moment of contact there occurs a bifurcation of formation of the special closed heteroclinic *contour connecting two saddle–focuses* (Fig. 2.23b). Numerical experiments show, that such separatrix contour is not a rough formation and that in any its neighbourhood a complex irregular dynamics of trajectories can be observed with presence of infinite number of unstable and stable limit cycles (see in detail in Chapter 3).

2.4.2.4   *Point–cycle bifurcation*

This bifurcation also is possible only in phase space of dimension $m > 2$. Let there be two singular points of a saddle–node and a saddle–focus type in the system (2.1). Let the saddle–focus have a one-dimensional stable W^s

and $(m-1)$-dimensional unstable W^u manifolds, and the saddle–node, on the contrary, have a one-dimensional unstable W^u and $(m-1)$-dimensional stable W^s manifolds. Let at $\mu < 0$ in the system there be an unstable cycle having one coil in a neighbourhood of a saddle–node and a number of coils n around a saddle–focus, increasing with growth of values of the parameter μ, so $n \to \infty$ when $\mu \to 0$ (Fig. 2.24a). Then at $\mu = 0$ the following bifurcation is possible: the cycle merges into a saddle–node and simultaneously a saddle or semi-stable cycle is born around the saddle–focus. In other words, a separatrix contour consisting of two separatrices of a saddle–node is born from an original cycle at $\mu = 0$. One of the separatrices tends to a new cycle which was born around a saddle–focus at $t \to +\infty$, and another at $t \to -\infty$ (Fig. 2.24b).

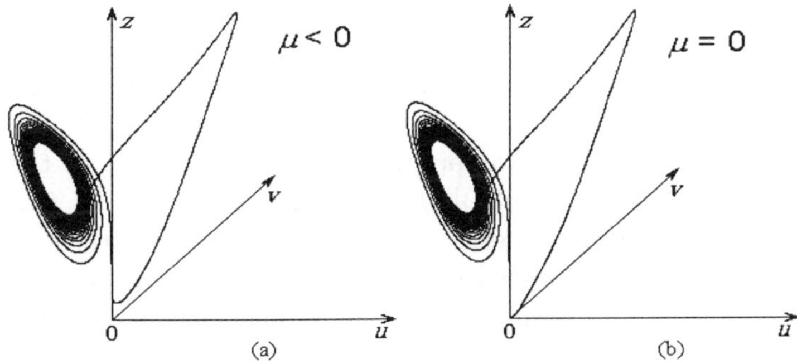

Fig. 2.24   A point–cycle bifurcation at $\mu < 0$ (a) and at $\mu = 0$ (b).

This bifurcation exists in the Lorenz system at some values of the system parameters. The numerical experiments spent with this system, show, that the described above separatrix contour is not a rough formation and that in any of its neighbourhood there are periodic, but there are no stable trajectories. More in detail about this see in Chapter 3.

Except for the separatrix contours in families of systems of type (2.1) considered above an existence of a heteroclinic contour connecting a saddle–node with two saddle focuses is possible. Such heteroclinic contour exists in the Lorenz system. Moreover, a heteroclinic contour connecting two saddle–nodes with one-dimensional stable and one-dimensional unstable manifolds is also possible. However, examples of such systems are not known to authors.

2.4.3   *Approximate method for finding bifurcation points of separatrix contours of singular points*

Homoclinic and heteroclinic contours of singular points play an important role in formation of irregular attractors of nonlinear systems of ordinary differential equations. From this point of view the solution of a problem of finding bifurcational surfaces and curves of existence of homoclinic separatrix loops and heteroclinic contours of singular points of system (2.1) in the space of parameters has, undoubtedly, a great value. It was possible to solve a similar problem analytically only for a separatrix loop of piecewise linear Chua system [Gribov and Krischenko (2001)]. In papers [Leonov (1988); Chen (1996)] it was found a necessary and sufficient condition (*Leonov–Chen inequality*)

$$3\sigma > 2b + 1$$

for existence of the third parameter r in the system of Lorenz Eqs. (2.3) at which the system with three parameters (σ, b, r) has the homoclinic saddle-node separatrix loop.

A new original method was proposed by N. Magnitskii in the paper [Magnitskii and Sidorov (2003b)], *for approximate finding of bifurcation surfaces (curves)* in the parameter space and for definition of their codimensions for the following separatrix contours of singular points: a heteroclinic contour connecting two saddle–nodes, two saddle–focuses or saddle–node and saddle–focus; homoclinic separatrix loop of a saddle–focus; homoclinic separatrix loop of a saddle–node. It was supposed, that at all values of parameters from some area $U \subset \mathbb{R}^k$ the conditions of the Grobman–Hartman theorem are satisfied for singular points of the family (2.1). So separatrices, entering into singular points along their stable manifolds and leaving them along their unstable manifolds, exist. The problem consists in the approximate finding of those values of parameters at which these separatrices form the closed contours in a phase space. In Chapter 3 the results of this section will be illustrated by examples of finding of all mentioned contours for the Lorenz system (2.3).

2.4.3.1   *Heteroclinic contours of saddle–nodes and saddle–focuses*

Let the singular point $\bar{x}(\mu)$ of the family (2.1) be a saddle–node or a saddle–focus, having one-dimensional stable W^s and $(m-1)$-dimensional unstable W^u manifolds, and the singular point $\tilde{x}(\mu)$ of the family (2.1) be a saddle–node or a saddle–focus, having, on the contrary, one-dimensional unstable

W^u and $(m-1)$-dimensional stable W^s manifolds. Thus Jacobi matrix of the right part of system (2.1) has exactly one negative real eigenvalue λ at point $\overline{x}(\mu)$ and exactly one positive real eigenvalue ν at point $\widetilde{x}(\mu)$. The problem consists in a finding of bifurcational surface S in the space of parameters $\mathbb{R}^k$ such, that at any value $\mu \in S$ separatrix, entering into a singular point $\widetilde{x}(\mu)$ of the family (2.1) along its one-dimensional unstable manifold when $t \to -\infty$, enters also into a singular point $\overline{x}(\mu)$ of the family (2.1) along its one-dimensional stable manifold at $t \to +\infty$.

Let us write down the equations of system (2.1) in the form of

$$\dot{x}_j(t) = f_j(x, \mu), \quad j = 1, \ldots, m. \tag{2.29}$$

Let us choose one of the coordinates, having most simple right part in a neighbourhood of the singular points $\overline{x}(\mu)$ and $\widetilde{x}(\mu)$, for example, x_1. We shall calculate derivatives dx_j/dx_1 at a singular point, for example, $\overline{x}(\mu)$, having used for this purpose by expansion of the right parts of the system (2.29) at this point

$$\frac{dx_j(\overline{x})}{dx_1} = \lim_{x \to \overline{x}} \frac{f_j(x, \mu)}{f_1(x, \mu)} = \frac{\displaystyle\sum_{l=1}^{m} \frac{\partial f_j(\overline{x}, \mu)}{\partial x_l} \cdot \frac{dx_l(\overline{x})}{dx_1}}{\displaystyle\sum_{l=1}^{m} \frac{\partial f_1(\overline{x}, \mu)}{\partial x_l} \cdot \frac{dx_l(\overline{x})}{dx_1}}, \quad j = 1, \ldots, m. \tag{2.30}$$

It is easy to see, that if the system (2.29) is reduced to a kind of $f_1(x, \mu) = x_2$ (that is $\dot{x}_1 = x_2$) the vector y with coordinates $y_j = dx_j/dx_1$, $j = 1, \ldots, m$, is the solution of system of the nonlinear equations

$$y_2 y_j = \sum_{l=1}^{m} a_{jl} y_l, \quad j = 1, \ldots, m, \tag{2.31}$$

where a_{jl} are elements of the Jacobi matrix of the right part of the system (2.29), calculated at a singular point. It follows from (2.31) , that

$$Ay = y_2 y$$

and, hence, value $y_2 = dx_2/dx_1$ is an eigenvalue for a matrix of linearization of system (2.29) at a singular point. We shall accept $\overline{y}_2 = \lambda$ at the point $\overline{x}(\mu)$ and $\widetilde{y}_2 = \nu$ at the point $\widetilde{x}(\mu)$. Other values $\widetilde{y}_j$, $j = 3, \ldots, n$, we shall define uniquely from system of the Eqs. (2.31). We shall notice, that if any value y_j will appear equal to zero at some singular point it will demand an additional expansion of the right parts in Taylor's series at this singular point before obtaining of derivatives of some order distinct from zero.

Let us issue from a neighbourhood of a singular point $\tilde{x}(\mu)$ of the system (2.29) a trajectory along its unstable manifold so, that in each projection (x_1, x_j) a tangent of an angle of inclination to an axis x_1 of a line, connecting initial and singular points, will be equal to $\tilde{y}_j$. With this purpose for any as much as small value $\varepsilon > 0$ we shall consider value $x_1(0) = \tilde{x}_1(\mu) + \varepsilon$ of coordinate x_1, as much as close to value of this coordinate in a singular point $\tilde{x}(\mu)$. Other coordinates of an initial point we shall accept equal to $x_j(0) = \tilde{x}_j(\mu) + \tilde{y}_j \varepsilon$. In other words, solving system (2.29) with the specified initial conditions we obtain a trajectory $x^+(t)$ of the system, as much as close to separatrix, starting from the singular point $\tilde{x}(\mu)$. Similarly, let us issue from a neighbourhood of a singular point $\overline{x}(\mu)$ a trajectory $x^-(t)$, as much as close to separatrix of the singular point, solving the system (2.29) in return time with initial conditions $x_1(0) = \overline{x}_1(\mu) + \varepsilon$, $x_j(0) = \overline{x}_j(\mu) + \overline{y}_j \varepsilon$, $j = 2, 3, \ldots, m$. Now it is necessary to join the trajectories $x^+(t)$ and $x^-(t)$. The point of joint can be chosen any way enough. For example, we can take points of crossing of axes x_1 in the plane (x_1, x_2) by projections of trajectories $x^+(t)$ and $x^-(t)$, i.e. that is points $x_1^+(t^+)$ and $x_1^-(t^-)$ such, that $x_2^+(t^+) = 0$ and $x_2^-(t^-) = 0$. Thus the moments t^+ and t^- of crossings of an axis x_1 by projections of these trajectories and the ends of these projections are defined uniquely. Then it is necessary and sufficient to solve a system of $m - 1$ equations with k parameters for finding the curve approximately coincident with a heteroclinic contour of the system (2.29)

$$\begin{aligned}
\lim_{\varepsilon \to 0} x_1^+(t^+, \mu, \varepsilon) &= \lim_{\varepsilon \to 0} x_1^-(t^-, \mu, \varepsilon), \\
\lim_{\varepsilon \to 0} x_j^+(t^+, \mu, \varepsilon) &= \lim_{\varepsilon \to 0} x_j^-(t^-, \mu, \varepsilon), \quad j = 3, \ldots, m.
\end{aligned} \tag{2.32}$$

Theorem 2.3 *If $k < m - 1$, then in case of general position the system (2.29) has no values in a space of parameters corresponding to heteroclinic contours of singular points connecting their one-dimensional manifolds. Otherwise a bifurcational surface of heteroclinic contours of the singular points, connecting their one-dimensional manifolds, has codimension $m - 1$ in a space of parameters and is defined with any accuracy at the numerical solution of system of the Eqs. (2.32) for some $\varepsilon > 0$.*

Thus, at $k = m - 1$ there is only the unique point in a space of parameters corresponding to a heteroclinic contour of the indicated kind, at $k = m$ there is a part of a line and at $k \geq m + 1$ — a part of a surface of dimension $k - m + 1$.

2.4.3.2 *Homoclinic separatrix loop of a saddle–focus*

Let the singular point $\overline{x}(\mu)$ of the family (2.1) be a saddle–focus, having two-dimensional unstable manifold W^u and $(m-2)$-dimensional stable manifold W^s. Thus Jacobi matrix of the right part of the system (2.1) has two complex conjugate eigenvalues λ_2 and λ_3 at a singular point with positive real parts and the others $m-2$ real negative eigenvalues $\lambda_1, \lambda_4, \ldots, \lambda_m$. The problem consists in finding of bifurcational surface S in the space of parameters $\mathbb{R}^k$ such, that at any $\mu \in S$ separatrix, entering at twisting into a singular point $\overline{x}(\mu)$ of the family (2.1) along its two-dimensional unstable manifold when $t \to -\infty$, enters also into the same singular point $\overline{x}(\mu)$ of the family (2.1) along its $(m-2)$-dimensional stable manifold when $t \to +\infty$.

First let us consider a case $m = 3$, that is the case of one-dimensional stable manifold of a saddle–focus $\overline{x}(\mu)$. To a similar case considered above, we shall find a vector of derivatives at a singular point $y_j = dx_j/dx_1$, $j = 1, 2, 3$ so, that $y_2 = \lambda_1 < 0$. Let us issue from a neighbourhood of a singular point $\overline{x}(\mu)$ in return time a trajectory $x^-(t)$, as much as close to separatrix so, that it will have in projections (x_1, x_2) and (x_1, x_3) a tangent of an angle of inclination at an initial point as much as close to the value $y_2 = \lambda_1$. For this purpose we solve the system (2.29) in return time with initial conditions $x_1(0) = \overline{x}_1(\mu) + \varepsilon$, $x_j(0) = \overline{x}_j(\mu) + y_j\varepsilon$, $j = 2, 3$, where ε is any as much as small positive number. We stop calculation of the trajectory $x^-(t)$ at the moment of time t^- when its projection in plane (x_1, x_2) will cross the axis x_1, that is when $x_2^-(t^-) = 0$.

To find a trajectory $x^+(t)$, as much as close to a separatrix, proceeding from a saddle–focus $\overline{x}(\mu)$ in direct time, we shall reduce a matrix of linearization of the system (2.29) to a canonical form by a non-degenerate transformation

$$B = C^{-1}AC = \begin{pmatrix} \alpha & -\beta & 0 \\ \beta & \alpha & 0 \\ 0 & 0 & \lambda_1 \end{pmatrix},$$

where $\lambda_1 < 0$, $\lambda_2 = \alpha + i\beta$, $\lambda_3 = \alpha - i\beta$, $\alpha > 0$.

In coordinates $z = C^{-1}(x - \overline{x}(\mu))$ the singular point $z = 0$ has one-dimensional stable manifold, tangent to an axis z_3, and two-dimensional unstable manifold, tangent to a plane (z_1, z_2). For any $\varepsilon > 0$ there is a family of initial conditions $z_{10} = \varepsilon \cos\varphi$, $z_{20} = \varepsilon \sin\varphi$, $z_{30} = 0$, depending on any phase φ, but lying in a distance of ε from a singular point. To these initial conditions there corresponds a family of trajectories $x^+(t, \varphi)$, issued in direct time from points

$$x(0) = \overline{x}(\mu) + C \begin{pmatrix} \varepsilon \cos \varphi \\ \varepsilon \sin \varphi \\ 0 \end{pmatrix}.$$

Each of such trajectories approximates as much as close one of separatrices of the saddle–focus $\overline{x}(\mu)$. We shall choose now a phase $\varphi(\varepsilon, \mu)$ so that $x_1^+(t^+) = x_1^-(t^-)$ where t^+ is the moment of time when the projection of a trajectory $x^+(t)$ in a plane (x_1, x_2) crosses an axis x_1, that is when $x_2^+(t^+) = 0$. Then for finding a separatrix loop of the saddle–focus in the case of $m = 3$ it is necessary and sufficient to solve only one equation concerning a set of parameters, namely

$$\lim_{\varepsilon \to 0} x_3^+(t^+, \mu, \varepsilon) = \lim_{\varepsilon \to 0} x_3^-(t^-, \mu, \varepsilon). \tag{2.33}$$

Hence, in this case a bifurcation of formation of the saddle–focus separatrix loop has codimension 1, that means it is a $(k-1)$-dimensional surface in a space of parameters. Now it is easy to see, that codimension of bifurcation of formation of a saddle–focus separatrix loop is equal to 1 also at $m > 3$, as there exists $(m-3)$-dimensional family of separatrices, entering into a singular point along its $(m-2)$-dimensional stable manifold when $t \to +\infty$. In other words there are $m - 3$ free parameters, necessary for closing of trajectories $x^+(t)$ and $x^-(t)$ in all projections (x_1, x_j), $j = 4, \ldots, m$. Thus, the following theorem takes place.

Theorem 2.4 *Homoclinic separatrix loop of a saddle–focus has codimension 1 in a space of parameters, and its bifurcational surface is a part of a $(k-1)$-dimensional hypersurface and can be defined approximately by the numerical solution of the Eq. (2.33) with some $\varepsilon > 0$.*

2.4.3.3 *Homoclinic separatrix loop of a saddle–node*

Let the singular point $\widetilde{x}(\mu)$ of the family (2.1) is a saddle–node, having one-dimensional unstable manifold W^u and $(m-1)$-dimensional stable manifold W^s. Thus the Jacobi matrix of the right part of the system (2.1) has one positive eigenvalue ν_2 and $m - 1$ negative real eigenvalues $\nu_1, \nu_3, \ldots, \nu_m$ at a singular point. The problem consists in a finding of bifurcational surface S in a space of parameters $\mathbb{R}^k$ such, that at any value $\mu \in S$ separatrix, entering into a singular point $\widetilde{x}(\mu)$ of the family (2.1) along its one-dimensional unstable manifold when $t \to -\infty$, enters also into the same singular point $\widetilde{x}(\mu)$ of the family (2.1) along its $(m-1)$-dimensional stable manifold when $t \to +\infty$. First let us consider a case $m = 2$, that is a case of one-dimensional stable manifold of a saddle–node $\tilde{x}(\mu)$ which in this case is simply a saddle. Similarly stated above we shall find values of a derivative $y_2 = dx_2/dx_1$ at a singular point. Thus, obviously, there are two values $y_2^+ = \nu_2 > 0$ and $y_2^- = \nu_1 < 0$, being real eigenvalues of the Jacobi matrix of the right part of the system (2.1) at a singular point.

Let us issue a trajectory $x^+(t)$ from a neighbourhood of a singular point $\tilde{x}(\mu)$ in direct time as much as close to separatrix, that a tangent of an angle of its inclination to the axis x_1 of a plane (x_1, x_2) at an initial point will be as much as poorly differing from value $y_2^+ = \nu_2 > 0$. For this purpose we solve the system (2.29) in direct time with initial conditions

$$x_1(0) = \tilde{x}_1(\mu) + \varepsilon, \quad x_2(0) = \tilde{x}_2(\mu) + y_2^+ \varepsilon,$$

where ε is as much as small positive number. Calculation of a trajectory $x^+(t)$ is stopped at the moment t^+, at which $x_2^+(t^+) = 0$. Similarly, we shall issue a trajectory $x^-(t)$ from a singular point $\tilde{x}(\mu)$ in return time as much as close to separatrix, that it will have a tangent of an angle of inclination to the axis x_1 of a plane (x_1, x_2) at initial point as much as poorly differing from value $y_2^- = \nu_1 < 0$. For this purpose we solve the system (2.29) in return time with initial conditions

$$x_1(0) = \tilde{x}_1(\mu) + \varepsilon, \quad x_2(0) = \tilde{x}_2(\mu) + y_2^- \varepsilon,$$

where ε is as much as small positive number. Calculation of trajectory $x^-(t)$ is stopped at the moment t^-, at which $x_2^-(t^-) = 0$. A necessary and sufficient condition of existence of a separatrix loop is the realization of only one equality

$$\lim_{\varepsilon \to 0} x_1^+(t^+, \mu, \varepsilon) = \lim_{\varepsilon \to 0} x_1^-(t^-, \mu, \varepsilon) \tag{2.34}$$

concerning the family of parameters μ. Thus, in case of $m = 2$ bifurcation of formation of a separatrix loop of a saddle has the codimension 1, that means it is a $(k-1)$-dimensional hypersurface in a space of parameters.

It is easy to see, that at $m > 2$ the codimension of bifurcation of formation of a separatrix loop of a saddle–node is also equal to 1, as there exists a $(m-2)$-dimensional family of separatrices, entering into a singular point along its $(m-1)$-dimensional stable manifold at $t \to +\infty$. In other words, there are else $m - 2$ free parameters necessary for closing trajectories $x^+(t)$ and $x^-(t)$ in all projections (x_1, x_j), $j = 3, \ldots, m$. Thus, the following theorem takes place.

Theorem 2.5   *A homoclinic separatrix loop of a saddle–node has codimension 1 in a space of parameters, and its bifurcational surface is a part of a $(k-1)$-dimensional hypersurface and it can be defined with any accuracy by the approximate solution of the Eq. (2.34) at some $\varepsilon > 0$.*

2.4.4   *Cascades of bifurcations, scenarios of transition to chaos*

In nonlinear dynamical systems described by both ordinary and partial differential equations and differential equations with delay arguments there can exist cascades of bifurcations, leading to occurrence of complex, chaotic regimes of behaviour. They have received the name of scenarios of transition to chaos. We shall consider the most important and typical of these scenarios.

2.4.4.1   *Cascade of period doubling bifurcations, Feigenbaum scenario*

This scenario is the universal and most widespread scenario of transition to chaos in nonlinear dynamical systems. The infinite *cascade of period doubling bifurcations* of stable limit cycles corresponds to this scenario. As already it was noticed above, it can be found out in many nonlinear dynamical systems having chaotic behaviour, both in mappings with discrete time, and in systems described by the differential equations. It is present, for example, at Lorenz hydrodynamical model (2.3) and in hypothetical Rossler models of chemical reactions (2.5), in Chua electrotechnical model (2.6) and in Mackey–Glass model of haemopoiesis (see Chapter 5), in Magnitskii macroeconomic model (2.7) and in models of various biological and ecological systems [Sviregev (1987)]. This cascade leading to occurrence of irregular Feigenbaum attractor (Fig. 2.25), is an initial stage of other, more complex cascades of bifurcations, leading to occurrence of more complex irregular attractors. Besides, while only for *Feigenbaum scenario* it was possible to prove some universal properties of sequence of values of bifurcation parameter μ_n at which next period doubling bifurcations of stable cycles occur and which converge to the value $\mu_\infty = \lim \mu_n$ when $n \to \infty$ (see Chapter 4). The first simplest irregular attractor, Feigenbaum attractor is born in any system at the parameter value μ_∞.

Numerous examples of cascades of the period doubling bifurcations in concrete systems of ordinary differential equations are considered in Chapter 3. The theory of transition to chaos through the Feigenbaum cascade is stated in Chapter 4.

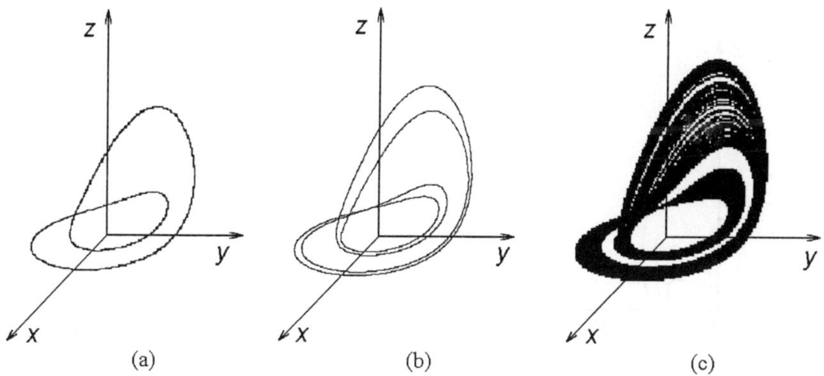

Fig. 2.25 Stable cycles of the period two (a), the period four (b) and Feigenbaum attractor (c) in the Rossler system.

2.4.4.2 *Subharmonic cascade of bifurcations of stable cycles, Sharkovskii scenario*

Sometimes in the scientific literature the Feigenbaum cascade of period doubling bifurcations is called the subharmonic cascade [Berger *et al.* (1984)]. We shall characterize by this term a much more complex cascade of bifurcations of occurrence of stable cycles of any period with the purpose to separate the concepts of subharmonic cascade and the period doubling cascade of bifurcations. In such definition the Feigenbaum cascade is only an initial stage of *the subharmonic cascade.* As a rule, the period doubling cascade of bifurcations, described above, has a continuation. At further increasing of values of bifurcation parameter $\mu > \mu_\infty$ a birth of stable limit cycles of any period takes place in any system in compliance with the scenario found in [Sharkovskii (1964)]. It has been proved by him, that there is an ordering relationship (*Sharkovskii order*) which orders cycles of continuous one-dimensional mappings on size of their period as follows:

$$1 \lhd 2 \lhd 2^2 \lhd 2^3 \lhd \cdots \lhd 2^2 \cdot 7 \lhd 2^2 \cdot 5 \lhd 2^2 \cdot 3 \lhd \cdots$$

$$\cdots \lhd 2 \cdot 7 \lhd 2 \cdot 5 \lhd 2 \cdot 3 \lhd \cdots \lhd 7 \lhd 5 \lhd 3. \qquad (2.35)$$

The first relation in this order means, that if one-dimensional continuous mapping have a cycle of the double period it has also a simple cycle. The

    New Methods for Chaotic Dynamics

cycle of the period three is the most complex cycle in this order. Existence of such cycle means as well the existence of any cycle of any period from the Sharkovskii order (2.35).

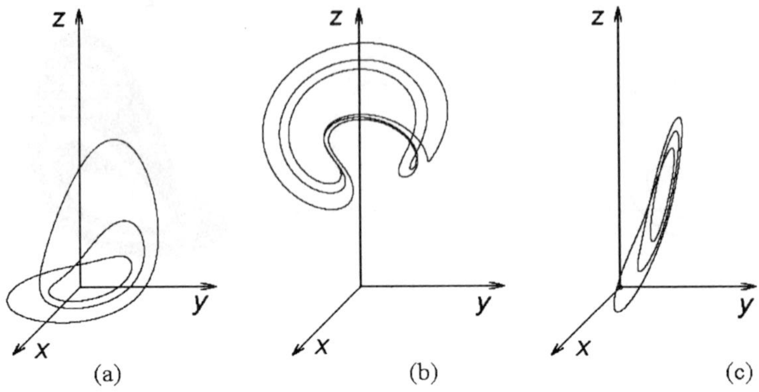

Fig. 2.26   Stable cycles of the period three in the Rossler (a), Lorenz (b) and Chua (c) systems.

It turned out, that not only Feigenbaum order, but also Sharkovskii order and even some more complex homoclinic order (see Sec. 2.4.4.3) take place for the typical cascade of bifurcations of stable cycles of nonlinear dynamical systems described by differential equations. An application to cascade of bifurcations in compliance with the Sharkovskii order the term subharmonic is quite justified as frequencies of born periodic orbits are subharmonics of the basic main frequency of an original cycle. The subharmonic cascade of bifurcations is present at the Lorenz, Rossler, Chua, Mackey–Glass, Magnitskii systems and many other dissipative systems of nonlinear differential equations. The main characteristic feature of cascade is the presence in a system of a stable limit cycle of the period three. Such cycles for some chaotic systems are shown in Fig. 2.26.

The subharmonic cascade of bifurcations of stable cycles generates infinite number of *cyclic subharmonic singular attractors* (in the sense of definition of Sec. 1.4.3), being much more complex attractors, than the elementary Feigenbaum attractor. Each of such attractors is generated by the cascade of the period doubling bifurcations of some stable cycle from the order (2.35), born as a result of corresponding saddle–node bifurcation. And so each of such singular attractors is an incomplete cyclic subharmonic attractor.   Complete cyclic subharmonic singular attractor arises after the cascade of period doubling bifurcations of a cycle of the period three. Complete cyclic subharmonic attractors of the Rossler, Lorenz and Chua systems are shown in Fig. 2.27.

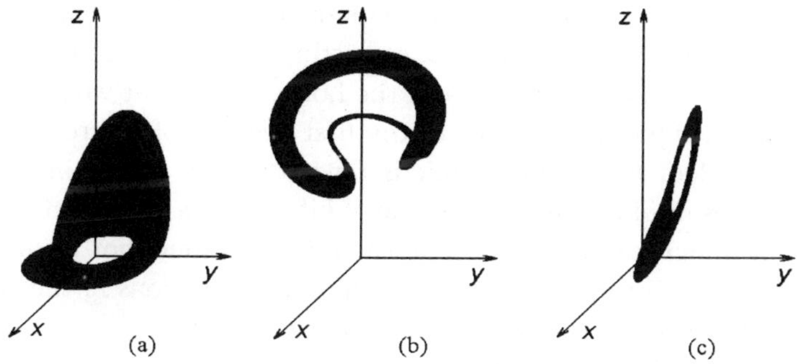

Fig. 2.27   Cyclic subharmonic singular attractors in the Rossler (a), Lorenz (b) and Chua (c) systems.

2.4.4.3   *Homoclinic cascade of bifurcations, Magnitskii scenario*

In many nonlinear dynamical systems described by the ordinary differential equations, process of complication of regimes of behaviour of trajectories does not come to the end with a birth of a complete cyclic subharmonic attractor. And it proceeds by the cascade of bifurcations of birth of stable homoclinic cycles tending to a homoclinic contour existing in the system, a separatrix loop of a saddle–focus, as a rule.

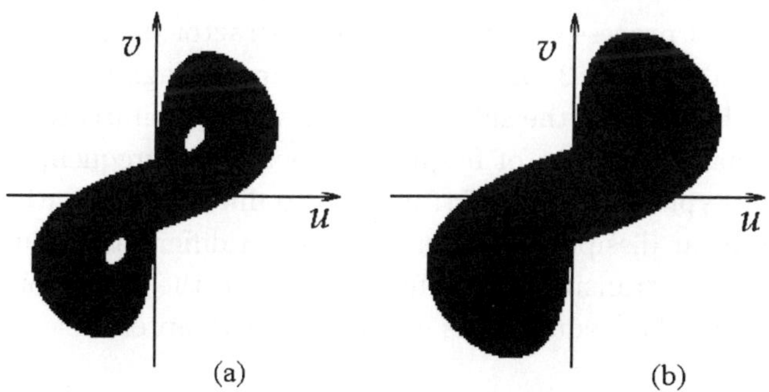

Fig. 2.28   Projections of classical incomplete double homoclinic Lorenz attractor (a) and complete double homoclinic singular attractor (b) in the Lorenz system.

The sequence of such cycles generating a double homoclinic cascade of bifurcations has been found out for the first time by authors in the Lorenz system (2.3) in the paper [Magnitskii and Sidorov (2001c)]. It has been shown, that as *incomplete* double *homoclinic cascade* of bifurcations, consisting of finite number of pairs of stable homoclinic cycles, as *complete* double *homoclinic cascade* of bifurcations, consisting of infinite number of pairs of stable homoclinic cycles, can exist in the Lorenz system at various values of system parameters. As a result of the first cascade of bifurcations the classical incomplete double homoclinic Lorenz attractor, possessing "eyes", is born. As a result of the second cascade of bifurcations a complete double *homoclinic singular attractor* is born in the Lorenz system, not having eyes (Fig. 2.28). In detail the homoclinic cascades of bifurcations in the Lorenz system and in other systems are considered in Chapter 3.

2.4.4.4 *Subharmonic cascade of bifurcations of stable two-dimensional tori*

It has been shown in [Magnitskii and Sidorov (2002)] at studying behaviour of solutions of complex system of the Lorenz Eqs. (2.20), that after formation of stable two-dimensional torus T^2 the further process of complication of solutions of the system (2.20) can occur through *period doubling and subharmonic* (on one frequency) *cascades of bifurcations of two-dimensional tori* at reduction of parameter values r_2. The basic characteristic feature of this cascade is the birth of stable two-dimensional torus of the period three, which is understood as a direct product of a simple cycle on a cycle of the period three. At the further reduction of values of bifurcation parameter r_2 a *complete subharmonic toroidal singular attractor* is born in the system. In detail process of a birth of such attractor in the system (2.20) is considered in Chapter 3. In the author's paper [Magnitskii and Sidorov (2005c)] it is shown, that the subharmonic cascade of bifurcations of stable two-dimensional tori at one of frequencies or at both frequencies simultaneously is the typical scenario of transition to diffusion or spatio-temporal chaos in nonlinear dissipative systems of partial differential equations. In detail a process of transition to diffusion chaos in the Kuramoto–Tsuzuki (or Ginzburg–Landau) equation is considered in Chapter 5.

2.4.4.5 *Cascade of Andronov–Hopf bifurcations, scenarios of Landau and Ruelle–Takens.*

In the middle of the XX century L. Landau and irrespective of him E. Hopf have put forward a hypothesis according to which chaotic dynamics of dissipative systems is a motion on an invariant torus of a large dimension, born as a result of a cascade of Andronov–Hopf bifurcations. As a result of the first such bifurcation there is a birth of a stable limit cycle, as a result of the second bifurcation there is a birth of an invariant two-dimensional torus. The further change of values of bifurcation parameter can lead to the sequence of bifurcations as a result of which a stable invariant tori of growing dimensions arise in the phase space. Finally a transition to complex quasiperiodic motion with k incommensurable frequencies is realized. Such motion looks like a chaotic motion if k is a big number.

However, *the Landau–Hopf scenario* has not found experimental confirmation. A sharp transition to chaotic motion is usually observed after a small number of bifurcations. For the first time Ruelle and Takens have paid attention in [Ruelle and Takens (1971)] to an opportunity of destruction of invariant torus of small dimension and birth of chaotic attractor. It has been shown by them together with Newhouse [Newhouse *et al.* (1979)], that three-frequency quasiperiodic motion, arising after three Andronov–Hopf bifurcations, is, as a rule, unstable and easily collapses, and a strange attractor can appear in place of destroyed three-dimensional torus. Such sequence of bifurcations, leading to formation of an irregular (strange) attractor, is called *a Ruelle–Takens scenario* of transition to chaos.

Ruelle–Takens scenario, as well as the Landau scenario is more likely a theoretically possible scenario of transition to chaos in nonlinear systems of differential equations. Only various modelling examples using, as a rule, iterations of two-dimensional mappings satisfy this scenario [Berger *et al.* (1984); Anishchenko *et al.* (1999)]. Any concrete examples of systems of differential equations in which transition to chaos would be carried out in a similar way are unknown to authors. In our opinion the situation is more natural when in real systems described by differential equations, first the occurrence of a stable two-dimensional torus T^2 takes place, and after that the further transition to chaos is carried out through a subharmonic cascade of bifurcations of two-dimensional tori on one or two frequencies. Such scenario takes place in the system of complex Lorenz Eqs. (2.20) and in systems of partial differential equations (see Chapters 3 and 5).

2.4.4.6 *Transition to chaos through an intermittency, the Pomeau–Manneville scenario*

The Pomeau–Manneville scenario of transition to chaos through an intermittency is one of three scenarios of occurrence of chaotic regimes of behaviour considered in the classical literature on nonlinear dissipative dynamical systems and synergetics along with the Feigenbaum and Ruelle–Takens scenarios [Berger *et al.* (1984); Loskutov and Mikhailov (1990); Haken (1983); Malinetskii and Potapov (2000); Schuster (1984)].

The phenomenon of intermittency, found out how it seemed, in the Lorenz system [Manneville and Pomeau (1980)], has no strictly formal definition. Its essence consists that the stable periodic solution of a system, disappearing at change of parameter, nevertheless leaves "memory" about itself. So trajectories of a system begin to make oscillations almost corresponding to oscillations of a disappeared stable periodic trajectory, but interrupted from time to time by abnormal chaotic fluctuations. The opportunity of existence of such phenomenon is illustrated mathematically only with examples of the elementary one-dimensional mappings. Therefore, after revealing the fact, that actually there are no intermittency and transition to chaos through an intermittency in the Lorenz system [Magnitskii and Sidorov (2001c)], and only owing to errors in calculations a jump of trajectory takes place from one fragment of a stable cycle to its another fragment close located to them in phase space, so, it follows to speak with care in each concrete case about the possibility of transition to chaos in complex nonlinear systems of differential equations through an intermittency. Any concrete examples of real systems of differential equations in which transition to chaos would be carried out in a similar way are unknown to the authors. The phenomenon of "intermittency" in the Lorenz system is in detail considered in Chapter 3.

2.4.5 *Bifurcations of irregular (singular) attractors*

As it was already repeatedly noticed above, except for regular attractors — stable singular points, cycles and tori, families of nonlinear systems of ordinary differential equations of a kind (2.1) can have other attractors, different from regular attractors by some features. In scientific literature there are various approaches to explanation of the nature of such attractors and the reasons of their occurrence. Uniform terminology is also absent (see Sec. 1.4.3). In Chapter 3 it is shown, that typical irregular attractor is not a rough formation which is born as a result of subharmonic or homoclinic cascade of soft bifurcations of regular attractors. Such attractor was named by us in Sec. 1.4.3 as singular attractor. The theory of singular attractors, named as *Feigenbaum–Sharkovskii–Magnitskii (FSM) theory* is stated in Chapter 4. The FSM-theory explains the presence of singular attractors in a wide class of systems of a kind (2.1) and completely coincides with results of numerical experiments. In Chapter 5 it is shown, that the FSM-theory is naturally transferred on many-dimensional and infinitely-dimensional systems of differential equations which also possess singular attractors.

Let us consider some possible bifurcations of singular attractors. As an infinite number of points of existence of various regular attractors exist in any neighbourhood of a point of existence of a singular attractor in a space of parameters, and complexity of singular attractors grows during cascades of bifurcations, then the following bifurcations are typical bifurcations for singular attractors.

2.4.5.1 *Bifurcation of formation of a regular attractor (a stable cycle)*

Let the parameter value $\mu = 0$ be a point of accumulation of some infinite cascade of parameter values $\mu_n < 0$ in which bifurcations of stable cycles of family (2.1) occur so, that $\mu_n \to 0$ when $n \to \infty$ (in occasion of cascades of bifurcations of stable cycles and tori look through Sec. 2.4.4 and Chapters 3 and 4). Then at $\mu = 0$ the system (2.1) has a not rough singular attractor. It is obvious, that there can be one of an infinite number of bifurcations of formation of some stable cycle at small perturbations of parameter μ in the area of negative values. At small perturbation of parameter in the area of positive values there can be also one of infinite number of bifurcations of formation of some stable cycle of some other infinite cascade of bifurcations. The elementary singular attractor of such type is the Feigenbaum attractor (see Chapters 3 and 4).

2.4.5.2 *Bifurcation of formation of a regular attractor (a stable torus)*

Let the parameter value $\mu = 0$ be a point of accumulation of some infinite cascade of parameter values $\mu_n < 0$ in which the period doubling bifurcations of stable tori of the family (2.1) occur so, that $\mu_n \to 0$ when $n \to \infty$. Then at $\mu = 0$ the system (2.1) has a not rough singular toroidal attractor. It is obvious, that there can be one of infinite number of bifurcations of formation of some stable torus of some period at small perturbation of parameter μ in an area of negative values. A singular attractor of such type is described in Sec. 3.2.

2.4.5.3   *Bifurcation of formation of singular attractor*

Let us in the first case, the parameter value $\mu = 0$ be a point of accumulation of some infinite cascade of parameter values $\mu_n < 0$ in which bifurcations of stable cycles of the family (2.1) occur so, that $\mu_n \to 0$ when $n \to \infty$. Then at $\mu = 0$ the system (2.1) has non-rough singular attractor (for example, a singular Feigenbaum attractor). At a small perturbation of parameter μ in an area of positive values there can be also one of bifurcations of formation of some other singular attractor alongside with a bifurcation of formation of some stable cycle. It can happen if the perturbed parameter value will be a point of accumulation of some other infinite cascade of bifurcations of stable cycles.

2.4.5.4   *Bifurcation of coherence*

This bifurcation consists in unification of parts of a singular attractor of the family of systems (2.1), attended by the everywhere dense trajectory in regular order. It is realized an infinite number of times in any subharmonic cascade of bifurcations, following the Feigenbaum period doubling cascade of bifurcations. That is, the complication of structure of each subsequent singular attractor, since the elementary singular Feigenbaum attractor, occurs as a result of unification of parts of the previous singular attractor (see Chapter 3). At that unification of parts of a singular attractor occurs by the return way to the order of the cascade of the period doubling bifurcations of stable cycles, which led to the birth of the Feigenbaum attractor.

2.4.5.5   *Bifurcation of unification of singular attractors*

Bifurcation of unification of two various singular attractors can occur also in the family of systems (2.1) similarly to unification of parts of one singular attractor. Such bifurcation is characteristic for systems with several singular points of saddle–focus type. In particular it can be observed at some values of system parameters in the system of Lorenz equations when there occurs a merging of two tapes of two subharmonic attractors, formed around two saddle–focuses of the system, in one singular attractor, covering spatial area around both singular points (see in detail Chapter 3).

Possible bifurcations of singular attractors of the family of systems of nonlinear ordinary differential Eqs. (2.1) considered above base on the large actual material resulted in Chapters 3 and 5, and on the strict mathematical theory of singular attractors stated in Chapter 4.

Chapter 3

Chaotic Systems of Ordinary Differential Equations

We shall name a nonlinear system of ordinary differential equations having irregular attractors as *chaotic system*. In this chapter examples of both classical, and less known chaotic systems are considered. It is shown, that all of them without exception have singular (see Sec. 1.4.3 and Chapter 4) attractors, and transition to chaos in all these systems is carried out either through subharmonic or homoclinic cascades of bifurcations.

3.1 System of the Lorenz Equations

The system of three nonlinear ordinary differential equations

$$\dot{x} = \sigma(y - x),$$
$$\dot{y} = x(r - z) - y, \qquad\qquad (3.1)$$
$$\dot{z} = xy - bz,$$

known as the Lorenz system [Lorenz (1963)], historically is the first dynamical system for which the existence of an irregular attractor (namely, the Lorenz attractor for $\sigma = 10$, $b = 8/3$, and $24.06 < r < 28$) was proved. For many years Lorenz system was been the subject of study by numerous authors, whose results were included in textbooks and university courses (see Refs. in Sec. 2.4.1 and Sec. 2.4.4). However, it did not lead to clearness in the question that Lorenz attractor represents from itself.

3.1.1 *Classical scenario of birth of the Lorenz attractor*

We shall briefly present in the form of theses the contemporary point of view on the structure of the Lorenz attractor and causes for its appearance.

107

(1) The origin $O(0,0,0)$ is a singular (stationary) point in system (3.1) for any r, σ and b. It is stable for $r < 1$.

(2) If $r > 1$, then in the system, there are two more stationary points

$$O_1\left(\sqrt{b(r-1)},\ \sqrt{b(r-1)},\ r-1\right),$$
$$O_2\left(-\sqrt{b(r-1)},\ -\sqrt{b(r-1)},\ r-1\right),$$

which are stable up to the parameter value $r_c = \dfrac{\sigma(\sigma+b+3)}{\sigma-b-1}$, ($r_c = 24.7368$ for $\sigma = 10$ and $b = 8/3$). For all $r > 1$, the point O is a saddle–node. It has a two-dimensional stable manifold W^s and a one-dimensional unstable manifold W^u.

(3) If $1 < r < r_1 \approx 13.926$, then separatrices Γ_1 and Γ_2 issuing from the point O along its one-dimensional unstable manifold W^u are attracted by their nearest stable points O_1 and O_2, respectively.

(4) If $r = r_1$, then each of the separatrices Γ_1 and Γ_2 becomes a closed homoclinic loop. In this case, two homoclinic loops are tangent to each other and the z-axis at the point O and form a figure referred to as a homoclinic butterfly. It is assumed that generation of an unstable homoclinic butterfly is one of the two bifurcations leading to the appearance of Lorenz attractor.

(5) If $r_1 < r < r_2 \approx 24.060$, then a saddle periodic trajectory bifurcates from each of the closed homoclinic loops (these trajectories will be denoted by L_1 and L_2, respectively). In this case, separatrices Γ_1 and Γ_2 tend to the stable points O_2 and O_1, respectively. It is usually assumed that stable manifolds of the saddle periodic trajectories L_1 and L_2 are the boundaries of attraction domains of points O_1 and O_2. A curve issuing from the exterior of these domains can make oscillations from the neighborhood of L_1 into a neighborhood of L_2 and conversely until it enters the attraction domain of the attractor O_1 or O_2; the closer is parameter r to the value r_2, the larger is the number of oscillations. This behavior of the system is referred to as metastable chaos.

(6) If $r = r_2$, then separatrices Γ_1 and Γ_2 do not tend to the points O_2 and O_1 but wind around the limit saddle cycles L_2 and L_1, respectively. Here the second bifurcation leading to the appearance of the Lorenz attractor takes place.

(7) If $r_2 < r < r_3 = r_c$, then points O_1 and O_2 are still stable. In addition, in the phase space, there is an attracting set B referred to as the Lorenz attractor; it is a set of integral curves moving from L_1 to L_2 and vice

versa. The saddle point O, together with its separatrices Γ_1 and Γ_2, belongs to the attractor.

(8) If $r \to r_3 = r_c$, then the saddle limit cycles L_1 and L_2 shrink to points O_1 and O_2; for $r = r_3$ they vanish and coincide with these points as a result of the Andronov–Hopf subcritical bifurcation.

(9) If $r_3 < r < r_4 \approx 30.1$, then the Lorenz attractor is the unique stable limit set of system (3.1). It is usually assumed that this set is a branching surface S lying near the plane $x - y = 0$ and consisting of infinitely many sheets tied together and infinitely close to each other. A phase trajectory issuing on the left from the z-axis comes untwisted along a spiral around the point O_1 until the transition to the right of the z-axis, after which it becomes untwisted along a spiral around the point O_2 in the opposite direction. The number of rotations around the points O_1 and O_2 varies irregularly; thus the motion looks chaotic. It is assumed that the attractor is not a two-dimensional manifold and has a fractal structure; its fractal dimension equals to 2.05 ± 0.01 for $\sigma = 10$, $b = 8/3$, and $r = 28$ [Schuster (1984)].

(10) If $r_4 < r \lesssim 313$, then the structure of solutions of the system of Lorenz equations becomes extremely complicated with alternation of chaotic and periodic modes. Several periodicity windows are known in this range of parameter r, for example, $99.524 < r < 100.795$, $145 < r < 166$, and $214.364 < r < 313$. It is usually assumed that there may be infinitely many periodicity windows, and each of such windows is a direct subharmonic cascade of bifurcations, which terminates with a basic stable limit cycle. For further growth of r, each of such cycles is destroyed by an intermittency of the first type, and appearance of periodicity window is preceded by the inverse cascade of bifurcations (for example, an inverse cascade exists for $197.4 < r < 214.364$, and an intermittency of the first type exists for $r > 166.07$ [Berger *et al.* (1984)]).

(11) If $r > 313$, then the unique stable limit cycle is an attractor in the Lorenz system.

Thus, items (1)–(11) contain basic commonly accepted assertions dealing with the Lorenz attractor and the scenarios of its appearance (and vanishing). Note that all these assertions are based only on computer experiments and speculative arguments rather than any analytic proofs. Some of these assertions can readily be verified, and their validity is not brought into question anywhere (for example, the assertions of items (1), (2), and (11)).

However, other assertions are difficult to verify, and they always look quite dubious. Thus, for example, while saddle periodic cycles L_1 and L_2 are really generated from homoclinic loops for $r = r_1$, and they are those determining "eyes" of the Lorenz attractor for $r = r_2$, but why are these eyes observed in the attractor in the case $r > r_3$ as well in which the cycles L_1 and L_2 already vanished? The only possible conclusion is the following: the eyes of attractor are not determined by the saddle cycles L_1 and L_2 even if they exist. But if they also exist, at all it is unessential that they are born at $r = r_1$, as a result of bifurcation of a homoclinic butterfly. Also, assertions on the structure of attractor and its dimension found on computer with incredible accuracy have always been questionable. Finally, the phenomenon of intermittency did not find its logic explanation.

It had been shown by authors in the paper [Magnitskii and Sidorov (2001c)], that actually in Lorenz system absolutely another scenario of transition to chaos is realized, through a double homoclinic cascade of bifurcations. We shall consider this scenario in more detail.

3.1.2 *Scenario of birth of the Lorenz attractor through an incomplete double homoclinic cascade of bifurcations*

An opportunity of existence in the Lorenz system (3.1) of various homoclinic and heteroclinic contours of singular points and the cascades of bifurcations connected with them follows from the presence in the system of a saddle–node and two saddle–focuses.

Definition 3.1 We shall name as *complete (incomplete) homoclinic cascade of bifurcations* in a system of nonlinear differential equations the cascade of bifurcations of stable limit cycles, generated by a unique original stable limit cycle at change of values of bifurcation parameter along a straight line which is passing in space of parameters through a point (near to a point) of existence of the homoclinic contour.

Definition 3.2 Any singular attractor, born as a result of complete (incomplete) homoclinic cascade of bifurcations of limit cycles, we shall name as *complete (incomplete) homoclinic attractor.*

If the straight line, along which the values of bifurcation parameter change, passes in space of parameters through a point (near to a point) of simultaneous existence of two homoclinic contours, then we shall name the formed cascade of bifurcations of stable cycles as *complete (incomplete) double homoclinic cascade.* And we shall name any singular attractor born as a result of such cascade as *complete (incomplete) double homoclinic attractor.*

3.1.2.1 *Method for investigating the Lorenz attractor*

It turns out that all cycles from infinite family of unstable cycles, generating Lorenz attractor, have crossing with one-dimensional unstable not invariant manifold V^u of the point O (do not confuse with the invariant unstable manifold W^u). This result though is not strictly proved directly for the Lorenz system, but it follows from the theory of dynamical chaos stated below in Chapter 4. Consequently, after the derivation of analytic formulas for the manifold V^u it becomes possible to reduce the problem of establishing and proving the existence of unstable cycles in the Lorenz system to the one-dimensional case, namely, to finding stable points of the one-dimensional first return mapping defined on the unstable manifold. Note that the method described below also makes it possible to find arbitrary stable cycles taking part in the formation of the attractor. Therefore, if the initial point lies on the one-dimensional manifold V^u, then one can find an arbitrary (either stable or not) cycle existing in the Lorenz system. It is shown below by this method that items (4) and (5) of the above-represented classical scenarios of transition to chaos in the Lorenz system are invalid. Some assertions of items (6)–(10) fail, while other require a more detailed investigation.

For each given parameter set $(\sigma,\ b,\ r > 1)$, we reduce the system of Eqs. (3.1) with the use of the nondegenerate change of variables

$$x = u + v,\ \sigma(y - x) = \lambda_1 u + \lambda_2 v,\ z = z \tag{3.2}$$

to principal axes on the plane $z = 0$. Since

$$\begin{pmatrix} x \\ y \end{pmatrix} = A \begin{pmatrix} u \\ v \end{pmatrix} = \begin{pmatrix} 1 & 1 \\ \dfrac{\lambda_1 + \sigma}{\sigma} & \dfrac{\lambda_2 + \sigma}{\sigma} \end{pmatrix} \begin{pmatrix} u \\ v \end{pmatrix},$$

$$A^{-1} = \frac{\sigma}{\lambda_2 - \lambda_1} \begin{pmatrix} \dfrac{\lambda_2 + \sigma}{\sigma} & -1 \\ -\dfrac{\lambda_1 + \sigma}{\sigma} & 1 \end{pmatrix},$$

we have

$$\begin{pmatrix} \dot{x} \\ \dot{y} \\ \dot{z} \end{pmatrix} = \begin{pmatrix} -\sigma & \sigma & 0 \\ r & -1 & 0 \\ 0 & 0 & -b \end{pmatrix} \begin{pmatrix} A & 0 \\ 0 & 1 \end{pmatrix} \begin{pmatrix} u \\ v \\ z \end{pmatrix} + \begin{pmatrix} 0 \\ -xz \\ xy \end{pmatrix},$$

or

$$\begin{pmatrix} \dot{u} \\ \dot{v} \\ \dot{z} \end{pmatrix} = \begin{pmatrix} \Lambda & 0 \\ 0 & -b \end{pmatrix} \begin{pmatrix} u \\ v \\ z \end{pmatrix} + \begin{pmatrix} \sigma xz/(\lambda_2 - \lambda_1) \\ \sigma xz/(\lambda_2 - \lambda_1) \\ xy \end{pmatrix},$$

where $\Lambda = A^{-1} \begin{pmatrix} -\sigma & \sigma \\ r & -1 \end{pmatrix} A =$

$$\begin{pmatrix} \dfrac{\lambda_1 \lambda_2 + (\sigma + 1)\lambda_1 - \sigma(r - 1)}{\lambda_2 - \lambda_1} & \dfrac{\lambda_2^2 + (\sigma + 1)\lambda_2 - \sigma(r - 1)}{\lambda_2 - \lambda_1} \\ \dfrac{-\lambda_1^2 - (\sigma + 1)\lambda_1 + \sigma(r - 1)}{\lambda_2 - \lambda_1} & \dfrac{-\lambda_1 \lambda_2 - (\sigma + 1)\lambda_1 + \sigma(r - 1)}{\lambda_2 - \lambda_1} \end{pmatrix}.$$

Consequently, if the quantities λ_1 and λ_2 are roots of the equation

$$\lambda^2 + (\sigma + 1)\lambda - \sigma(r - 1) = 0,$$

i.e.

$$\lambda_1 = -\frac{\sigma + 1}{2} - \sqrt{\frac{(\sigma + 1)^2}{4} + \sigma(r - 1)} < 0,$$

$$\lambda_2 = -\frac{\sigma + 1}{2} + \sqrt{\frac{(\sigma + 1)^2}{4} + \sigma(r - 1)} > 0,$$

then $\Lambda = \mathrm{diag}(\lambda_1, \lambda_2)$. Therefore, system (3.1) acquires the form

$$\begin{aligned} \dot{u} &= \lambda_1 u + \frac{\sigma z(u + v)}{\lambda_2 - \lambda_1}, \\ \dot{v} &= \lambda_2 v - \frac{\sigma z(u + v)}{\lambda_2 - \lambda_1}, \\ \dot{z} &= -bz + (u + v)^2 + \frac{(u + v)(\lambda_1 u + \lambda_2 v)}{\sigma}. \end{aligned} \tag{3.3}$$

The notion of Lorenz equations in the variables (u, v, z) is more convenient than the corresponding notion in the variables (x, y, z), since for any r, the stable manifold W^s of the origin (that is, the point O) is tangent to the u-axis and the unstable manifold W^u is tangent to the v-axis. This fact allows to see clearly some new important characteristics of the attractor, which are invariant with respect to the parameter r.

Note that projections of the fixed points

$$O_1(\lambda_2\sqrt{b(r-1)}/(\lambda_2 - \lambda_1),\ -\lambda_1\sqrt{b(r-1)}/(\lambda_2 - \lambda_1),\ r-1),$$
$$O_2(-\lambda_2\sqrt{b(r-1)}/(\lambda_2 - \lambda_1),\ \lambda_1\sqrt{b(r-1)}/(\lambda_2 - \lambda_1),\ r-1)$$

of system (3.3) on the plane $z = 0$ lie symmetrically around the z-axis in the first and third quadrants of the plane (u, v). If we simultaneously look at projections of an arbitrary trajectory Γ of system (3.3) on the planes (u, v) and (u, z) for a value of the parameter r corresponding to the chaotic behavior of the system (for example, $r = 28$), then we can see that the trajectory belongs to two sets, which are sheets of two dimensional surface, described in Chapter 4. The first of these sets S, has the form of lateral surfaces of two deformed truncated cones with vertices the points O_1 and O_2. Each cone lies on its generatrix $z = z_s(u, v)$, $0 < z_{s\,\min} \leq z \leq z_{s\,\max} < r - 1$, on which local minima of various trajectories with respect to z are achieved under their conditional rotation around the point O_1 (respectively, O_2). In the projection on the plane $z = 0$, rotation takes place in the clockwise direction. The top part of the cone is deformed and curved. The smaller face of the cone determines an eye of the attractor, inside which a trajectory does not enter, and the larger face determines the maximal size of the attractor for which the trajectory passes from rotation around the point O_1 (respectively, O_2) to rotation around the point O_2 (respectively, O_1). This passage is performed along two symmetric parts of the second set G on which the trajectory also has points of local minimum with respect to z, which lie on some curve $z = z_g(u, v)$, where $0 < z_{g\,\min} \leq z \leq z_{g\,\max} < r - 1$. As it was mentioned above, it is usually assumed that the attractor eye is determined by a saddle cycle appearing from a homoclinic contour for $r = r_1$. But if $r = 28$, then the cycle is already absent, but the eye exists, although it is smaller than in the case of $r = r_2 \approx 24.06$. The causes of this phenomenon are considered below, and now we note that it is natural to assume that the attractor eyes are determined not by saddle cycles appearing around the stationary points O_1 and O_2 but by some other cycles surrounding simultaneously both points O_1 and O_2. Then the trajectories of these cycles must have local minima on the curve $z_g(u, v)$. Thus, finding an equation of that curve, we virtually reduce the problem of finding cycles to the one-dimensional case. To this end, in system (3.1), we perform a substitution similar to (3.2) but depending on z

$$x = \widetilde{u} + \widetilde{v}, \ \sigma(y - x) = \lambda_1(z)\widetilde{u} + \lambda_2(z)\widetilde{v}, \ z = z.$$

Then

$$\begin{pmatrix} x \\ y \end{pmatrix} = A(z) \begin{pmatrix} \widetilde{u} \\ \widetilde{v} \end{pmatrix} = \begin{pmatrix} 1 & 1 \\ \dfrac{\lambda_1(z) + \sigma}{\sigma} & \dfrac{\lambda_2(z) + \sigma}{\sigma} \end{pmatrix} \begin{pmatrix} \widetilde{u} \\ \widetilde{v} \end{pmatrix},$$

$$A^{-1}(z) = \frac{\sigma}{\lambda_2(z) - \lambda_1(z)} \begin{pmatrix} \dfrac{\lambda_2(z) + \sigma}{\sigma} & -1 \\ -\dfrac{\lambda_1(z) + \sigma}{\sigma} & 1 \end{pmatrix},$$

and consequently,

$$\begin{pmatrix} \dot{A}(z) & 0 \\ 0 & 0 \end{pmatrix} \begin{pmatrix} \widetilde{u} \\ \widetilde{v} \\ z \end{pmatrix} + \begin{pmatrix} A(z) & 0 \\ 0 & 1 \end{pmatrix} \begin{pmatrix} \dot{\widetilde{u}} \\ \dot{\widetilde{v}} \\ \dot{z} \end{pmatrix} = \begin{pmatrix} -\sigma & \sigma & 0 \\ r - z & -1 & 0 \\ 0 & 0 & -b \end{pmatrix} \begin{pmatrix} A(z) & 0 \\ 0 & 1 \end{pmatrix} \begin{pmatrix} \widetilde{u} \\ \widetilde{v} \\ z \end{pmatrix} + \begin{pmatrix} 0 \\ 0 \\ xy \end{pmatrix}.$$

Therefore,

$$\begin{pmatrix} \dot{\widetilde{u}} \\ \dot{\widetilde{v}} \\ \dot{z} \end{pmatrix} = \begin{pmatrix} \Lambda(z) & 0 \\ 0 & -b \end{pmatrix} \begin{pmatrix} \widetilde{u} \\ \widetilde{v} \\ z \end{pmatrix} + \begin{pmatrix} 0 \\ 0 \\ xy \end{pmatrix},$$

where

$$\Lambda(z) = A^{-1}(z) \begin{pmatrix} -\sigma & \sigma \\ r - z & -1 \end{pmatrix} A(z) - A^{-1}(z)\dot{A}(z)$$

$$= \frac{1}{\lambda_2(z) - \lambda_1(z)} \left[\begin{pmatrix} \Lambda_{11} & \Lambda_{12} \\ \Lambda_{21} & \Lambda_{22} \end{pmatrix} - \begin{pmatrix} -\dot{\lambda}_1(z) & -\dot{\lambda}_2(z) \\ \dot{\lambda}_1(z) & \dot{\lambda}_2(z) \end{pmatrix} \right],$$

$$\Lambda_{11} = \lambda_1(z)\lambda_2(z) + (\sigma + 1)\lambda_1(z) - \sigma(r - z - 1),$$
$$\Lambda_{12} = \lambda_2^2(z) + (\sigma + 1)\lambda_2(z) - \sigma(r - z - 1),$$
$$\Lambda_{21} = -\lambda_1^2(z) - (\sigma + 1)\lambda_1(z) + \sigma(r - z - 1),$$
$$\Lambda_{22} = -\lambda_1(z)\lambda_2(z) - (\sigma + 1)\lambda_1(z) + \sigma(r - z - 1).$$

Consequently, if the quantities $\lambda_1(z)$ and $\lambda_2(z)$ are roots of the equation

$$\lambda^2 + (\sigma + 1)\lambda - \sigma(r - z - 1) = 0,$$

i.e.

$$\lambda_1(z) = -\frac{\sigma+1}{2} - \sqrt{\frac{(\sigma+1)^2}{4} + \sigma(r - z - 1)} < 0,$$

$$\lambda_2(z) = -\frac{\sigma+1}{2} + \sqrt{\frac{(\sigma+1)^2}{4} + \sigma(r - z - 1)} > 0,$$

then

$$\Lambda(z) = \mathrm{diag}(\lambda_1(z), \lambda_2(z)) + \frac{\sigma\dot{z}}{(\lambda_2(z) - \lambda_1(z))^2} \begin{pmatrix} 1 & -1 \\ -1 & 1 \end{pmatrix}.$$

In this case, system (3.1) can be reduced to the form

$$\dot{\widetilde{u}} = \lambda_1(z)\widetilde{u} + \frac{\sigma\dot{z}(\widetilde{u} - \widetilde{v})}{(\lambda_2(z) - \lambda_1(z)^2},$$

$$\dot{\widetilde{v}} = \lambda_2(z)\widetilde{v} - \frac{\sigma\dot{z}(\widetilde{u} - \widetilde{v})}{(\lambda_2(z) - \lambda_1(z))^2}, \tag{3.4}$$

$$\dot{z} = -bz + (\widetilde{u} + \widetilde{v})^2 + \frac{(\widetilde{u} + \widetilde{v})(\lambda_1(z)\widetilde{u} + \lambda_2(z)\widetilde{v})}{\sigma}.$$

Obviously, system (3.4) is solvable only for $z < r$ rather than for all initial conditions and hence is not equivalent to system (3.1). However, the points of the curve $z_g(u, v)$ lie in this range and hence must satisfy system (3.4). It is also obvious that, at least in some neighborhood of the origin (the point O), the curve $z_g(u, v)$ must be related in some way to the one-dimensional unstable manifold V^u of the point O of system (3.3) as well as system (3.4). It is remarkable that the manifold V^u can be explicitly found from system (3.4). Indeed, since each point of manifold V^u should not move in a plane parallel to stable manifold W^s of the point O, it follows that the conditions $\dot{z} = 0$ and $\dot{\widetilde{u}} = 0$ must necessarily be satisfied on V^u. Moreover, it follows from system (3.4) that V^u simply coincides with the direction of $\widetilde{v}$. By setting $\dot{z} = 0$ and $\dot{\widetilde{u}} = 0$ (and hence $\widetilde{u} = 0$ in (3.4)), we obtain

$$\widetilde{v} = \pm\sqrt{\frac{b\sigma z}{\sigma + \lambda_2(z)}}. \tag{3.5}$$

Further, since

$$\begin{pmatrix} x \\ y \end{pmatrix} = A \begin{pmatrix} u \\ v \end{pmatrix} = A(z) \begin{pmatrix} \widetilde{u} \\ \widetilde{v} \end{pmatrix},$$

then

$$\begin{pmatrix} u \\ v \end{pmatrix} = A^{-1}A(z)\begin{pmatrix} \widetilde{u} \\ \widetilde{v} \end{pmatrix} = \frac{1}{\lambda_2 - \lambda_1}\begin{pmatrix} \lambda_2 - \lambda_1(z) & \lambda_2 - \lambda_2(z) \\ \lambda_1(z) - \lambda_1 & \lambda_2(z) - \lambda_1 \end{pmatrix}\begin{pmatrix} \widetilde{u} \\ \widetilde{v} \end{pmatrix}.$$

On the other hand,

$$\begin{pmatrix} \widetilde{u} \\ \widetilde{v} \end{pmatrix} = A^{-1}(z)A\begin{pmatrix} u \\ v \end{pmatrix} = \frac{1}{\lambda_2(z) - \lambda_1(z)}\begin{pmatrix} \lambda_2(z) - \lambda_1 & \lambda_2(z) - \lambda_2 \\ \lambda_1 - \lambda_1(z) & \lambda_2 - \lambda_1(z) \end{pmatrix}\begin{pmatrix} u \\ v \end{pmatrix}.$$

Consequently, the point $(0, \widetilde{v}(z), z)$ of the $(\widetilde{u}, \widetilde{v}, z)$-space corresponds to the point $(u(z), v(z), z)$ of the space (u, v, z), lying on the unstable manifold V^u of the point O of system (3.3) and satisfying the equations

$$u(z) = \frac{\lambda_2 - \lambda_2(z)}{\lambda_2 - \lambda_1}\widetilde{v}(z), \ \ v(z) = \frac{\lambda_2(z) - \lambda_1}{\lambda_2 - \lambda_1}\widetilde{v}(z). \tag{3.6}$$

Obviously, the curve described by Eqs. (3.5) and (3.6) is tangent to the v-axis as $z \to 0$, since

$$\frac{dv}{du} = \frac{\lambda_2(z) - \lambda_1}{\lambda_2 - \lambda_2(z)} \to \infty$$

along this curve as $z \to 0$.

Since $\dot{z} = 0$ on the resulting curve, it follows that this curve (the manifold V^u) is the desired curve $z = z_g(u(z), v(z))$ if $0 < z < r - 1$, and the passage of a trajectory in the Lorenz attractor from one part of the set S into the other (from one half-space into another) takes place through the unstable one-dimensional manifold V^u of the point O. The problem of finding unstable trajectories of the Lorenz attractor and, in particular, all unstable cycles in this attractor is thereby reduced to the one-dimensional case and can be solved by the method of returns to the one-dimensional unstable manifold V^u of the point O.

We define the first return mapping $f_1(z)$ as follows. For every z_0, $0 < z_0 < r - 1$, we draw a trajectory of system (3.3) issuing from the point $(z_0, u(z_0), v(z_0)) \in V^u$, which lies, for example, in the right half-space. This trajectory makes some number of rotations around the point O_1 in the right half-space, then passes into the left half-space, and again makes some number of rotations around the point O_2. After that the trajectory returns (for the first time) to the original right half-space; before this, it intersects the plane $\widetilde{u} = 0$ of system (3.4) at a point of the manifold V^u but, possibly, for a different value of the variable $z = f_1(z_0)$. The point of intersection of the trajectory with the plane $\widetilde{u} = 0$ is found from the equation

$$\widetilde{u}(t) = \frac{\lambda_2(z) - \lambda_1}{\lambda_2(z) - \lambda_1(z)} u(t) + \frac{\lambda_2(z) - \lambda_2}{\lambda_2(z) - \lambda_1(z)} v(t) = 0. \qquad (3.7)$$

Eq. (3.7) must be solved rather precisely. Therefore, in a neighborhood of the plane $\widetilde{u} = 0$, whose position in the plane (u, v) depends on the value of the variable z, system (3.3) must be integrated with a small step h (for example, $h \sim 10^{-7}$ in the fourth-order Runge–Kutta method). Associated with the solution of Eq. (3.7) is the time T of first return to the unstable manifold V^u, and the value $z(T) = f_1(z_0)$ of the variable z determines the first return mapping. In a similar way, we can define the k-th return mapping $f_k(z_0)$ of the manifold V^u.

Since the return mapping of arbitrary order is a continuous one-dimensional mapping, it follows that the above-described method allows one not only to find any unstable cycle in the Lorenz system with arbitrary accuracy as a fixed point of the corresponding return mapping but also to prove the existence of such a cycle beyond doubt. From this viewpoint, the method of returns to the one-dimensional unstable manifold has an obvious advantage over the well-known Poincare section method. The accuracy of the initial value z_0 was $10^{-10} - 10^{-15}$ in some cases to avoid ill-posedness of the solved problem.

3.1.2.2 *Scenario of transition to chaos*

Now we present a scenario of the appearance of the Lorenz attractor obtained with the use of the above-described method.

Items (1)–(3) and (11) of the scenario given at the beginning of the present chapter remain the same.

(4) If $r = r_1 \approx 13.926667$, then the separatrices Γ_1 and Γ_2 do not form two separate homoclinic loops. Here we have a bifurcation with the generation of a single closed contour surrounding both stationary points O_1 and O_2; the end of the separatrix Γ_1 enters the beginning of the separatrix Γ_2, and vice versa, the end of Γ_2 enters the beginning of Γ_1. As r grows, from this contour, a closed cycle C_0 appears there first. It is an eight-shaped figure surrounding both points O_1 and O_2. It can be clearly observed for $r = 14$ ($z \approx 0.152293$, $h = 10^{-4}$). We note that it is this cycle that exists in the Lorenz system for all $r > r_1$ and is the unique stable attractor of the system for $r > 313$ (see Fig. 3.1).

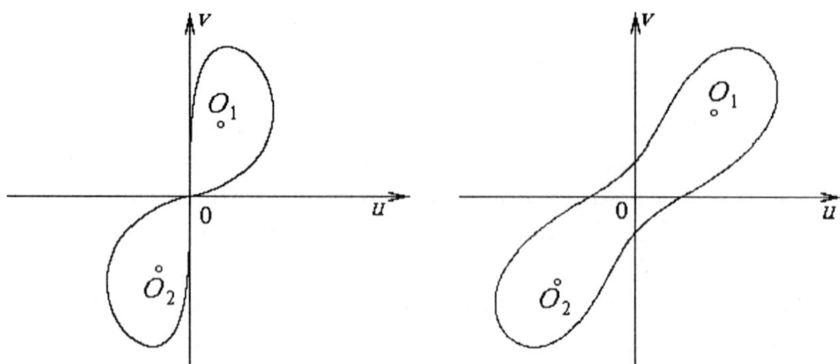

Fig. 3.1   Projections of the phase portrait of the cycle C_0 on the plane (u, v) for $r = 14$ (left) and $r = 350$ (right).

(5) If $r_1 < r < r_2 \approx 24.058$, then cycles L_1 and L_2 surrounding the points O_1 and O_2, respectively, do not appear; but with further growth of r, pairs of cycles C_n^+, C_n^-, $n = 0, 1, \ldots$, are successively generated. They determine the generation of the Lorenz attractor. The cycle C_n^+ makes n complete rotations in the half-space containing the point O_1 and one incomplete rotation around the point O_2. Conversely, the cycle C_n^- makes n complete rotations around the point O_2 and one incomplete rotation around the point O_1. The cycles C_1^+ and C_1^- are also clearly observable for $r = 14$.

For each r, $r_1 < r < r_2$, there exists the number $n(r)$ $(n(r) \to \infty$ as $r \to r_2)$ such that in the (u, v, z)-space, there are unstable cycles C_0, C_k^+, C_k^-, $k = 0, \ldots, n$, and cycles C_{km}^+, C_{km}^-, $k, m < n$, which make k rotations around the point O_1 and m rotations around the point O_2 and are various combinations of the cycles C_n^+ and C_n^-, and many other cycles generated by bifurcations of the cycles C_n^+ and C_n^- (see item (10) below). Points of intersection of all these cycles with the manifold V^u have the following arrangement on the curve V^u for $0 \le z_{min} \le z \le z_{max} < r - 1$. The point z_{min} corresponds to the right large single loop of the cycle C_n^-. This loop is the larger face of the right truncated cone of the set S. Further, the trajectory of the cycle passes into the left half-plane and makes n clockwise rotations around the point O_2. The smallest first loop around the point O_2 is the smaller face of the truncated cone of the set S. The point z_{max} corresponds to the smallest loop of the cycle C_n^+ around the point O_1. This loop is the smaller face of the right truncated cone. Further, the trajectory of this cycle makes n rotations around the point O_1 clockwise, passes into the left half-plane, and make one large rotation around the point O_2. This rotation is the larger face of the left truncated cone. Between the points $z_{\min}$ and $z_{\max}$ there is a point z_0 corresponding to the main cycle C_0. Note that all cycles C_{km}^+ obtained for $z > z_0$ are symmetric to the cycles C_{km}^- formed for $z < z_0$ in the z-axis and can be found by the change of sign of the initial point in (3.5). Therefore, to find all cycles for given r, it suffices to find the pairs of cycles for the interval $z_{\min} \leq z \leq z_0$. This is also advisable for the following reason. The cycles found for $z_{\min} \leq z \leq z_0$ make a small number of rotations around the point O_1 in the right half-plane and then pass into the left half-plane. These rotations cannot be mistaken for a multiple rotation around some single saddle cycle. Conversely, the cycles observed for $z_0 < z \leq z_{\max}$ make a very large number of rotations around the point O_1 in the right half-plane. This looks like a rotation around a single limit cycle, the mythical saddle cycle L_1 in the conventional scenario. This illusion deepens as r approaches r_2 and the number of rotations of the cycle C_n^+ grows infinitely. The cycles C_{237}^+ and C_{237}^- with $r = 23.5$ are shown in Fig. 3.2.

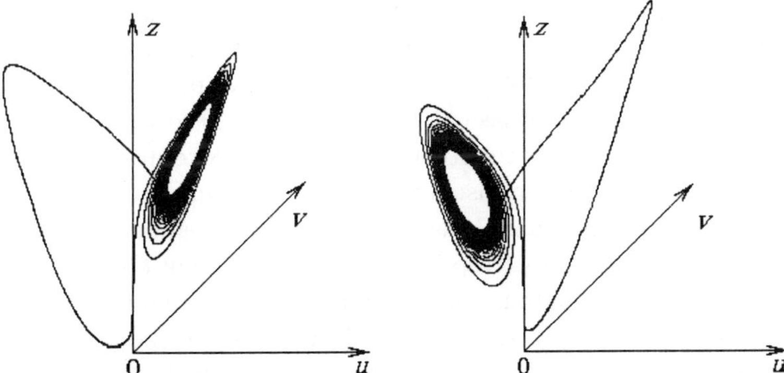

Fig. 3.2   Trajectories of cycles C_{237}^+ (left) and C_{237}^- (right) in the phase space (u, v, z).

Boundaries of the attraction domains of the stable points O_1 and O_2 are given by the smallest loops of the cycles C_n^+ and C_n^-, whose size decay as r grows. Therefore, for some $r = r_a$, the attraction domain of the set $B = S \cup G$ no longer intersects the attraction domains of points O_1 and O_2, and set B becomes an attractor. Numerical computations show that this is not a result of bifurcation at point $r = r_2$ but happens somewhat earlier, for $r = r_a \approx 23.9$, i.e. for $r_a < r_2$. This has a logical explanation if we use the concept by which the Lorenz attractor is formed as a result of bifurcations of finitely many stable cycles $C_k^{\pm}$, $k = 0, \ldots, l$, as r decreases starting from $r \approx 313$ (see item (10) below). Since the attraction domain of the attractor (and its eyes) is formed by cycles $C_l^{\pm}$ and a set of cycles generated by their bifurcations, it follows that the smaller loops of cycles $C_n^{\pm}$ lie inside the attractor eyes for some $n > l$ $(r_a < r < r_2)$, i.e. the attractor does not intersect attraction domains of points O_1 and O_2.

Therefore, in the Lorenz system ($\sigma = 10$, $b = 8/3$), metastable chaos exists only in the interval $r_1 < r \leq r_a$, and in the interval $r_a < r < r_2$, the system has three stable limit sets, namely, the points O_1 and O_2 and the Lorenz attractor.

(6) If $r \to r_2$, then the eye size decreases as the number of rotations of cycles C_n^{+} and C_n^{-} around points O_1 and O_2, respectively, grows. The value $z_{\max}$ grows, and $z_{\min}$ decays; moreover, $z_{\min} \to 0$ as $r \to r_2$. The lengths of generatrices of truncated cones grow, since additional rotations are added to the cone vertex and diminish the size of the smaller face. Conversely, the larger face grows. If $r = r_2$, then $z_{\min} = 0$, but $z_{\max} < r - 1$; thus, the larger face of each cone achieves its maximal size, while the smaller face is not contracted into a point, the cone vertex. The following bifurcation takes place. In the limit as $n \to \infty$, each set of cycles C_n^{+} (respectively, C_n^{-}) forms a point–cycle heteroclinic structure consisting of two separatrix contours of the point O. The first contour consists of a separatrix issuing from the point O along its unstable manifold and spinning on the appearing (only for $r = r_2$) saddle cycle L_1 (respectively, L_2) of the point O_1 (respectively, O_2). The second contour consists of the separatrix spinning out from the saddle cycle L_1 (respectively, L_2) and entering the point O along its stable manifold. This procedure can readily be imagined from Fig. 3.2 on which an almost limit situation is shown but without the cycles L_1 and L_2 necessary for separation of closed contours shown in the figure in two parts.

As mentioned in the previous item (5), the described bifurcation does not lead to generation of the Lorenz attractor for $r = r_2$. It is more correct to say that it is only a prerequisite of destruction of the attractor as r decays. The attractor itself, existing in the system for $r = r_2$, is formed from finitely many stable cycles $C_k^{\pm}$, $k = 0, \ldots, l$, for $r < 313$. It contains neither separatrices Γ_1 and Γ_2 of the point O nor infinitely many unstable cycles $C_n^{\pm}$ existing in the neighborhood of the point–cycle heteroclinic structure.

(7) If $r_2 < r < r_3$, then points O_1 and O_2 are still stable, and their attraction domains are bound by the appearing limit cycles L_1 and L_2 contracting to points as $r \to r_3$. But the Lorenz attractor B is not a set of integral curves going from L_1 to L_2 and back, and separatrices Γ_1 and Γ_2 of the saddle point O do not belong to the attractor. Cycles L_1 and L_2 have already made their job at $r = r_2$ and no longer have anything to do with the attractor. If $r_2 < r < r_3$, then, just as in the case of $r_1 < r < r_2$, the cycles C_n^+ and C_n^- appear again from separatrix contours. For example, one can observe cycles C_{201}^+ and C_{201}^- for $r = 24.06$ and cycles C_{36}^+ and C_{36}^- for $r = 24.5$. The attractor is determined by finitely many such cycles (see item (10)).

(8) For $r = r_3$, saddle cycles L_1 and L_2 disappear. In the system, there is a unique limit set, namely, the Lorenz attractor.

(9) There exist one more important value r_4, of the parameter r, which affect the formation of the Lorenz attractor. By our results, this is a point $r_4 \approx 30.485$. If r grows from r_3 to r_4, then the number of rotations of cycles C_n^+ and C_n^- first rapidly decays, then grows again, and attains a value of 36 at the point r_4. In this case, eyes formed by separatrices of the point O are much smaller than attractor eyes and begin to grow as r increases. Therefore, point r_4 is the point of minimum distance from the line ($\sigma = 10$, $b = 8/3$) in the space of parameters (σ, b, r) to the curve of heteroclinic contours joining the point O with points O_1 and O_2. Separatrices of the point O approach one-dimensional stable manifolds of points O_1 and O_2 by the minimal distance but do not hit these points! As it will be shown in Sec. 3.1.4, the value r_4 is also the point of minimum distance from the line ($\sigma = 10$, $b = 8/3$) in the space of parameters (σ, b, r) to halfsurface of homoclinic contours of points O_1 and O_2 (see Sec. 3.1.4 and 3.1.5). Therefore, almost heteroclinic and almost homoclinic contours exist in system (3.3) at point r_4. These contours are shown in Fig. 3.3. Obviously, in system (3.3) (and system (3.1)), there exist more complicated attractors for values of parameters σ and b such that the r-axis passes exactly through the points of heteroclinic or homoclinic contours. Since homoclinic contours intersect the manifold V^u, it follows

that the attractor is most complicated for the values of parameters σ^* and b^* for which the homoclinic contours coincide with the heteroclinic ones in the limit as $r \to r^*$ (see Secs. 3.1.3–3.1.5). In case of the Lorenz attractor ($\sigma = 10$, $b = 8/3$), situation is somewhat different. Attractor is formed only by finitely many stable cycles $C_n^{\pm}$ generated around homoclinic contours of points O_1 and O_2 as r decreases and by all of their bifurcations for $r > r_4$ (see the next item). Therefore, the Lorenz attractor, as well as other well-known irregular attractors, appears from stable limit sets, which is quite natural and is in good accordance with the intuitive treatment of the attractors nature. It is formed not as r grows and not due to bifurcations at points r_1 and r_2 but, on the opposite, as r decreases from $r \approx 313$ through infinitely many bifurcations at the points considered in the following item. The last of these points is r_4. If $r < r_4$, then eyes of the attractor grow, the domain filled by the attractor gradually decays, and the attractor itself is destroyed through bifurcations at the points r_2, r_a, and r_1.

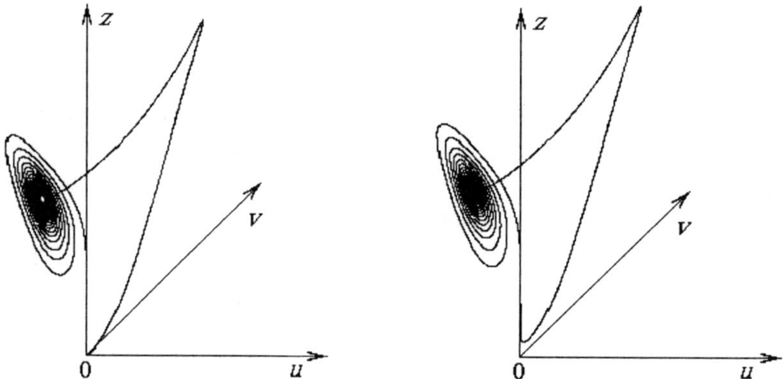

Fig. 3.3   An almost heteroclinic contour (left) and an almost homoclinic contour (right) for $r \approx 30.485$.

(10) The process of generation of the Lorenz attractor ($\sigma = 10$, $b = 8/3$) in the system (3.3) (system (3.1)) as r decays from 313 up to r_4 is referred to as the incomplete double homoclinic cascade. The complete cascade occurs if the r-axis passes exactly through the point of existence of two homoclinic contours. Note that in systems with a single homoclinic contour, there can be a simple complete or incomplete homoclinic cascade of bifurcations of transition to chaos. Let

us give a detailed description of transition to chaos through the double homoclinic (complete or incomplete) cascade of bifurcations. As was already mentioned, if $r > 313$, then in the system, there exists a unique stable limit cycle C_0 surrounding both points O_1 and O_2. If $r \approx 313$, then the cycle C_0 becomes unstable and generates two stable cycles C_0^+ and C_0^-, which also surround the points O_1 and O_2 but have deflections in the direction of corresponding halves of the unstable manifold V^u of the point O. This is the point where the double homoclinic cascade of bifurcations really begins. In case of an incomplete cascade, it consists of finitely many stages of appearance of stable cycles $C_k^\pm$, $k = 0, \ldots, l$, and their infinitely many further bifurcations. But in case of a complete cascade, the number of stages is infinite, and at the limit of $l \to \infty$, cycles tend to homoclinic contours of the points O_1 and O_2, respectively. At the k-th stage of the cascade, originally stable cycles $C_k^\pm$ undergo a subharmonic cascade of bifurcations and form two band-form attractors that consist of infinitely many unstable limit cycles intersecting the respective domains of the unstable manifold V^u of the point O. Then these two bands merge and form a single attractor surrounding both points O_1 and O_2, after which there is a cascade of bifurcations of cycles generated as a result of the merger and making rotations separately around points O_1 and O_2 and simultaneously around both points. The last cascade of bifurcations has the property of self-organization, since it is characterized by simplification of the structure of cycles and the generation of new stable cycles with a smaller number of rotations around the points O_1 and O_2 as r decays. Each cycle of *the cascade of self-organization bifurcations* undergoes its own subharmonic cascade of bifurcations, after which all cycles formed during infinitely many bifurcations of all subharmonic cascades and cascades of self-organization bifurcations of cycles become unstable and form some set B_k. After an incomplete homoclinic cascade of bifurcations, we obtain a set $B = \cup B_k$ consisting of infinitely many possible unstable cycles appearing at all l stages of the cascade. These cycles generate an incomplete double homoclinic attractor, that is the classical Lorenz attractor.

Let us describe the first stage of the double homoclinic cascade in more detail. Each of the cycles C_0^+ and C_0^- undergoes a subharmonic cascade of bifurcations consisting of the appearance of stable cycles with an arbitrary period in a neighborhood of an original cycle and a further period doubling cascade for these cycles. For example, one can observe double-period cycles for $r = 222$, quadruple-period cycles for $r = 216$, cycles of period 6 for $r = 214$, cycles of period 3 for $r = 209$, and so on. If $r = 203$, then, after termination of the subharmonic cascade of bifurcations, there appear two stable sets (respectively, two attractors) each of which consists of infinitely many unstable cycles formed due to bifurcations of cycles C_0^+ (respectively, C_0^-) and passing through the manifold V^u. Visually, these sets look like two wide bands. If $r < 203$, then two attractors formed by cycles C_0^+ and C_0^- merge and form a single attractor, which for the moment lies at some distance from the z-axis and hence has no eye. As a result, there may appear stable cycles making rotations around both points O_1 and O_2; moreover, the number of rotations in such cycles decreases as r decreases (see Fig. 3.4), and a cascade of self-organization bifurcations of cycles begins. If r further decreases, and consequently, the cycles forming the attractor approach the z-axis, then trajectories of the system begin to twist around points O_1 and O_2; and in the attractor, there appear eyes for $r = 197.6$. This is caused by the appearance of stable cycles like $C_{km}^{\pm}$, which make rotations separately around each of the points O_1 and O_2 as well as around both of them. These cycles undergo self-organization and appear in pairs in a huge number as a result of saddle–node bifurcation in the range of $170 < r < 197.6$. All of them successively undergo subharmonic cascades of bifurcations, also form narrower bands, and also pass through the manifold V^u.

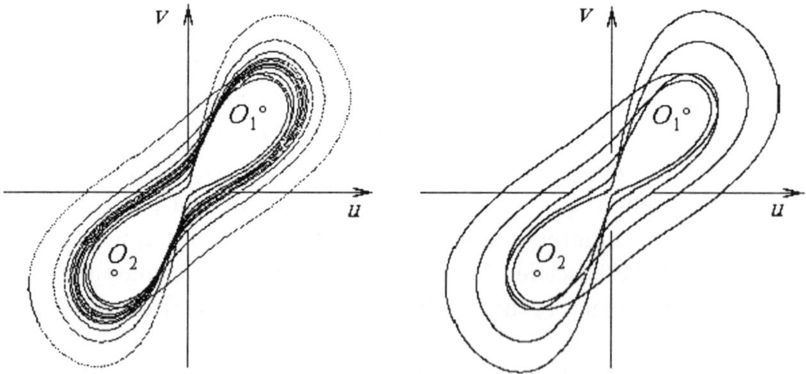

Fig. 3.4   Projections of the phase portrait of stable cycles appearing as a result of a self-organization bifurcation on the (u, v)-plane for $r = 202.384$ (left) and $r = 198.986$ (right).

Note that one should evaluate the cycle bifurcations for $r < 203$ very carefully. The matter is that, as a rule, in this interval, cycles are quite close to the z-axis and have large values of gradient in its neighborhood. We face a typical ill-posed problem of evaluating the derivative of a function that is not known precisely. Therefore, for a wrongly chosen integration step, a jump of trajectory from one cycle to another in a neighborhood of the z-axis can be mistaken for some property of the attractor, although such a jump is explained only by instability of the computational procedure. This error explains the "detection" of the so-called intermittency of the first kind in the Lorenz system (for details see below).

If $r \approx 170$, then there appears a stable cycle C_{11}, which undergoes the same bifurcations as the original cycle C_0 but in the interval of $100.795 < r < 170$. This completes the first stage of the double homoclinic cascade, and for $r \approx 100.795$, there appears a pair of stable cycles $C_1^{\pm}$. If $r \approx 71.52$, then the third stage begins with generation of a pair of stable cycles $C_2^{\pm}$ and goes on till the value $r \approx 59.25$ for which there appears a pair of stable cycles $C_3^{\pm}$, and so on. As was mentioned above, the incomplete homoclinic cascade finishes at the value of $r = r_4$. For smaller values of r, the system has a unique stable limit set $\overline{B}$ that is the Lorenz attractor in traditional sense. The problem as to whether $\overline{B}$ is fractal and has a fractional dimension remains open. The solution of this problem is directly related to the answer to the question as to what the structure of the set of points of all unstable limit cycles inducing the attractor on the curve V^u is and whether it is dense everywhere on V^u or has the structure of a Cantor set. In the first case, the attractor cannot be a fractal and its dimension is equal to two. Numerous numerical experiments performed by the authors show that the first case is more probable, i.e. points of the cycles composing each band of a homoclinic cascade are dense everywhere in their domain of the manifold V^u; consequently, intersection of the attractor with V^u is a segment on which $0 < z_{\min} \leq z \leq z_{\max} < r - 1$. We have not faced any reason to consider that the Lorenz attractor is a fractal and has a fractional dimension. These conclusions are completely agreed with *the theory of a birth of singular attractors* on smooth submanifolds of phase space (see Chapter 4). According to this theory the dimension of such attractors can not be more than two, and that actually in the interval of change of parameter values $r_a < r < r_4$ there is not one structurally stable attractor, but there are infinite number of structurally unstable singular attractors at points of accumulation of parameter values r, corresponding to various cascades of period doubling bifurcations.

More interesting is the situation of the "intermittency" phenomenon discovered by Pomeau and Manneville in the Lorenz system at the parameter value of $r = 166.1$ [Manneville and Pomeau (1980)]. It is claimed that the stable cycle existing in the Lorenz system for $148.5 < r < 166.07$ vanishes as r grows, and the system enters the intermittency interval in which the motion in the neighborhood of the former cycle is interrupted by irregular chaotic splashes. The system allegedly keeps memory of the cycle earlier existing in it. For further growth of r, there appears chaos in the system. The transition to chaos through intermittency is one of the three widespread scenarios for appearance of chaotic regimes in behavior of nonlinear dynamical systems.

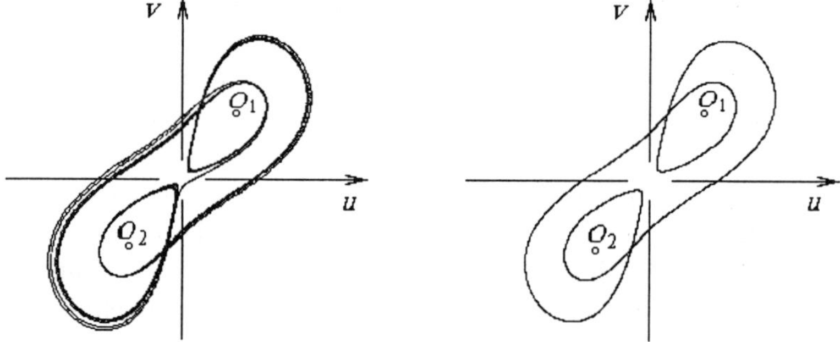

Fig. 3.5   Projections of the phase portrait of the cycle C_{11} with $r = 166.1$, $h = 0.003$ (left) and $r = 170$, $h = 0.03$ (right) on the plane (u, v).

Indeed, a chaotic splash takes place in the Lorenz system for $r = 166.1$ if we use the Runge–Kutta method of the fourth order with $h \sim 10^{-3} - 10^{-5}$ for integration of the system. But if we set $h = 2 \cdot 10^{-2}$, then we do not observe any chaotic splash for $r = 166.1$ as well as for $r = 166.8$. Moreover, we can readily see that, in the system, there exists a stable limit cycle up to the value of $r = 170$ if $h = 3 \cdot 10^{-2}$. This is the same cycle C_{11} as that observed for $r < 166.07$ (see Fig. 3.5). Therefore, in the Lorenz system there is neither intermittency nor transition to chaos through intermittency. This effect is solely due to the numerical error caused by the ill-posedness of the problem of evaluating a derivative in the neighborhood of the z-axis. The fact that multipliers of the cycle with $r = 166.06$ (that is, the case in which intermittency is still absent) have the values $(1; 0.91; 0)$ also shows that this value of parameter r lies still very far from the boundary of domain of a cycle generation.

3.1.3 *Scenario of birth of a complete double homoclinic attractor in the Lorenz system*

Let us consider the problem of finding parameters (σ^*, b^*) for which decrease of the parameter r results in transition to chaos in the Lorenz system (3.1) by a complete double homoclinic bifurcation cascade. This corresponds to the case when line $\sigma = \sigma^*$, $b = b^*$ passes in the parameter space (σ, b, r) through the existence point (σ^*, b^*, r^*) of two homoclinic contours of points O_1 and O_2. We shall show in Sec. 3.1.4. that such points form a halfsurface in the parameter space (σ, b, r). The attractor created in this case for $r = r^*$ has no "eyes" and occupies the upper part $(0 < z^* < z < r-1)$ of the manifold V^u, where z^* is the value of coordinate z on the manifold V^u corresponding to the homoclinic contour. The boundary of the homoclinic halfsurface is the locus of points where homoclinic contours of fixed points O_1 and O_2 of the Lorenz system coalesce with the heteroclinic contours of the points (O, O_1) and (O, O_2). Intersection of the homoclinic halfsurface with the plane $b = 8/3$ in the parameter space (σ, b, r) is presumably a curvilinear ray issuing from the point $(\sigma^* \approx 10.1668,\ r^* \approx 30.868)$ and passing, in particular, through the point $(\sigma^* \approx 10.5, r^* \approx 33.2189)$. A detailed description of transition to chaos in the Lorenz system at exact point $(b = 8/3, \sigma = 10.1668)$ is impossible to obtain because the problem is strongly ill-posed. Our purpose is therefore to illustrate the scenario of transition to chaos in the Lorenz system for $b = 8/3$, $\sigma = 10.5$ as parameter r decreases from $r = 350$ to $r \approx 33.2189$. In this scenario, the transition to chaos is via a complete double homoclinic bifurcation cascade and it involves the creation of a complete double homoclinic attractor, which in general is different from the classical Lorenz attractor. Method for investigating the attractor is similar to the method described in Sec. 3.1.2.

Integrating the system (3.1) by the fourth-order Runge–Kutta method with initial conditions (3.6), we can compute with an arbitrary accuracy every stable or unstable cycle by the method of return to the one-dimensional manifold V^u.

Now let us describe the transition to chaos via a complete double homoclinic bifurcation cascade. For $r > 340$ the system (3.1) has a unique stable limit cycle C_0 that encircles both equilibrium states O_1 and O_2. It is also an unstable cycle of system (3.1) for $13.958 < r < 340$, and it disappears as a result of the homoclinic butterfly bifurcation (Fig. 3.6). For $r \approx 340$ the cycle C_0 becomes unstable and generates two stable cycles C_0^+ and C_0^-, which also encircle the equilibrium states O_1 and O_2 but now have deflections in the direction of their halves of unstable manifold V^u of the point O (Fig. 3.7). This is essentially where the double homoclinic bifurcation cascade starts.

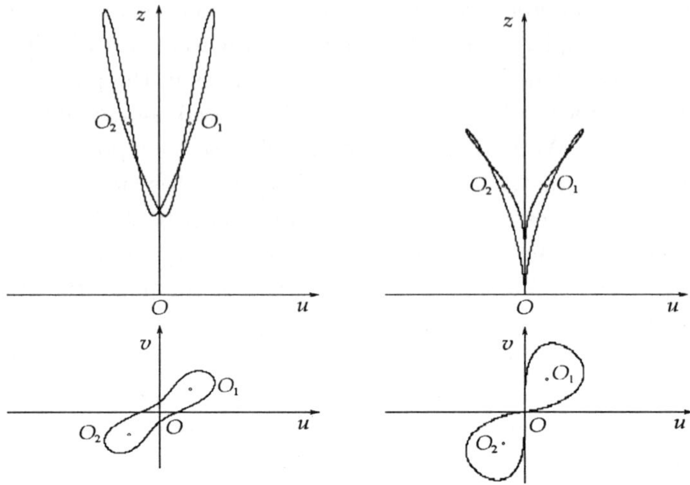

Fig. 3.6   Projections of the cycle C_0 for $r = 350$ (left) and $r = r_1 \approx 13.958$ (right).

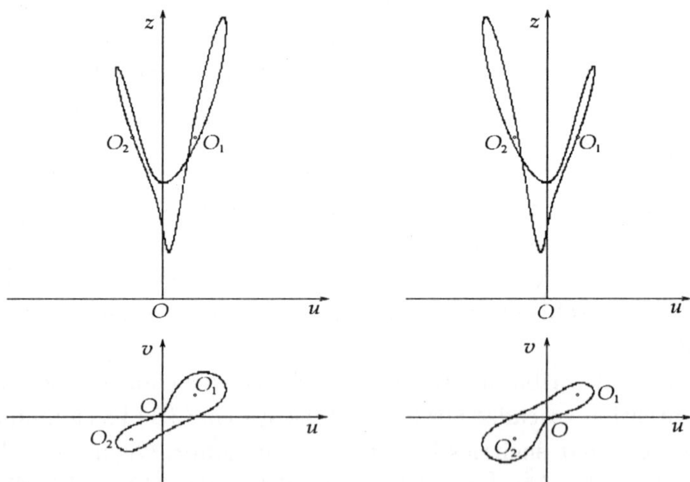

Fig. 3.7   Projections of the cycles C_0^- (left) and C_0^+ (right) for $r = 300$.

An incomplete cascade consists of finitely many stages generating stable cycles $C_k^{\pm}$, $k = 0, 1, \ldots, l$, followed by infinitely many bifurcations. A complete cascade consists of infinitely many stages, and in the limit of $l \to \infty$ the cycles tend to the homoclinic contours of points O_1 and O_2 respectively. The cycle C_k^+ makes k complete rotations around the point O_1 in the half-space containing this point and one incomplete rotation around the point O_2. The cycle C_k^-, conversely, makes k complete rotations around the point O_2 in the half-space containing this point and one incomplete rotation around the point O_1.

Note that at each stage of the cascade, the pair of stable cycles $C_k^{\pm}$ is simultaneously accompanied by a pair of identical unstable cycles. In stage k of the cascade, initially stable cycles $C_k^{\pm}$ undergo a subharmonic bifurcation cascade producing two attractors. These attractors are in the shape of bands consisting of infinitely many unstable limit cycles intersecting their regions of unstable manifold V^u of the point O. Then these bands coalesce forming a single attractor around both points O_1 and O_2. This is followed by a bifurcation cascade of the cycles produced by coalescing bands; these cycles make turns around each of points O_1 and O_2 separately and about both points simultaneously. The last cascade is self-organizing: its cycles show progressive simplification of structure, i.e. as r decreases, new stable cycles are created with fewer turns around points O_1 and O_2. Each cycle in the self-organizing bifurcation cascade undergoes its own subharmonic bifurcation cascade; after that all the cycles created as a result of infinitely many bifurcations in all the subharmonic cascades and in the self-organizing cascades become unstable and lie with their bands in their own regions of the manifold V^u, creating in this way some set B_k. The complete homoclinic bifurcation cascade produces the set $B = \cup B_k$, $k = 0, 1, \ldots$, which consists of infinitely many unstable cycles created in all stages of the cascade. These cycles generate a complete double homoclinic attractor in the Lorenz system.

Let us consider in more detail the first stage of the double homoclinic cascade. Each of the cycles C_0^+ and C_0^- undergoes a subharmonic bifurcation cascade, which involves the appearance in the neighborhood of the initial cycle of stable cycles of arbitrary period and a further period doubling cascades of these cycles. For instance, double-period cycles are created for $r \approx 251$ (Fig. 3.8); quadruple-period cycles are created for $r \approx 237$; cycles of period 5 are created for $r \approx 229.5$ (Fig. 3.9); cycles of period 3 are created for $r \approx 226.9$ (Fig. 3.10), and so on. For $r \approx 225.5$, when the subharmonic cascade ends, there are two stable sets (two subharmonic attractors), created by bifurcations of the cycles C_0^+ (C_0^-). These sets look like two wide bands (Fig. 3.11). For $r \leq 219.9$, the two attractors created by the cycles C_0^+ and C_0^- coalesce into one irregular attractor, which meanwhile remains at some distance from the z-axis and therefore has no "eyes" (Fig. 3.12). This coalescence leads to the possibility of formation of stable cycles that make turns around both points O_1 and O_2, number of turns in these cycles decreases with the decrease of r (Fig. 3.13) and a cascade of self-organizing bifurcations begins.

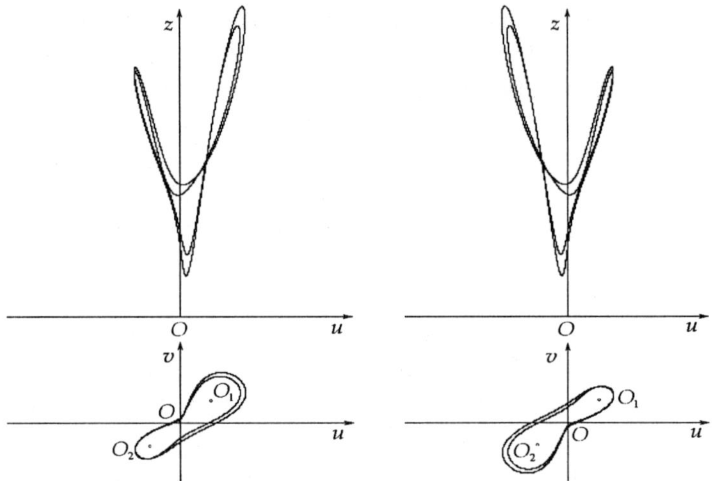

Fig. 3.8   Projections of double-period cycles C_0^- (left) and C_0^+ (right) for $r = 244$.

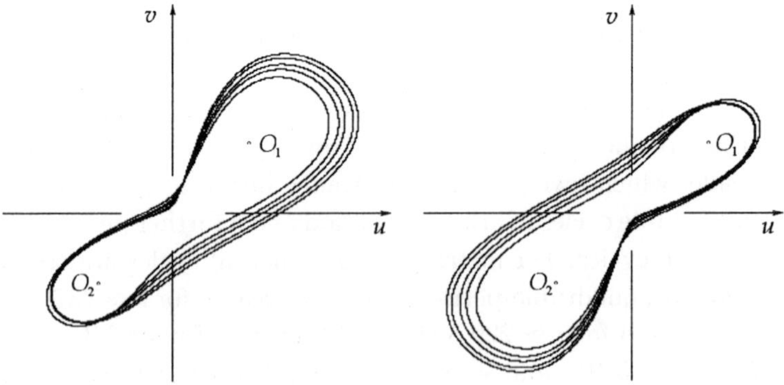

Fig. 3.9   Projections of period 5 cycles C_0^- (left) and C_0^+ (right) for $r = 229.5$.

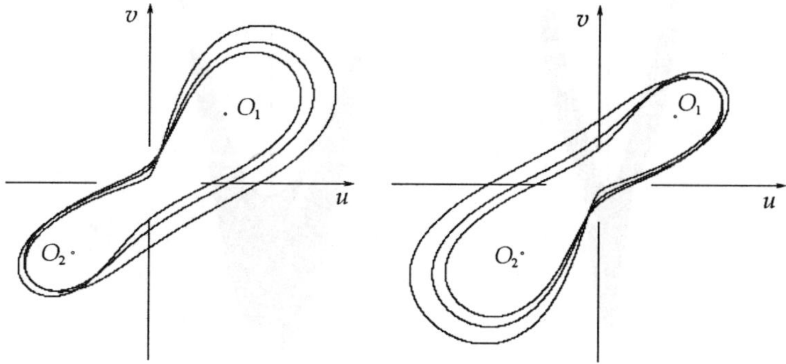

Fig. 3.10 Projections of period 3 cycles C_0^- (left) and C_0^+ (right) for $r = 226.7$.

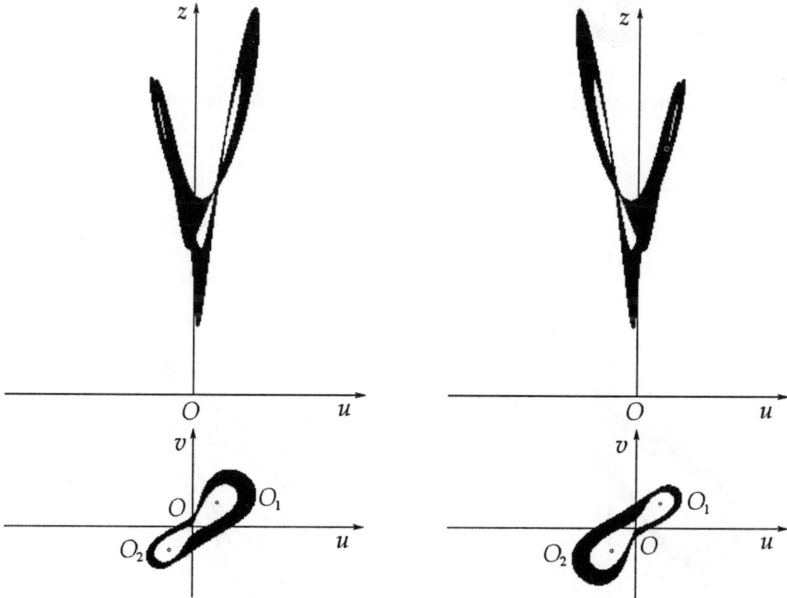

Fig. 3.11 Projections of the subharmonic attractors generated by the cycles C_0^- (left) and C_0^+ (right) for $r = 220$.

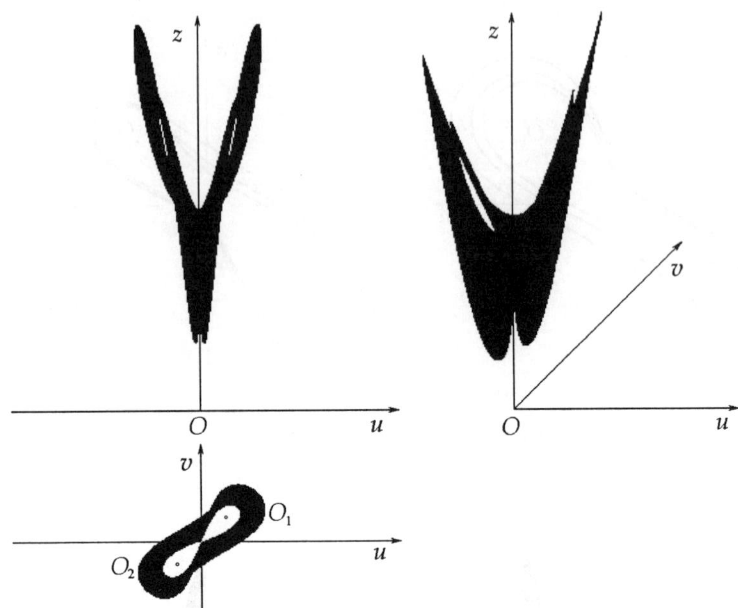

Fig. 3.12   Irregular attractor created by the coalescence of two subharmonic attractors for $r = 219.9$.

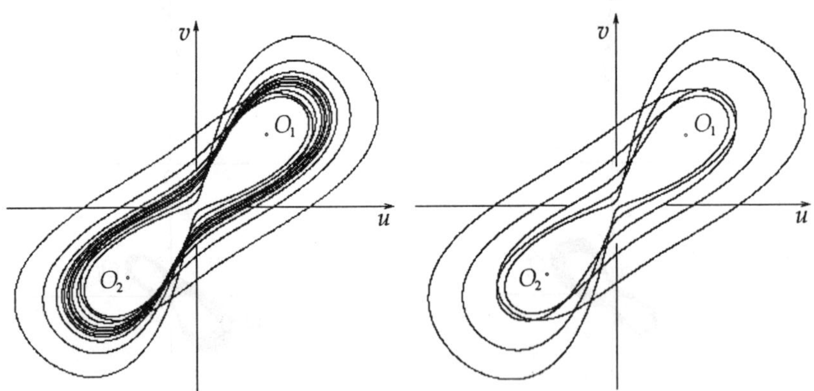

Fig. 3.13   Projections of period 11 and period 5 cycles in the self-organizing cascade for $r = 219.263$ and $r = 215.260$, respectively.

Further reduction of r drives the cycles closer to the z-axis and the system trajectories begin to wind around points O_1 and O_2. The self-organizing cycles thus acquire "eyes" for $r \approx 213.55$. Stable cycles of the type $C_{km}^{\pm}$ appear, making turns around each of points O_1 and O_2 separately (precisely one turn) and around both points simultaneously. These self-organizing cycles are created in pairs in huge numbers in the interval $183 < r < 213.55$ (Figs. 3.14–3.16). They all subsequently undergo subharmonic bifurcation cascades forming their narrower bands, and also lie on the manifold V^u.

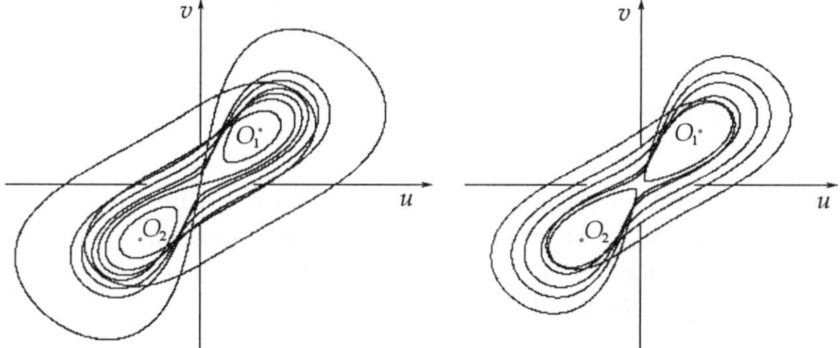

Fig. 3.14   Projections of self-organizing cascade cycles for $r = 209.090$ (left) and $r = 199.412$ (right).

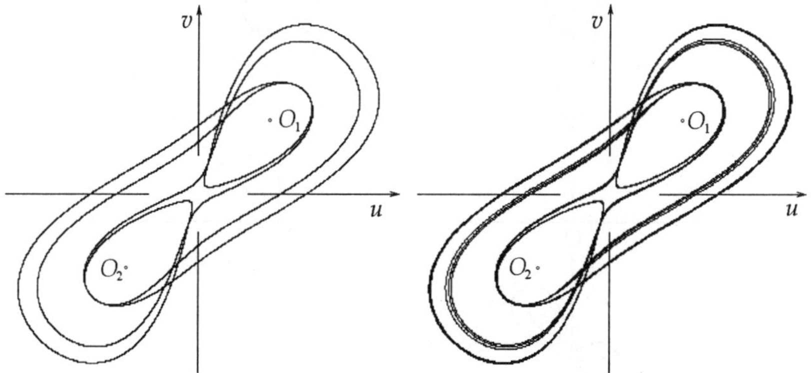

Fig. 3.15   Projections of self-organizing cascade cycles for $r = 194.75$ (left) and $r = 193.9175$ (right).

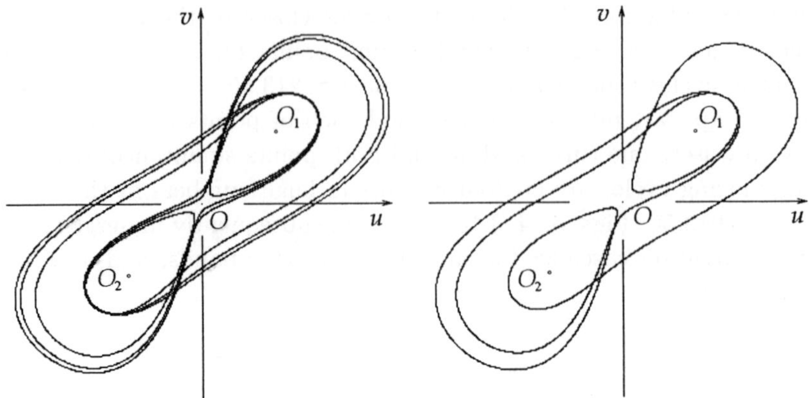

Fig. 3.16　Projections of self-organizing cascade cycles for $r = 190.561$ (left) and $r = 184.943$ (right).

For $r \approx 183$ self-organizing bifurcation cascade ends with creation of a stable cycle C_{11}, which subsequently (with decreasing r) undergoes the same bifurcations as the original cycle C_0, but in a different interval $108.339 < r < 183$ (Figs. 3.17–3.20).

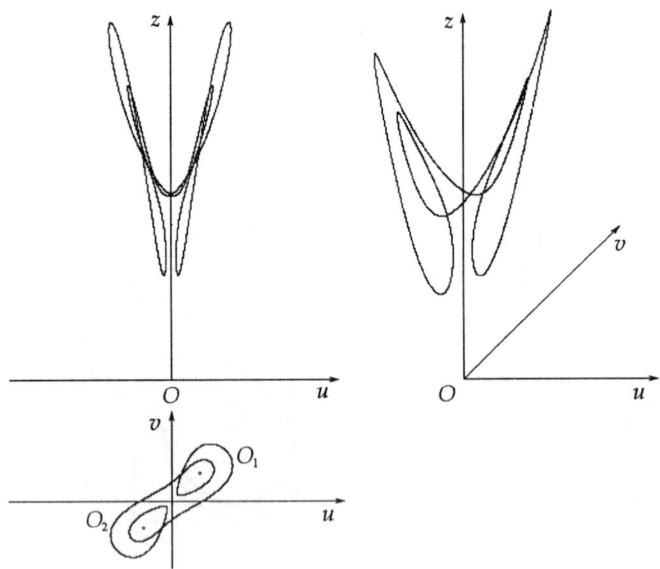

Fig. 3.17　Projections and three-dimensional visualization of the cycle C_{11} for $r = 176$.

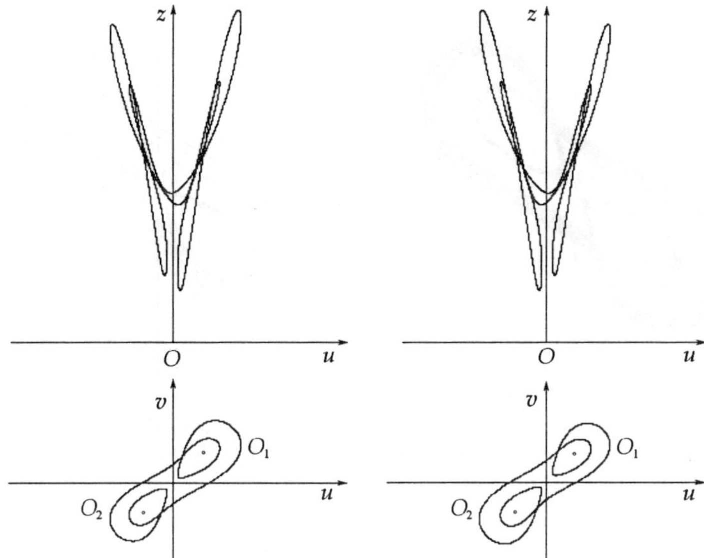

Fig. 3.18 Projections of the stable cycles C_{11}^- (left) and C_{11}^+ (right) for $r = 164$.

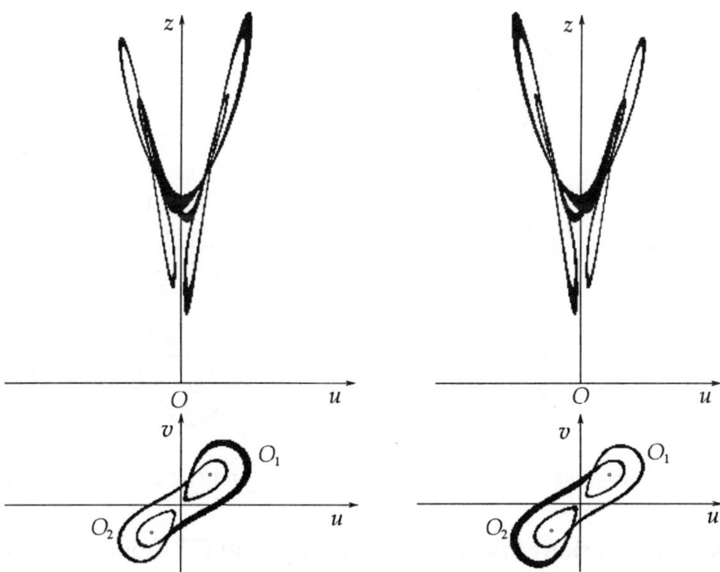

Fig. 3.19 Projections of the subharmonic attractors generated by the cycles C_{11}^- (left) and C_{11}^+ (right) for $r = 155.42$.

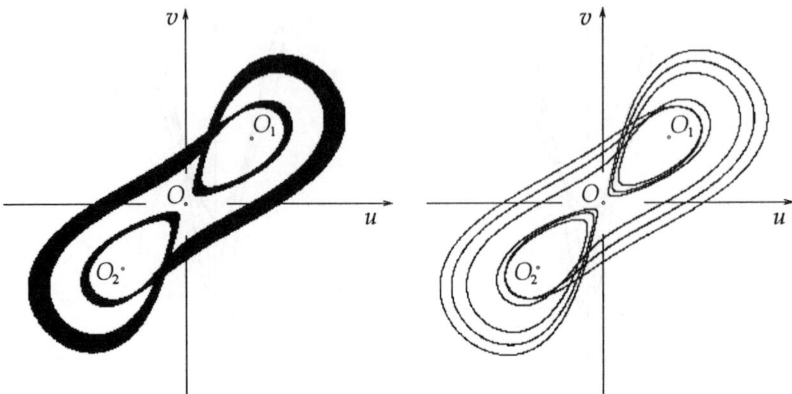

Fig. 3.20   Projection of the singular attractor formed by the coalescence of two subharmonic attractors for $r = 155.41$ (left) and projection of one of the self-organizing cycles for $r = 153.35$.

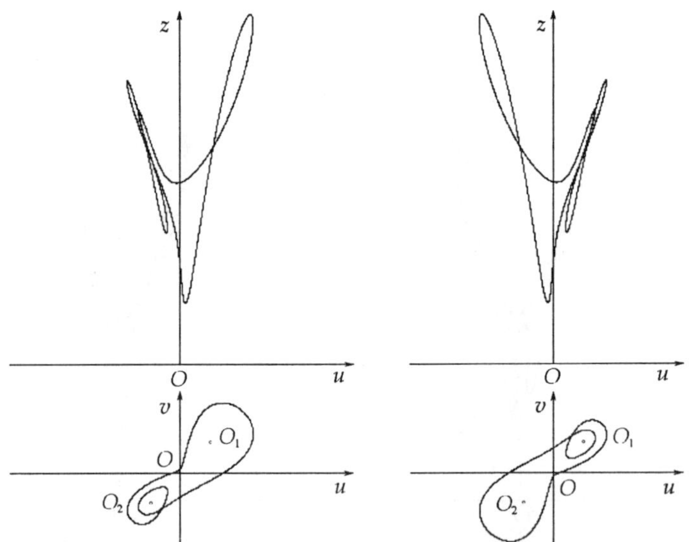

Fig. 3.21   Projections of the stable cycles C_1^- (left) and C_1^+ (right) for $r = 107.75$.

Note that creation of two stable cycles C_{11}^- and C_{11}^+ from one stable cycle C_{11} occurs at $r \approx 165.2$. This ends the first stage of a double homoclinic cascade, and a pair of stable cycles $C_1^{\pm}$ is created for $r \approx 108.339$ (Fig. 3.21). For $r \approx 76.281$ creation of a pair of stable cycles $C_2^{\pm}$ signals the beginning of a third stage, which extends to $r \approx 63$, when a pair of stable cycles $C_3^{\pm}$ is created. The fourth stage of the cascade extends to $r \approx 55.67$, when a pair of stable cycles $C_4^{\pm}$ is created; fifth stage extends to $r \approx 51.01$, sixth stage extends to $r \approx 47.78$, when a pair of stable cycles $C_6^{\pm}$ is created (Fig. 3.22).

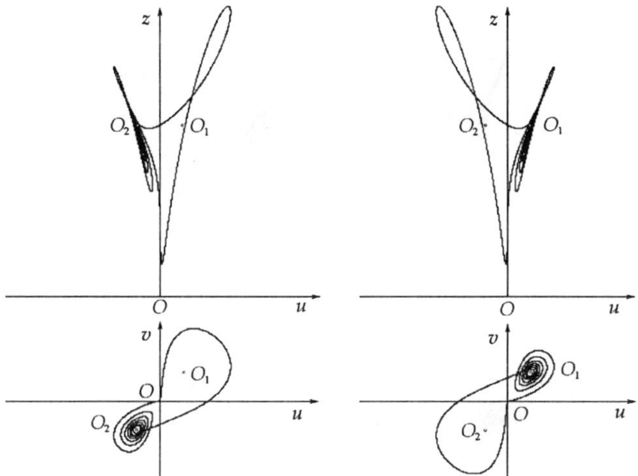

Fig. 3.22   Projections of the stable homoclinic cycles C_6^- (left) and C_6^+ (right) for $r = 47.7788$, $z_0 = 13.72$.

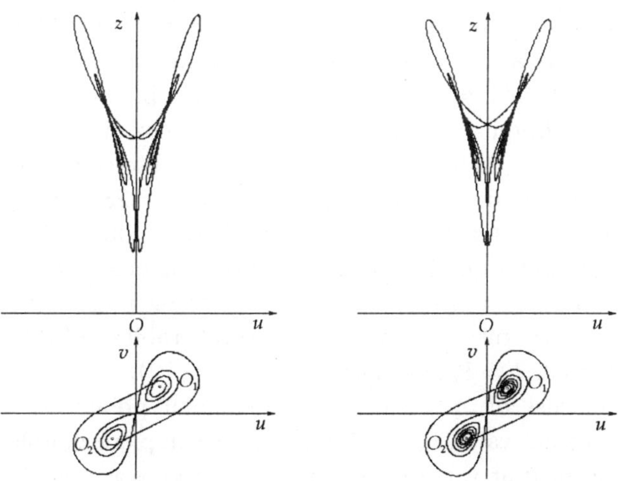

Fig. 3.23   Projections of the stable cycles C_{33} and C_{66} for $r = 74.09$ and $r = 50.8682$ respectively.

Inside each stage k of the cascade we observe creation of stable homoclinic cycles C_{kk} and $C_{kk}^{\pm}$, as well as creation of all possible stable cycles of type C_{km} and C_{mk}, $m = 1, \ldots, k - 1$.

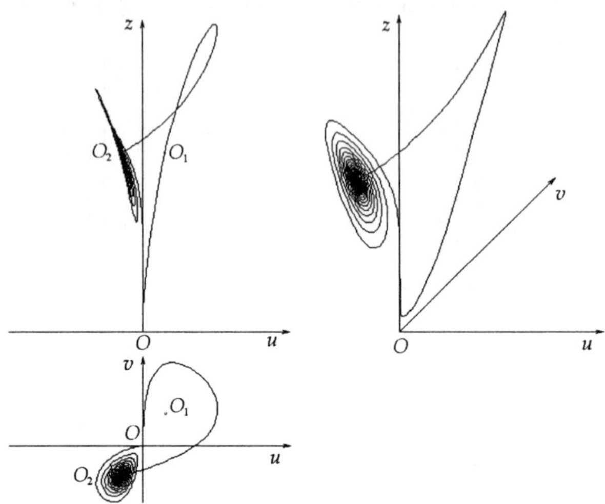

Fig. 3.24   Projections and three-dimensional visualization of the cycle C_{121}^{-} for $r \approx$ 33.21896.

Cycles C_{33} and C_{66} are shown in Fig. 3.23. Fig. 3.24 shows homoclinic cycle C_{121}^{-} created for $r \approx 33.2189$. Creation of this cycle has been numerically detected for $r = 33.21895995031$, $z_0 = 2.3784027452$ with integration step $h = 0.01$. As a result, when the complete double homoclinic bifurcation cascade ends at $r \approx 33.2189$, a complete double homoclinic attractor is created in the Lorenz system. This is not a Lorenz attractor, which in itself is an incomplete double homoclinic attractor. For comparison Fig. 3.25 shows the Lorenz attractor ($b = 8/3$, $\sigma = 10$, $r = 28$) and the complete double homoclinic attractor ($b = 8/3$, $\sigma = 10.5$, $r = 33.2189$).

Further investigation of a complete double homoclinic attractors in the Lorenz system at $\sigma = 10$, $b = 0.5$ allows strong reasons for doubt not only in fractal structure, but also in structural stability of classical Lorenz attractor ($\sigma = 10$, $b = 8/3$) at all values $r \in (r_a, r_4)$.

Indeed, in case of $\sigma = 10$, $b = 0.5$ singular points O_1 and O_2 lose stability at critical value $r_3 \approx 15.882$, and a complete double homoclinic attractor is formed at value $r^* \approx 234.177$. At values $r < r^*$, in the Lorenz system, there appear features which are distinct from the scenario with classical values of parameters. So, with reduction of values of parameter

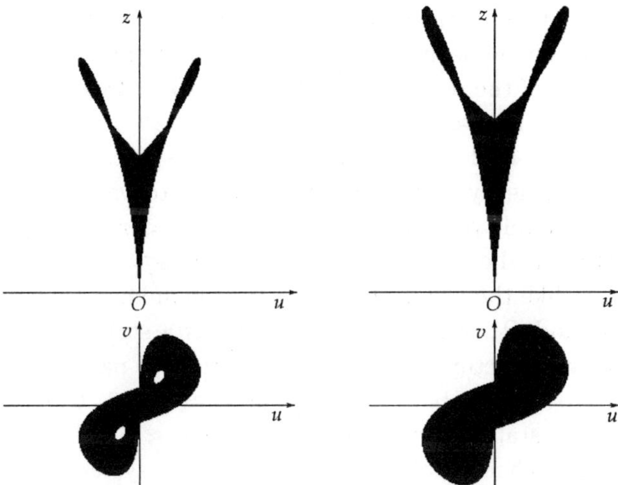

Fig. 3.25   Projections of the Lorenz attractor for $b = 8/3$, $\sigma = 10$, $r = 28$ (left) and the complete double homoclinic attractor for $b = 8/3$, $\sigma = 10.5$, $r = 33.2189$ (right).

r a return double homoclinic cascade is observed, i.e. there is a dump of coils of stable cycles $C_n^{\pm}$ and $C_{nn}^{\pm}$. We shall note, that it is not observed at classical values of parameters. At value $r \approx 212.32$ there is a stable cycle C_{44} in system which earlier was observed also stable at $r \approx 265.4$, greater than r^* (Fig. 3.26). The obtained results give the basis to confirm, that most likely, in any neighbourhood of the value r^* there are systems with stable limit cycles which we cannot observe exclusively as a result of computing errors owing to an ill-posedness of a solved problem.

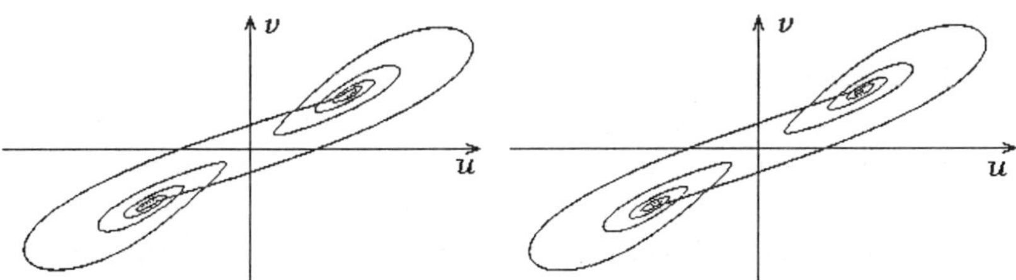

Fig. 3.26   Projections of the stable cycles C_{44} for $r = 265.4$ (left) and for $r = 212.32$ (right).

Let us note also, that at the fixed values of parameters $\sigma = 10$, $b = 0.5$ stable cycles $C_0^{\pm}$ are observed in an interval $19 < r < 38.5$, and at the further reduction r the next double homoclinic bifurcation cascade starts to be formed, beginning by the cascade of the period doubling bifurcations and proceeding by the subharmonic cascade of bifurcations. So, for example, stable cycles $C_0^{\pm}$ of the period 4 are observed at $r \approx 18.1$, and stable cycles of the period 3 — at $r \approx 17.76$.

For values $r < 16.2$ it is possible to note one more interesting feature, creation of a stable cycle C_{22} which exists up to value $r \approx 15.1$ where it generates a double-period cycle. Hence, at $r < r_3 \approx 15.882$ there are three attractors in the system: two stable focuses O_1 and O_2 and a stable cycle. Thus, unlike the classical scenario, at the moment of loss of stability by points O_1 and O_2, we have a limit cycle as a stable limit set, instead of an irregular attractor. At further reduction of parameter values r it was possible to find stable cycles $C_2^{\pm}$ at $r \approx 13.07$, C_{33} at $r \approx 12.95$ and some other stable cycles down to value $r \approx 12.29$. At $r < r_a \approx 11.78$ metastable chaos is already observed in the system. Thus, irregular attractor, if possible, exists only in an interval of change of parameter $11.78 < r < 12.29$. Though more plausible the hypothesis seems that stable cycles exist almost everywhere in this interval of change of values of bifurcation parameter r, except for points of accumulation (limiting values of separate cascades of bifurcations). This, however, will completely agree with the theory of singular attractors, stated in Chapter 4.

At transition r through value r_a the system passes into a state of metastable chaos. At $r = r_2 \approx 10.103$ there occurs a point–cycle bifurcation in the system. Trajectories leaving the neighbourhood of the zero do not tend to points O_1 and O_2, and are reeled on boundaries of their attraction domains. However in a classical case the trajectory, having come off a boundary and not getting to a stable point at once, will eternally move along an attractor ($r_a \approx 23.9 < r_2 \approx 24.06$ for $b = 8/3$ and $\sigma = 10$). In a considered case the trajectory, not having got to a stable point at once, anyway sooner or later will be attracted by points O_1 or O_2 as their areas of stability are great enough, and the system continues to be in the domain of metastable chaos ($r_2 \approx 10.103 < r_a \approx 11.78$ for $b = 0.5$ and $\sigma = 10$). Thus, this difference from the classical scenario once again confirms the conclusion of paper [Magnitskii and Sidorov (2001c)] that the point r_2 has no relation to formation of the Lorenz attractor. The metastable chaos in the system proceeds down to value $r = r_1 \approx 6.493$ at which the homoclinic butterfly bifurcation completely destroys the remainder of attractor, destroying the last unstable cycle C_0. For $1 < r < r_1$ any trajectory tends to the nearest stable point O_1 or O_2.

Thus, contrary to the traditional scenario, a classical (and any another) irregular attractor of system of the Lorenz equations is born from stable limit sets (cycles) as a result of subharmonic and (or) homoclinic cascades of soft bifurcations. The theory of such (singular) attractors is presented in Chapter 4.

3.1.4 *Bifurcations of homoclinic and heteroclinic contours in the Lorenz system*

We shall apply the Magnitskii method presented in Sec. 2.4.3 for finding bifurcation surfaces and curves of homoclinic separatrix loops and heteroclinic contours of singular points for the system of Lorenz Eqs. (3.1) in the space of parameters (σ, b, r).

3.1.4.1 *Heteroclinic contours connecting the saddle–node with the saddle–focus*

We shall consider the Lorenz system (3.1) in the domain of parameter values $\sigma > 0, b > 0, r > 1$ and we shall rewrite it in a form convenient for the method application

$$
\begin{aligned}
\dot{x} &= y, \\
\dot{y} &= -(\sigma + 1)y - \sigma(z - r + 1)x, \\
\dot{z} &= x^2 + xy/\sigma - bz.
\end{aligned}
\tag{3.8}
$$

If $r > 1$, then system (3.8) has the saddle–node $O(0,0,0)$ with real eigenvalues of the linearization matrix

$$
\nu_1 = -\frac{\sigma + 1}{2} - \sqrt{\frac{(\sigma + 1)^2}{4} + \sigma(r - 1)} < 0,
$$

$$
\nu_2 = -\frac{\sigma + 1}{2} + \sqrt{\frac{(\sigma + 1)^2}{4} + \sigma(r - 1)} > 0, \qquad \nu_3 = -b < 0.
$$

In addition, system (3.8) has two saddle–focuses

$$
O_{1,2}\left(\pm\sqrt{b(r - 1)},\ 0,\ r - 1\right)
$$

with one negative real eigenvalue λ_1 and two complex conjugate eigenvalues $\lambda_{2,3}$, which have positive real parts for

$$
r > r_c = \sigma(\sigma + b + 3)/(\sigma - b - 1)
$$

and satisfy the characteristic equation

$$\lambda^3 + (\sigma + b + 1)\lambda^2 + b(\sigma + r)\lambda + 2b\sigma(r - 1) = 0. \tag{3.9}$$

The problem is to find values of parameters b, σ, r for which in system (3.8), there exist heteroclinic contours connecting the saddle–node O with the saddle–focus O_1. Note that for these values of parameters, the system simultaneously has heteroclinic contours connecting the saddle–node O with the saddle–focus O_2.

In accordance with the above-presented theory, we reduce system (3.8) to the form

$$\frac{dy}{dx} = -\frac{(\sigma + 1)y - \sigma(z - r + 1)x}{y}, \quad \frac{dz}{dx} = \frac{x^2 + xy/\sigma - bz}{y}. \tag{3.10}$$

System (3.10) has singularities at points O and $O_{1,2}$, since $y = 0$ here. By using the Taylor expansion at point O, we obtain

$$y'(0) = -\frac{(\sigma + 1)y'(0) - \sigma(r - 1)}{y'(0)}, \quad z'(0) = -\frac{bz'(0)}{y'(0)},$$

which implies that $y'(0) = \nu_2 > 0$, $z'(0) = 0$. Next, we obtain

$$z''(0) = \frac{2 + 2y'(0)/\sigma - bz''(0)}{2y'(0)} = \frac{2(1 + \nu_2/\sigma)}{2\nu_2 + b} > 0.$$

Likewise, at point O_1, we obtain

$$y'\left(-\sqrt{b(r - 1)}\right) = \lambda_1 < 0, \ z'\left(-\sqrt{b(r - 1)}\right) = -\frac{\sqrt{b(r - 1)}(2 + \lambda_1/\sigma)}{b + \lambda_1} > 0,$$

where λ_1 is a negative real root of the characteristic equation (3.9).

Further, for small $\varepsilon > 0$, we numerically solve system (3.8) in direct time with initial conditions

$$x(0) = \varepsilon, \ y(0) = y'(0)\varepsilon = \nu_2\varepsilon, \ z(0) = \frac{z''(0)\varepsilon^2}{2} = \frac{1 + \nu_2/\sigma}{2\nu_2 + b}\varepsilon^2$$

and obtain a trajectory $(x^+(t), \ y^+(t), \ z^+(t))$, that is arbitrarily close to the separatrix issuing from the saddle–node O. Then we numerically solve system (3.8) in reverse time with the initial conditions

$$x(0) = -\sqrt{b(r - 1)}+\varepsilon, \quad y(0) = \lambda_1\varepsilon, \quad z(0) = r-1-\frac{\sqrt{b(r - 1)}(2 + \lambda_1/\sigma)}{b + \lambda_1}\varepsilon$$

and obtain a trajectory $(x^-(t),\ y^-(t),\ z^-(t))$ that is arbitrarily close to the separatrix entering the saddle–focus O_1. The trajectories are sewn on the x- axis at the instants t^+ and t^- for which $y^+(t^+)=y^-(t^-)=0$. To satisfy the last condition, it is necessary and sufficient to solve the system of two equations

$$x^+(\sigma,b,r,t^+) = x^-(\sigma,b,r,t^-),$$
$$z^+(\sigma,b,r,t^+) = z^-(\sigma,b,r,t^-) \tag{3.11}$$

for three parameters. Therefore the heteroclinic contour bifurcation in the Lorenz system has codimension 2, i.e. the bifurcation surface is a part of line in the space of parameters.

In Fig. 3.27, one can see the (x,y)- and (x,z)-projections of the heteroclinic contour of system (3.8) found by the numerical solution of system (3.11) for the parameters σ and r for $b = 8/3$, $\varepsilon = 10^{-6}$. The represented results imply that, for $b = 8/3$, the contour exists only for $\sigma = 10.1672937 \pm 2 \cdot 10^{-7}$ and $r = 30.868108 \pm 2 \cdot 10^{-6}$. Hence it follows that, in the classical case (in which $\sigma = 10$, $b = 8/3$), the Lorenz system has no heteroclinic contour for any r, $1 < r < \infty$.

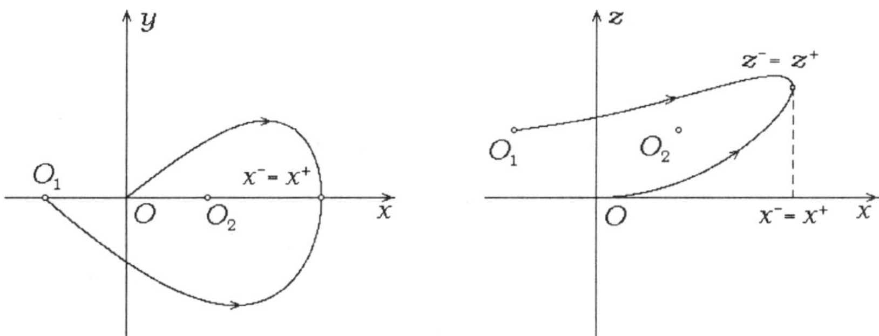

Fig. 3.27   The construction of heteroclinic contour connecting the saddle–node with the saddle–focus in the Lorenz system.

Second part of the heteroclinic contour represented in Fig. 3.27 is formed by the separatrix winding around the saddle–focus O_1 as $t \to -\infty$ and tending to the saddle–node O along its stable two-dimensional manifold W^s as $t \to +\infty$. This separatrix enters the point O in the (x,y)-projection at an angle whose tangent is equal to $y'(0) = \nu_1 < 0$. The possibility to close the contour in the (x,y)-projection uniquely is provided by the choice of the free parameter, that is, the initial phase φ of the separatrix during the winding around the saddle–focus O_1. The possibility to close the contour in the (x, z)-projection is provided by the choice of another free parameter, that is, the coefficient $\varkappa$ in the representation $z(x) \approx (\varkappa x)^{-b/\nu_1}$, $x \to 0$, which follows from the second equation in system (3.10).

The (x, y)- and (x, z)-projections of the complete heteroclinic contour of the Lorenz system for $b = 8/3$ ($\sigma = 10.1672937$, $r = 30.868108$) are shown in Fig. 3.28.

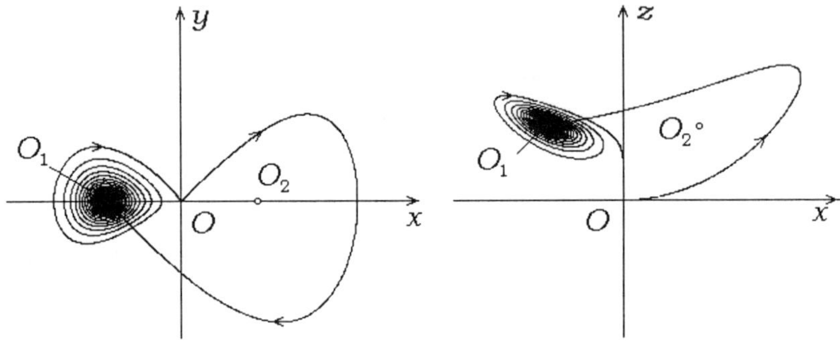

Fig. 3.28   Projections of the complete heteroclinic contour connecting the saddle–node with the saddle–focus in the Lorenz system (we have obtained $\sigma = 10.1672937 \pm 2 \cdot 10^{-7}$ and $r = 30.868108 \pm 2 \cdot 10^{-6}$) for $b = 8/3$.

3.1.4.2   *Homoclinic saddle–focus separatrix loop*

We shall apply the presented approach to finding a two-dimensional half-surface of homoclinic contours of saddle–focuses O_1 and O_2 of the Lorenz system (3.8) in the domain of parameter values $b > 0$, $\sigma > 0$, $r > r_c$. The linearization matrix

$$A = \begin{pmatrix} 0 & 1 & 0 \\ 0 & -(\sigma + 1) & \sigma\sqrt{b(r-1)} \\ -2\sqrt{b(r-1)} & -\sqrt{b(r-1)}/\sigma & -b \end{pmatrix}$$

at the point O_1 has a negative real eigenvalue λ_1 and two complex conjugate eigenvalues $\lambda_{2,3} = \alpha \pm i\beta$ with positive real part $\alpha > 0$, which satisfy the characteristic equation (3.9). We perform the change of variables $(x, y, z)^{\mathrm{T}} = C(u, v, w)^{\mathrm{T}}$. By a straightforward substitution, one can show that if matrix C has the form

$$C = \begin{pmatrix} \dfrac{1}{\alpha} & 0 & \dfrac{\sigma\sqrt{b(r-1)}}{\lambda_1(\sigma+1+\lambda_1)} \\[2ex] 1 & -\dfrac{\beta}{\alpha} & \dfrac{\sigma\sqrt{b(r-1)}}{\sigma+1+\lambda_1} \\[2ex] \dfrac{\beta^2-\alpha(\sigma+1+\alpha)}{-\alpha\sigma\sqrt{b(r-1)}} & \dfrac{\beta(\sigma+1+2\alpha)}{-\alpha\sigma\sqrt{b(r-1)}} & 1 \end{pmatrix},$$

then

$$C^{-1}AC = \begin{pmatrix} \alpha & -\beta & 0 \\ \beta & \alpha & 0 \\ 0 & 0 & \lambda_1 \end{pmatrix}.$$

Let us now take the initial conditions $u_0 = \varepsilon\cos\varphi$, $v_0 = \varepsilon\sin\varphi$, $w_0 = 0$ and, hence,

$$
\begin{aligned}
x(0) &= -\sqrt{b(r-1)} + \frac{\varepsilon\cos\varphi}{\alpha}, \\
y(0) &= \varepsilon\cos\varphi - \frac{\beta\varepsilon\sin\varphi}{\alpha}, \\
z(0) &= r-1 - \frac{\beta^2-\alpha(\sigma+1+\alpha)}{\alpha\sigma\sqrt{b(r-1)}}\varepsilon\cos\varphi - \frac{\beta(\sigma+1+2\alpha)}{\alpha\sigma\sqrt{b(r-1)}}\varepsilon\sin\varphi.
\end{aligned}
\tag{3.12}
$$

By solving system (3.8) in direct time with the initial conditions (3.12), we obtain a trajectory $(x^+(t),\, y^+(t),\, z^+(t))$ that is arbitrarily close to the separatrix issuing from the saddle–focus O_1. Then, by solving the system (3.8) in reverse time with the initial conditions

$$
\begin{aligned}
x(0) &= -\sqrt{b(r-1)} + \varepsilon, \\
y(0) &= \lambda_1\varepsilon, \\
z(0) &= r - 1 - \frac{\sqrt{b(r-1)}(2+\lambda_1/\sigma)}{b+\lambda_1}\varepsilon,
\end{aligned}
$$

we obtain a trajectory $(x^-(t),\, y^-(t),\, z^-(t))$ that is arbitrarily close to the separatrix entering the saddle–focus O_1. The trajectories are sewn at the time instants t^+ and t^- at which $y^+(t^+) = y^-(t^-) = 0$. One should choose the phase $\varphi(\varepsilon,\mu)$ so as to ensure that

$$x^+(\sigma,b,r,t^+,\varphi) = x^-(\sigma,b,r,t^-,\varphi).$$

The remaining condition of obtaining a homoclinic contour with arbitrary accuracy, that is

$$z^+(\sigma, b, r, t^+, \varphi) = z^-(\sigma, b, r, t^-, \varphi), \qquad (3.13)$$

defines the bifurcation half-surface in the space of parameters (σ, b, r).

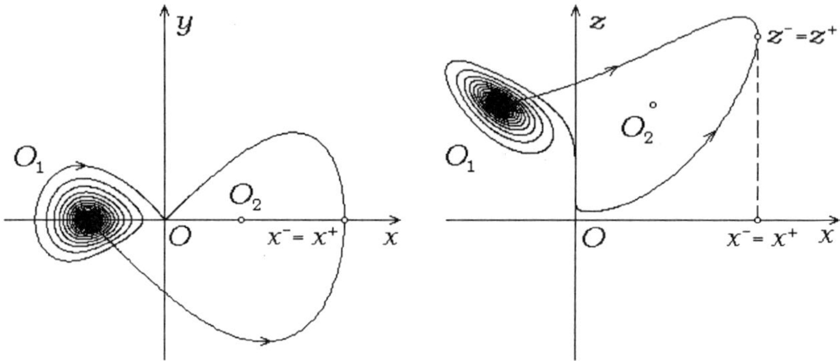

Fig. 3.29   Construction of the homoclinic separatrix loop of the saddle–focus O_1 in the Lorenz system (we have obtained $r = 33.21926 \pm 1 \cdot 10^{-5}$) for $b = 8/3$ and $\sigma = 10.5$.

Fig. 3.29 represents the (x, y)- and (x, z)-projections of homoclinic contour of system (3.8) found by the numerical solution of (3.13) for the parameter r with $b = 8/3$, $\sigma = 10.5$ and $\varepsilon = 10^{-5}$. Note that, in addition to homoclinic separatrix loop of the saddle–focus O_1, the Lorenz system has the symmetric (in the z-axis) homoclinic separatrix loop of the saddle–focus O_2. Cross-section of the bifurcation half-surface by the plane $b = 8/3$ is a half-ray in the (σ, r) -plane issuing from the point $\sigma = 10.1672937$, $r = 30.868108$ of existence of heteroclinic contour (see Fig. 3.28). Therefore, curve of heteroclinic contours of the Lorenz system in the space of parameters (σ, b, r) is the boundary (the limit case) of half-surface of homoclinic saddle–focus separatrix loops. In classical case, the line $\sigma = 10$, $b = 8/3$ does not intersect this half-surface. Therefore, in classical case, the Lorenz system can have only incomplete double homoclinic attractor. Note also that resulting value of parameter r for which there exists a homoclinic saddle–focus separatrix loop in the Lorenz system with $\sigma = 10.5$, $b = 8/3$ refines the value obtained in Sec. 3.1.3 by another method.

3.1.4.3 *Homoclinic saddle–node separatrix loop*

We apply our approach to finding the two-dimensional half-surface of homoclinic separatrix loops of the saddle–node O of the Lorenz system (3.8) in the domain $\sigma > 0$, $b > 0$, $r > 1$. Substitution $x^2 = v$, $y = xu$, $z = z$ reduces the system (3.8) to a more convenient form for application of the offered method

$$\dot{u} = -(u - \nu_1)(u - \nu_2) - \sigma z,$$
$$\dot{v} = 2uv, \tag{3.14}$$
$$\dot{z} = v(1 + u/\sigma) - bz.$$

Thus presence of a separatrix loop of the saddle–node O of the system (3.8) is equivalent to presence of a heteroclinic contour of the system (3.14) connecting its two singular points $P_1(\nu_1, 0, 0)$ and $P_2(\nu_2, 0, 0)$, where, as it was mentioned above, quantities

$$\nu_{1,2} = -\frac{\sigma + 1}{2} \mp \sqrt{\frac{(\sigma + 1)^2}{4} + \sigma(r - 1)}, \ \nu_1 < 0, \ \nu_2 > 0,$$

are the eigenvalues of linearization matrix of the Lorenz system (3.8) at the saddle–node $O(0, 0, 0)$. We choose a coordinate u on which the singular points P_1 and P_2 of the system (3.14) lie and, following the above considerations, reduce the system (3.14) to the form

$$\frac{dv}{du} = -\frac{2uv}{f(u) + \sigma z}, \qquad \frac{dz}{du} = -\frac{v(1 + u/\sigma) - bz}{f(u) + \sigma z}, \tag{3.15}$$

where $f(u) = (u - \nu_1)(u - \nu_2)$. By using the Taylor expansion of numerator and denominator of the right-hand side in the system (3.15) at the point $P_2(\nu_2, 0, 0)$, we obtain

$$v'(\nu_2) = \frac{2v'(\nu_2)\nu_2}{\nu_1 - \nu_2 - \sigma z'(\nu_2)}, \ z'(\nu_2) = \frac{v'(\nu_2)(1 + \nu_2/\sigma) - bz'(\nu_2)}{\nu_1 - \nu_2 - \sigma z'(\nu_2)}.$$

The last system of equations with $v'(\nu_2) \neq 0$ implies that $\nu_1 - \nu_2 - \sigma z'(\nu_2) = 2\nu_2$, and consequently,

$$z'(\nu_2) = \frac{\nu_1 - 3\nu_2}{\sigma} < 0,$$
$$v'(\nu_2) = \frac{z'(\nu_2)(2\nu_2 + b)}{1 + \nu_2/\sigma} = \frac{(\nu_1 - 3\nu_2)(2\nu_2 + b)}{\sigma + \nu_2} < 0. \tag{3.16}$$

Conversely, $v'(\nu_1) = 0$ at the point P_1. Therefore, $\nu_2 - \nu_1 - \sigma z'(\nu_1) = -b$, and hence, $z'(\nu_1) = (b + \nu_2 - \nu_1)/\sigma$.

For an arbitrarily small $\varepsilon > 0$, we solve numerically the system (3.14) in direct time with the initial conditions

$$u(0) = \nu_2 - \varepsilon, \ v(0) = -v'(\nu_2)\varepsilon, \ z(0) = -z'(\nu_2)\varepsilon,$$

where the values $v'(\nu_2)$ and $z'(\nu_2)$ are defined in accordance with (3.16). We obtain a trajectory $(u^+(t), v^+(t), z^+(t))$ that is arbitrarily close to the separatrix issuing from the singular point $P_2(\nu_2, 0, 0)$ of system (3.14). The trajectories are sewed for $u = 0$, that is we terminate the calculation of the trajectory issuing from a neighbourhood of the point P_2 at the time instant t^+ such that $u^+(t^+) = 0$. In this case, we have values $v^+(t^+)$ and $z^+(t^+)$.

Likewise, from a neighbourhood of the singular point $P_1(\nu_1, 0, 0)$ in reverse time, we draw a one-parameter family of trajectories $(u^-(t, \varkappa), v^-(t, \varkappa), z^-(t, \varkappa))$ arbitrary close to separatrices entering the point P_1. For this purpose, in reverse time, we solve numerically the system (3.14) with initial conditions

$$u(0) = \nu_1 + \varepsilon, \ v(0) = (\varkappa\varepsilon)^{-2\nu_1/b}, \ z(0) = z'(\nu_1)\varepsilon,$$

where $\varkappa$ is an arbitrary constant.

To clarify the nature of the one-parameter family of initial conditions at point P_1 for the function $v(t)$ we perform the change of variables $v = w^{-2\nu_1/b}$ in (3.14). Then $\dot{v} = -2w^{-2\nu_1/b-1}\dot{w}\nu_1/b$. On the other hand $\dot{v} = 2w^{-2\nu_1/b}u$. Hence,

$$\dot{w} = -\frac{b}{\nu_1}wu \qquad \text{and} \qquad w'(\nu_1) = -\frac{bw'(\nu_1)}{\nu_2 - \nu_1 - \sigma z'(\nu_1)}.$$

The last condition is satisfied for each $w'(\nu_1) = \varkappa = $ const, since $\nu_2 - \nu_1 - \sigma z'(\nu_1) = -b$. Therefore, the initial value of the function $w(t)$ has the form $w(0) = \varkappa\varepsilon$, which induces an initial condition for the function $v(t)$ at the point P_1 in the form $v(0) = (\varkappa\varepsilon)^{-2\nu_1/b}$.

Now we chose a time instant t^- and a value $\varkappa$ such that

$$u^-(t^-, \varkappa) = 0, \ v^-(t^-, \varkappa) = v^+(t^+).$$

That is we close projection of the heteroclinic contour of the system (3.14) in the (u, v)-plane. Then to obtain the entire heteroclinic contour with a given accuracy, it is necessary and sufficient to solve the single equation

$$z^+(\sigma, b, r, t^+) = z^-(\sigma, b, r, t^-, \varkappa), \tag{3.17}$$

which defines the bifurcation surface in the space of parameters (σ, b, r). Fig. 3.30 represents (u, v)- and (u, z)-projections of the heteroclinic contour of the system (3.14), which were obtained by numerical solution of equation (3.17) for parameter r for the given values $b = 8/3$, $\sigma = 10$ and $\varepsilon = 10^{-6}$. In this case we found the value $\varkappa = 0.293356 \pm 1 \cdot 10^{-6}$ of the constant and the value $r = 13.926560 \pm 5 \cdot 10^{-6}$ of the system parameter.

By expressions $x = \pm\sqrt{v}$, $y = xu$, this heteroclinic contour of the system (3.14) induces two saddle–node separatrix loops of the original Lorenz system (3.8) forming the so-called homoclinic butterfly. Fig. 3.31 represents (x, y)- and (x, z)-projections of the homoclinic butterfly of the original system (3.8) of the Lorenz equations corresponding to the classical parameter values $b = 8/3$, $\sigma = 10$. Thus as it has been already noted above, it was possible to refine the classical value r for which the homoclinic butterfly takes place.

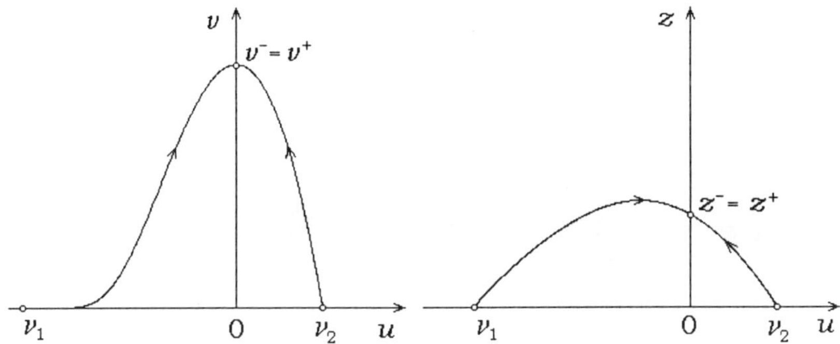

Fig. 3.30   The construction of a homoclinic saddle–node separatrix loop (homoclinic butterfly) in the Lorenz system.

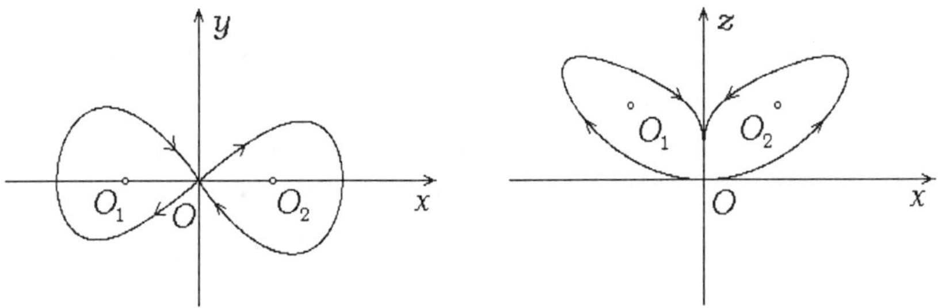

Fig. 3.31   Projections of the homoclinic butterfly of the Lorenz system (for $b = 8/3$ and $\sigma = 10$ we obtained $r = 13.926560 \pm 5 \cdot 10^{-6}$).

Let us pay attention to the almost fourteenth order of tangency of solution $v(u)$ of the system (3.14) to u-axis at the singular point P_1 (Fig. 3.30) and an approximate seventh order of tangency of solution $x(z)$ of the system (3.8) to the z-axis at the saddle–node O (Fig. 3.31). There is no wonder, that no numerical integration method can deal with calculation of such superunstable trajectories in a neighbourhood of saddle–node O in the Lorenz system.

3.1.4.4 *Heteroclinic contour connecting saddle–node with two saddle–focuses*

Applying a similar technique, we found the bifurcation curve C of existence in the space of parameters σ, b, r of heteroclinic contours connecting the saddle–node O with saddle–focuses O_1 and O_2. This contour connects three singular points (see Fig. 3.32), and it is a limit case of heteroclinic contours connecting two saddle–focuses O_1 and O_2 as well as homoclinic separatrix loops of saddle–focuses O_1 and O_2.

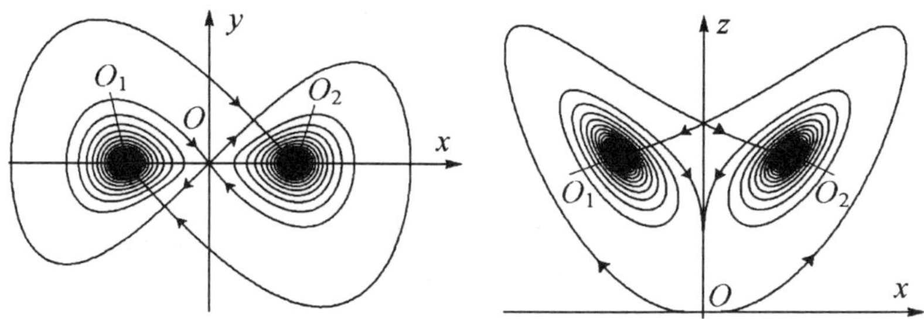

Fig. 3.32   Projections of the heteroclinic contour connecting the saddle–node with two saddle–focuses for $b = 8/3$, $\sigma = 10.167293$, $r = 30.8681$.

3.1.5 *Diagrams of nonlocal bifurcations in the Lorenz system*

The developed technique of finding homoclinic and heteroclinic contours of singular points of the Lorenz system enables to construct numerically bifurcation surfaces of bifurcations in the parameter space (σ, b, r) considered above.

First such construction of the bifurcation surface of existence of the homoclinic butterfly has been carried out in the paper [Kaloshin *et al.* (2003)]. It was verified the Leonov–Chen inequality [Leonov (1988); Chen

(1996)] $3\sigma > 2b + 1$, that is a necessary and sufficient condition of existence of a value r for which the Lorenz system has two homoclinic saddle–node separatrix loops (homoclinic butterfly).

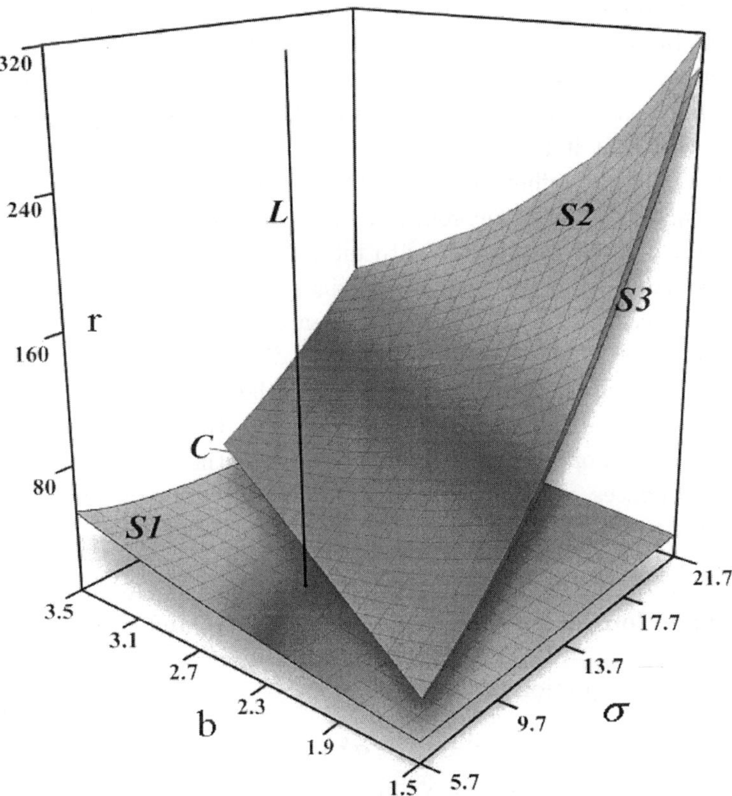

Fig. 3.33 S_1 is the bifurcation surface of existence of homoclinic butterflies; S_2 is the bifurcation surface of existence of homoclinic separatrix loops of saddle–focuses O_1 and O_2; S_3 is the bifurcation surface of existence of heteroclinic contours of saddle–focuses O_1 and O_2; C is the curve of existence of heteroclinic contours connecting the saddle–node O with saddle–focuses O_1 and O_2; L is a straight line of classical parameters of the Lorenz system.

The full bifurcation diagram of all nonlocal bifurcations existing in the Lorenz system is shown in Fig. 3.33. This is the representation of all bifurcation curves and surfaces of the Lorenz system on one diagram in the space of parameters (σ, b, r). The surface S_1 of existence of homoclinic butterflies is situated below all. The curve C of existence of heteroclinic contours connecting a saddle–focuses O_1 and O_2 with a saddle–node O is situated hardly above S_1, not touching and not crossing it. Two semi-surfaces S_2 and S_3 go away from the curve C, and S_2 lies above S_3 in the space of parameters. Semi-surface S_2 is the bifurcation surface of existence of homoclinic separatrix loops of saddle–focuses O_1 and O_2 in the Lorenz system, and S_3 is the bifurcation surface of existence of heteroclinic separatrix contours connecting saddle–focuses O_1 and O_2. Straight line L on the given diagram represents classical parameter values of the system $\sigma = 10, b = 8/3$. It is easy to see, that it intersects only the bifurcation surface S_1 of existence of homoclinic butterflies at $r \approx 13.926$ and passes very close to a curve C of existence of heteroclinic contours of saddle–focuses and a saddle–node. Moreover, L is not intersected with bifurcation semi-surfaces S_2 and S_3.

Once again it shows, that at classical values of parameters transition to chaos is carried out through an incomplete double homoclinic cascade of bifurcations. Thus the full bifurcation diagram shows, that behaviour of trajectories in the Lorenz system is not the most complex at values of parameters $\sigma = 10$, $b = 8/3$, as in the system there do not exist neither saddle–focuses homoclinic separatrix loops, neither heteroclinic contours of saddle–focuses, nor heteroclinic contours connecting a saddle–node with saddle–focuses.

3.2   The Complex System of Lorenz Equations

The system of equations

$$
\begin{aligned}
\dot{X} &= -\sigma X + \sigma Y, \\
\dot{Y} &= -XZ + rX - aY, \\
\dot{Z} &= -bZ + \frac{1}{2}(X^*Y + XY^*)
\end{aligned}
\tag{3.18}
$$

for two complex variables X and Y and one real variable Z, which has real parameters σ and b and complex parameters $r = r_1 + ir_2$ and $a = 1 - ie$, is referred to as the complex system of Lorenz equations [Gibbon (1981)].

We show, that a scenario of transition to chaos through a subharmonic (with respect to one frequency) cascade of bifurcations of two-dimensional tori takes place in the complex system of Lorenz equations. As was noted above, the specific feature of such a cascade should be the emergence of a stable two-dimensional torus of the triple period treated as the direct product of a simple cycle and a cycle of the triple period.

By introducing the new real variables $x_1 = \Re\{X\}$, $x_2 = \Im\{X\}$, $y_1 = \Re\{Y\}$, $y_2 = \Im\{Y\}$, and by setting $e = 0$, one can rewrite the system (3.18) in the form of a five-dimensional system of real ordinary differential equations

$$
\begin{aligned}
\dot{x}_1 &= -\sigma x_1 + \sigma y_1, \\
\dot{x}_2 &= -\sigma x_2 + \sigma y_2, \\
\dot{y}_1 &= -x_1 z + r_1 x_1 - y_1 - r_2 x_2, \\
\dot{y}_2 &= -x_2 z + r_1 x_2 - y_2 + r_2 x_1, \\
\dot{z} &= x_1 y_1 + x_2 y_2 - bz,
\end{aligned}
\tag{3.19}
$$

where $\sigma > 0$, $b > 0$, $r_1 > 0$, $r_2 > 0$. System (3.19) has a unique equilibrium, the origin, whose stability is determined by eigenvalues of the Jacobian $J(0)$ at this point:

$$
J(0) = \begin{pmatrix}
-\sigma & 0 & \sigma & 0 & 0 \\
0 & -\sigma & 0 & \sigma & 0 \\
r_1 & -r_2 & -1 & 0 & 0 \\
r_2 & r_1 & 0 & -1 & 0 \\
0 & 0 & 0 & 0 & -b
\end{pmatrix}.
$$

In the characteristic equation

$$
(\lambda + b)(\lambda^4 + a_1 \lambda^3 + a_2 \lambda^2 + a_3 \lambda + a_4) = 0,
$$

where

$$
a_1 = 2(\sigma + 1), \quad a_2 = (\sigma + 1)^2 + 2\sigma(1 - r_1),
$$
$$
a_3 = 2\sigma(\sigma + 1)(1 - r_1) + 2\sigma r_2, \quad a_4 = \sigma^2((1 - r_1)^2 + r_2^2),
$$

one eigenvalue $\lambda = -b$ is negative. Therefore, stability of the equilibrium depends on roots of a polynomial of degree 4. The Routh–Hurwitz conditions

$$
a_1 > 0, \quad a_1 a_2 - a_3 > 0, \quad a_1(a_2 a_3 - a_1 a_4) - a_3^2 > 0, \quad a_4 > 0,
$$

which determine the stability of that polynomial, are equivalent to the system of two inequalities

$$
a_1 a_2 - a_3 > 0, \qquad a_1(a_2 a_3 - a_1 a_4) - a_3^2 > 0.
$$

By substituting the above expressions for coefficients of the polynomial into this system, we find that the fixed point becomes unstable if

$$(r_1 - 1)(\sigma + 1)^2 + \sigma r_2^2 > 0, \tag{3.20}$$

and a limit cycle appears in the system (3.19). Let us show that this limit cycle has the form

$$x_1(t) = R\cos\omega t, \quad x_2(t) = R\sin\omega t. \tag{3.21}$$

Indeed, by substituting (3.21) into (3.19), we obtain

$$y_1(t) = -\frac{R\omega}{\sigma}\sin\omega t + R\cos\omega t, \quad y_2(t) = \frac{R\omega}{\sigma}\cos\omega t + R\sin\omega t.$$

Then

$$\dot{y}_1(t) = -\frac{R\omega^2}{\sigma}\cos\omega t - R\omega\sin\omega t$$

$$= R\cos\omega t(r_1 - z) - r_2 R\sin\omega t + \frac{R\omega}{\sigma}\sin\omega t - R\cos\omega t$$

for

$$-\frac{R\omega^2}{\sigma} = R(r_1 - z - 1) \qquad \text{and} \qquad R\omega = r_2 R - \frac{R\omega}{\sigma}.$$

In a similar way, we have

$$\dot{y}_2(t) = -\frac{R\omega^2}{\sigma}\sin\omega t + R\omega\cos\omega t$$

$$= R\sin\omega t(r_1 - z) + r_2 R\cos\omega t - \frac{R\omega}{\sigma}\cos\omega t - R\sin\omega t$$

for

$$-\frac{R\omega^2}{\sigma} = R(r_1 - z - 1) \qquad \text{and} \qquad R\omega = r_2 R - \frac{R\omega}{\sigma}.$$

Consequently, $\omega = \dfrac{\sigma r_2}{\sigma + 1}$, $z = \dfrac{\omega^2}{\sigma} + r_1 - 1 = z_0$. To find amplitude R, we use the fifth equation in (3.19) and set $z = z_0$ in this equation:

$$bz_0 = -\frac{R^2\omega}{\sigma}\sin\omega t\cos\omega t + R^2\cos^2\omega t + \frac{R^2\omega}{\sigma}\sin\omega t\cos\omega t + R^2\sin^2\omega t = R^2.$$

Consequently, $R^2 = bz_0 = b\left(\dfrac{\omega^2}{\sigma} + r_1 - 1\right) > 0$ by (3.20). If $r_2 = 0$, then the five-dimensional space of solutions of system (3.19) is partitioned into three-dimensional Lorenz spaces defined by given initial values of angles $\tan\varphi = x_2(0)/x_1(0)$ and $\tan\psi = y_2(0)/y_1(0)$. In the course of time, the trajectory is entirely drawn into one of Lorenz spaces and remains there forever. For example, if $\tan\varphi = \tan\psi = 1$, then the trajectory does not leave the Lorenz space $x_2 = x_1$, $y_2 = y_1$, which can be easily established by subtracting the second equation of the system (3.19) from the first, and the fourth equation from the third, and by taking account of the relation $r_2 = 0$. If $x_2(0) = y_2(0) = 0$, then the trajectory does not leave the three-dimensional Lorenz space defined by the variables x_1, y_1, and z. Therefore, one can assume that if $r_2 = 0$, then dynamics of system (3.19) is equivalent to dynamics of the three-dimensional system of Lorenz equations

$$\dot{x}_1 = -\sigma x_1 + \sigma y_1,$$
$$\dot{y}_1 = -x_1 z + r_1 x_1 - y_1, \tag{3.22}$$
$$\dot{z} = -bz + x_1 y_1.$$

We proved in Sec. 3.1 that, for parameter values $\sigma = 10.5$, $b = 8/3$, and $r_1^* \approx 33.2189$, the system (3.22) has the most complicated complete double homoclinic attractor, transition to which is through a complete double homoclinic cascade of bifurcations of stable cycles as parameter r_1 diminishes from $r_1 = 350$ to $r_1^* \approx 33.2189$. Therefore, it is most interesting to consider the transition to chaos in a five-dimensional system (3.19) as $r_2 \to 0$ for given values of the remaining parameters of the system, namely, $\sigma = 10.5$, $b = 8/3$, and $r_1 = 33.2189$. In what follows, we will show that such a transition to chaos is performed through *the Sharkovskii subharmonic cascade* of bifurcations of two-dimensional tori.

3.2.1 *Scenario of transition to chaos*

As follows from the preceding, inequality (3.20), which necessarily holds for the values of parameters σ, r_1 and r_2 given above, is a condition for existence of a limit cycle of the system (3.19). Therefore, for all r_2 exceeding some value r_2^*, the cycle (3.21) is a unique stable limit set of the system (3.19). If $r_2 = r_2^* \approx 1.93$, then, as a result of the Andronov–Hopf bifurcation, a stable torus T^2 is generated from the cycle (3.21); after that, two-frequency oscillation modes occur in the system (3.19). The multipliers μ_i, $i = 1,\ldots,5$, of the cycle (3.21) measured for system (3.19) with $r_2 = 2.00$, $r_2 = 1.95$, and $r_2 = 1.93$ are equal to $(0;0;1;0.683 \pm 0.635i)$ with $|\mu_4| = |\mu_5| = 0.932$; $(0;0;1;-0.166 \pm 0.966i)$ with $|\mu_4| = |\mu_5| = 0.971$ and

$(0; 0; 1; -0.570 \pm 0.812i)$ with $|\mu_4| = |\mu_5| = 0.992$, and justify the classical transition of two complex-conjugate multipliers through the circle of unit radius under the torus-generating bifurcation, which takes place in the case under consideration as r_2 diminishes. To analyze the so-generated torus T^2 and all of its subsequent transformations caused by related bifurcations, it is convenient to consider the projections of intersection between the torus and the four-dimensional subspace $x_2 = 0$ on the planes (x_1, y_1) and (x_1, z). Corresponding projections for the parameter value $r_2 = 1.001$ are shown in Fig. 3.34. Each projection is the projection of a pair of closed curves in which the torus T^2 intersects the four-dimensional subspace $x_2 = 0$.

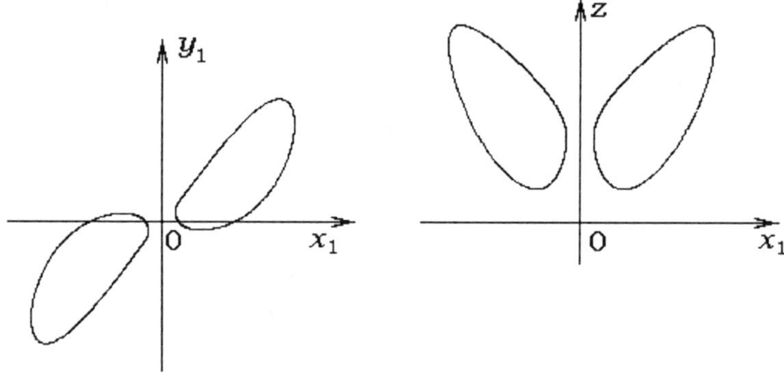

Fig. 3.34   Projections of a simple two-dimensional torus from the space $x_2 = 0$ on the planes (x_1, y_1) and (x_1, z) for $r_2 = 1.001$.

The stable torus T^2 exists in system (3.19) for the parameter values $0.7925 \lesssim r_2 \lesssim 1.93$. If $r_2 = 0.7925$, then in system (3.19), there is a period doubling bifurcation for closed curves, which are intersections of the torus and the subspace $x_2 = 0$ (see Fig. 3.35). In this case, torus remains two-dimensional, but its period is two, which corresponds to the doubling of the surface area inside a bounded volume of the five-dimensional phase space.

Further reduction in the parameter r_2 in system (3.19) leads to a subharmonic cascade of bifurcations of closed curves that are intersection of the torus with the subspace $x_2 = 0$, which reflects generation of new stable two-dimensional tori of various periods with increasing surface areas inside a bounded volume of phase space. Thus, system (3.19) has a stable two-dimensional torus of period 5 for $0.418 < r_2 < 0.421$ and a stable two-dimensional torus of period 3 (see Fig. 3.36) for $0.349 < r_2 < 0.352$. This implies that if $r_2 < 0.349$, then system (3.19) has infinitely many two-

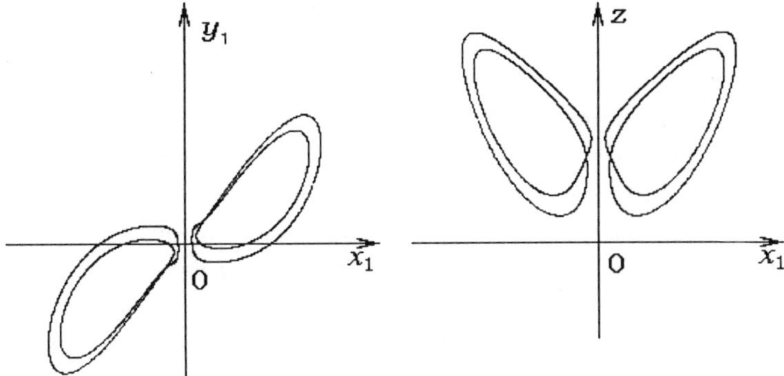

Fig. 3.35   Projections of a two-dimensional torus of period two from the space $x_2 = 0$ on the planes (x_1, y_1) and (x_1, z) for $r_2 = 0.6015$.

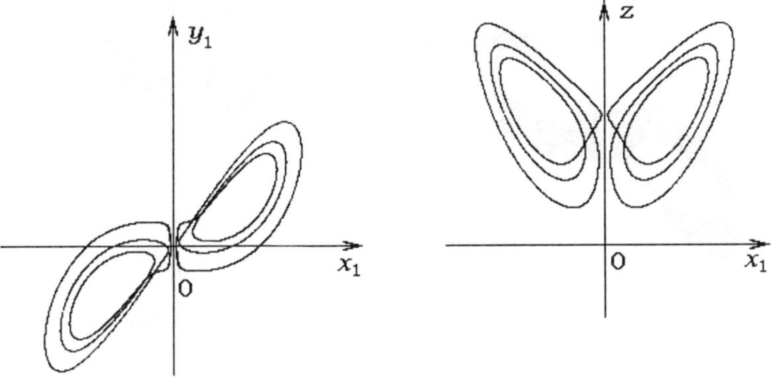

Fig. 3.36   Projections of a two-dimensional torus of period three from the space $x_2 = 0$ on the planes (x_1, y_1) and (x_1, z) for $r_2 = 0.351$.

dimensional unstable tori of all periods, in accordance with the Sharkovskii order. Therefore, if $r_2 < 0.349$, the subharmonic cascade of bifurcations generates in system (3.19) a *toroidal singular attractor* that is generated by the set of windings of all two-dimensional unstable tori, which were generated initially stable by the subharmonic cascade of bifurcations described above (see Fig. 3.37). The question of the attractor dimension has remained unanswered as yet. Our hypothesis is that the attractor generated in this way is not a fractal and has an integer dimension equal to 3. This is in accordance with the conjecture that complete homoclinic attractor of a three-dimensional real Lorenz system has an integer dimension equal to two, since the set of cycles taking part in its generation is assumed to be dense everywhere on some unstable one-dimensional not invariant manifold of the origin. In accordance with the suggested hypothesis, intersection of the attractor and a four-dimensional subspace should have dimension two, and its projections from the three-dimensional subspace $x_2 = 0$, $z = r_1 - 1$ on the planes (x_1, y_1), (x_1, y_2) and (y_2, y_1) should be one-dimensional curves, which is in agreement with results of computer simulations (see Fig. 3.38). Note that the set which is the intersection of attractor with subspace $x_2 = 0$ tends, as $r_2 \to 0$, to the supposedly two-dimensional complete homoclinic attractor of the Lorenz system, which lies in the three-dimensional space (x_1, y_1, z).

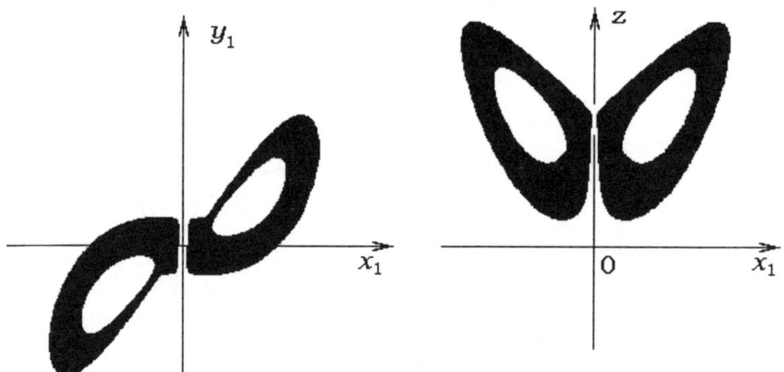

Fig. 3.37   Projections of the subharmonic singular attractor from space $x_2 = 0$ on planes (x_1, y_1) and (x_1, z) for $r_2 = 0.25$.

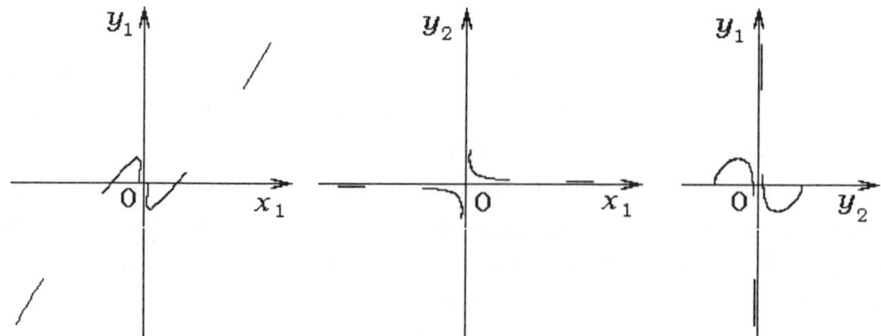

Fig. 3.38   Projections of the subharmonic singular attractor from the space $x_2 = 0$, $z = r - 1$ on the planes (x_1, y_1), (x_1, y_2) and (y_2, y_1) for $r_2 = 0.25$.

3.3 Systems of the Rossler Equations

Rossler offered a number of nonlinear systems of ordinary differential equations for modelling some hypothetical chemical reactions possessing chaotic behaviour [Rossler (1976); Gurel and Gurel (1983)]. Here we consider two systems, most known of which looks like

$$\begin{aligned}
\dot{x} &= -y - z, \\
\dot{y} &= x + ay, \\
\dot{z} &= b + z(x - \mu).
\end{aligned} \tag{3.23}$$

Qualitative analysis of the limit cycle birth bifurcation for this system is carried out in the paper [Magnitskii (1995)], and the domain of parameters in which the Andronov–Hopf bifurcation takes place is found. The scenario of transition to chaos, certainly, is connected with the limit cycle birth bifurcation which is also the beginning of the Feigenbaum cascade of bifurcations and the subharmonic cascade of bifurcations.

The system (3.23) is not dissipative everywhere in phase space since $\operatorname{div} F(x, y, z) = x + a - \mu$ depends on a variable x, and the plane $x + a - \mu = 0$ divides all phase space into two disjoint subspaces. Let us define positions of singular points $O_i(x_i^*, y_i^*, z_i^*)$, $i = 1, 2$ of the system (3.23), where

$$x_i^* = \frac{\mu}{2} + (-1)^i \sqrt{\frac{\mu^2}{4} - ab}, \ y_i^* = -\frac{x_i^*}{a}, \ z_i^* = -y_i^*.$$

The condition that point O_1 belongs to a dissipative domain is

$$a - \frac{\mu}{2} - \sqrt{\frac{\mu^2}{4} - ab} < 0.$$

This is equivalent to the following system of inequalities in view of $a, b, \mu > 0$

$$\begin{cases} \mu > 2a, \\ \mu \geq 2\sqrt{ab}, \end{cases} \qquad \begin{cases} \mu \leq 2a, \\ \mu > a + b. \end{cases}$$

According to [Magnitskii (1995)] a stable limit cycle is born in case of loss of stability of point O_1 under condition of $a < b$. As thus the second system is incompatible, then point $O_1 \in D = \{x \mid \operatorname{div} F(x) < 0\}$ for any values $\mu > 2\sqrt{ab}$.

The condition that point $O_2 \in D = \{x \mid \operatorname{div} F(x) < 0\}$, looks like

$$a - \frac{\mu}{2} + \sqrt{\frac{\mu^2}{4} - ab} < 0.$$

This is equivalent to the following system of inequalities

$$\mu \geq 2a, \qquad \mu > 2\sqrt{ab}, \qquad \mu < a + b,$$

which is compatible for $a < b$ and $2\sqrt{ab} < \mu < a + b$. It is shown in the paper [Magnitskii (1995)], that point O_1 loses stability as a result of Andronov–Hopf bifurcation for parameter value $\mu^* = p + ab/p$, where $\sqrt{ab} < p = a/2 + \sqrt{a^2/4 + (b-a)/a}$. At that it passes into an equilibrium state of a saddle–focus type with one-dimensional stable and two-dimensional unstable manifolds. It is easy to show, that value μ^* is not included into an interval $(2\sqrt{ab}, a + b)$ for $ab < 1$ and, thus, the point O_2 lies outside of the dissipative domain. Thus the type of an equilibrium state of the point O_2 is defined by the characteristic equation

$$\lambda^3 + \lambda^2(\mu - a - x^*) + \lambda(1 + z^* - a(\mu - x^*)) - \sqrt{\mu^2 - 4ab} = 0 \qquad (3.24)$$

for the Jacobi matrix of system (3.23), linearized at the point O_2. The numerical solution of the equation (3.24) depending on values of parameter μ for the fixed values of parameters $a = 0.5$ and $b = 0.75$ is shown in Fig. 3.39.

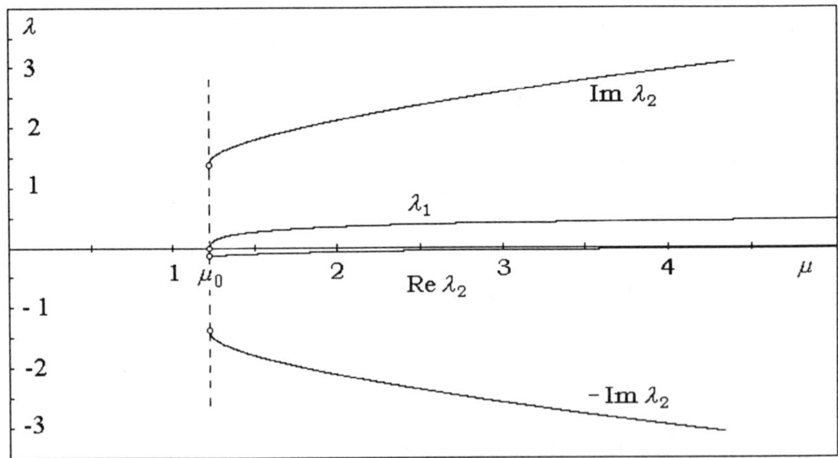

Fig. 3.39   Dependence of roots of the characteristic equation (3.24) from the parameter values μ for the singular point O_2.

It is visible from the figure, that point O_2 is a saddle–focus type singular point with two-dimensional stable manifold and one-dimensional unstable manifold for all parameter values $\mu > \mu_0 = 2\sqrt{ab}$. If point O_2 is situated in the dissipative domain of system (3.23), it would be possible to expect occurrence of heteroclinic contours connecting points O_1 and O_2. But as only point O_1 lies in the dissipative domain, there are only two possible scenarios of transition to chaotic behaviour: the Feigenbaum scenario (or more general subharmonic cascade of bifurcations), and the homoclinic cascade of bifurcations.

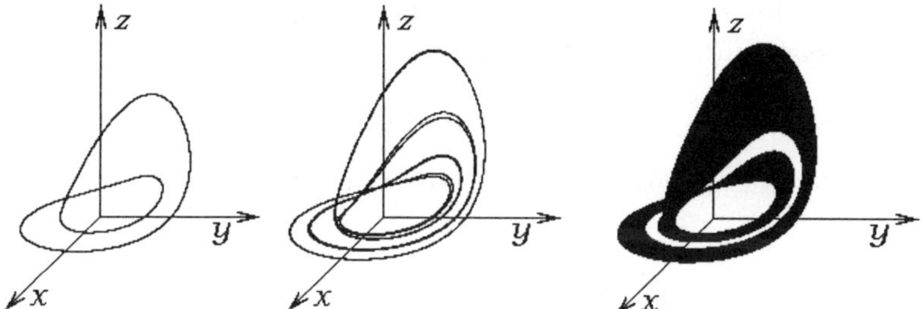

Fig. 3.40   Cycles of the periods 2 and 8 accordingly for the parameter values $\mu = 2.2$ and $\mu = 2.325$ and irregular singular attractor for $\mu = 2.35$.

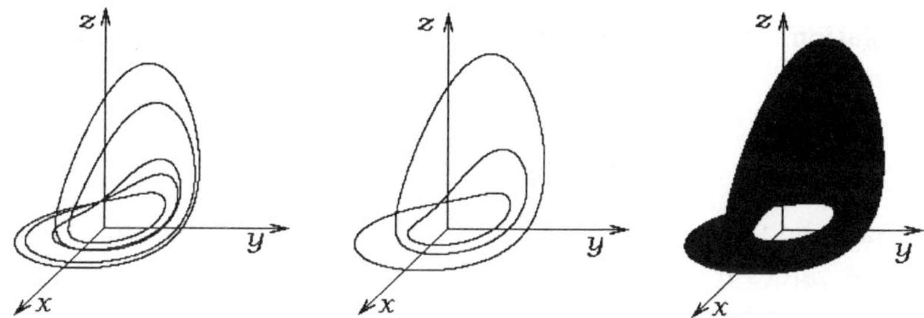

Fig. 3.41   Cycles of the periods 5 and 3 accordingly for the parameter values $\mu = 2.421$ and $\mu = 2.446$ and irregular singular attractor for $\mu = 2.5$.

Indeed, the Feigenbaum cascade of period doubling bifurcations is observed for specified above fixed values of parameters a and b with increase in parameter μ from the value $\mu = 1.375$ up to the value $\mu \approx 2.35$ (Fig. 3.40). And the cycles accordingly of periods 5 and 3 are observed for great values of parameter $\mu = 2.421$ and $\mu = 2.446$, that testifies the presence of subharmonic cascade of bifurcations in the scenario of transition to chaos (Fig. 3.41) in the Rossler system, i.e. this scenario is developed according to the Sharkovskii order. Irregular attractor in the Rossler system disappears for the value $\mu > 2.899$, and the solution of system becomes unlimited.

Other Rossler system considered by us looks like

$$\begin{aligned}
\dot{x} &= -y - z, \\
\dot{y} &= x + ay, \\
\dot{z} &= bx + z(x - \mu).
\end{aligned} \qquad (3.25)$$

The system (3.25), as well as previous, is not dissipative everywhere. Divergence of this system is negative in the domain $x < \mu - a$ of the phase space. The system (3.25) has two singular (fixed) points

$$O_1(0,0,0) \qquad \text{and} \qquad O_2(\mu - ab, b - \mu/a, \mu/a - b).$$

The characteristic equation for the Jacobi matrix at point O_1 looks as follows

$$\lambda^3 + \lambda^2(\mu - a) + \lambda(1 + b - a\mu) + \mu - ab = 0. \qquad (3.26)$$

According to the Routh–Hurwitz criterion the point O_1 is stable under conditions

$$\mu - a > 0, \ (\mu - a)(1 + b - a\mu) > \mu - ab > 0. \qquad (3.27)$$

The solution of system (3.27) in a case $b > 2a - a^2$ gives two critical values of parameter

$$\mu_{1,2}^* = \frac{a^2 + b \pm \sqrt{(a^2 + b)^2 - 4a^2}}{2a},$$

for which the loss of stability of point O_1 takes place. Point O_1 is a stable focus for values $\mu \in (\mu_1^*, \mu_2^*)$, and for values $\mu = \mu_1^*$ or $\mu = \mu_2^*$ the bifurcation of creation of a limit cycle occurs. In case $b < 2a - a^2$ the point O_1 is unstable for anyone $\mu > a$, and there can be both stable, and unstable periodic solutions in its neighbourhood. Existence of such solutions is influenced by the second singular point O_2. Its equilibrium state is defined by roots of the characteristic equation

$$\lambda^3 + \lambda^2 a(b - 1) + \lambda(1 - a^2 b + \mu/a) + ab - \mu = 0. \qquad (3.28)$$

Comparing expressions (3.26) and (3.28), we see, that instability of point O_2 follows from stability of point O_1, and point O_2 remains unstable even in case $b > 1$.

We shall carry out research of the scenario of transition to chaos in the given system for fixed values of parameters $a = 0.38$ and $b = 0.3$. For given values of parameters $b < 2a - a^2$, point O_1 is unstable saddle–focus, and there exists a cycle in system for the values $\mu \in (0.9, 1.933)$. For values $\mu < 0.9$ a trajectory of system passes into a domain of instability of point O_2 and tends to infinity along an unstable one-dimensional manifold of this point.

With growth of values of parameter μ, in the system (3.25), a cascade of the period doubling bifurcations of an original cycle is observed. So for the value $\mu \approx 1.934$ a cycle of the period 2 is born, for $\mu \approx 2.535$ a cycle of the period 4 is born, for $\mu \approx 2.680$ a cycle of the period 8 is born, *etc.* Feigenbaum cascade comes to the end for the value of parameter $\mu \approx 2.8$ with formation of an irregular singular attractor as a tape (Fig. 3.42). For the further increase in values of parameter μ, cycles of periods 7, 5 and 3 are observed accordingly for the values $\mu \in (2.8895, 2.8904)$, $\mu \in (2.954, 2.966)$ and $\mu \in (3.198, 3.532)$. This confirms the presence in the system of the subharmonic cascade of bifurcations. The cycles of this cascade are similar to the cycles represented in Fig. 3.41.

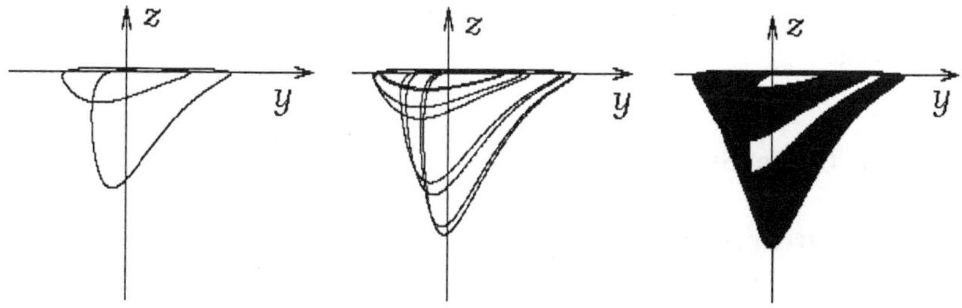

Fig. 3.42 Cycles of the periods 2 and 8 in system (3.25) accordingly for the values of parameter $\mu = 2.2$ and $\mu = 2.69$ and an irregular singular attractor for $\mu = 2.8$.

It is necessary to note, that in the given Rossler system the complete subharmonic cascade of bifurcations, smoothly passing into the simple homoclinic cascade in which the cycles tend to the homoclinic contour of the saddle–focus O_1, is observed for the fixed values of parameters $a = 0.38$ and $b = 0.3$. As well as in the Lorenz system a solution such as self-organizing cycle exists in the system for the parameter value $\mu = 3.668$, and the cycle C_4 of the homoclinic cascade of bifurcations exists for value $\mu = 4.367$. This cascade comes to the end by formation of irregular singular homoclinic attractor which is observed for value $\mu \approx 4.828$ (Fig. 3.43). The question of, whether the homoclinic cascade in system (3.25) is complete, demands special research in view of strong ill-posedness of a solved problem in a neighbourhood of axis z.

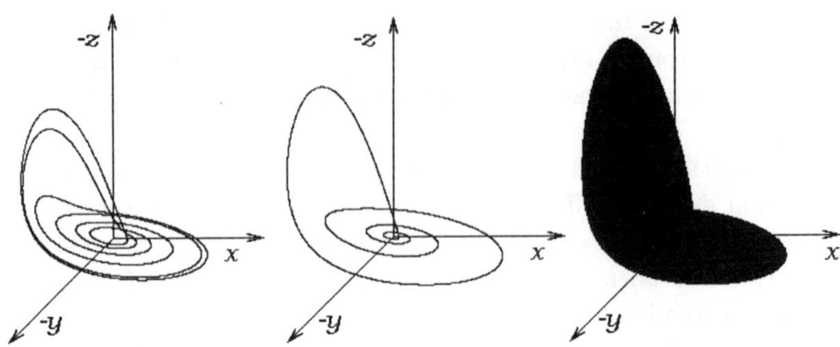

Fig. 3.43   A self-organizing cycle ($\mu = 3.668$), a cycle C_4 ($\mu = 4.367$) and a homoclinic attractor ($\mu = 4.828$) in system (3.25).

Thus, transition to chaos in Rossler systems of kinds (3.23) and (3.25) is carried out in the same way, as in the Lorenz system, namely: through *the Feigenbaum period doubling cascade* of bifurcations, *the Sharkovskii subharmonic cascade* of bifurcations and *the Magnitskii homoclinic cascade* of bifurcations.

3.4   The Chua System

The Chua system models some electric circuit offered by L.Chua for generation of chaotic oscillations [Chua *et al.* (1986)]. Behaviour of this electric circuit and the same system of ordinary differential equations was widely investigated as in numerous physical experiences and mathematical methods, including numerical experiments and analytical calculations [Anishchenko *et al.* (1999); Shil'nikov (1994)].

Let us consider this system in a following kind

$$\begin{aligned}
\dot{x} &= \mu[y - h(x)], \\
\dot{y} &= x - y + z, \\
\dot{z} &= -\beta y,
\end{aligned} \qquad (3.29)$$

where

$$h(x) = \begin{cases} bx + a + b, & x < -1, \\ -ax, & |x| \le 1, \\ bx - a - b, & x > 1, \end{cases}$$

a, b, β, μ are some positive parameters. We shall fix parameters $a =$

$1/7$, $b = 2/7$, $\beta > 1$ and investigate solutions of the system (3.29) depending on parameter μ. The system of equations (3.29) is symmetric concerning the point of origin, i.e. a vector field is invariant in relation to transformation

$$(x, y, z) \rightarrow (-x, -y, -z).$$

Fixed (singular) points of system (3.29) are defined from the conditions

$$h(x) = 0, \qquad y = 0, \qquad x + z = 0.$$

Using a special kind of function $h(x)$, we shall find, that the system (3.29) has one by one singular point in each of areas: $D_{-1} = \{x | x < -1\}$, $D_0 = \{x | |x| < 1\}$ and $D_1 = \{x | x > 1\}$. We shall designate the singular points corresponding to these areas as $O_{-1}(-\xi, 0, \xi)$, $O_0(0, 0, 0)$ and $O_1(\xi, 0, -\xi)$, where $\xi = 1 + a/b$. The system (3.29) is a linear system in each of areas D_{-1}, D_0, D_1, and the Jacobi matrix of this system looks like

$$J = \begin{pmatrix} -\mu c & \mu & 0 \\ 1 & -1 & 1 \\ 0 & -\beta & 0 \end{pmatrix},$$

where $c = -a$ for $x \in D_0$ and $c = b$ otherwise. We shall find the eigenvalues of the Jacobi matrix from the characteristic equation

$$\lambda^3 + \lambda^2(1 + \mu c) + \lambda(\mu c - \mu + \beta) + \mu\beta c = 0. \tag{3.30}$$

Using the Routh–Hurwitz conditions

$$1 + \mu c > 0, \qquad (1 + \mu c)(\mu c - \mu + \beta) > \mu\beta c, \qquad \mu\beta c > 0, \tag{3.31}$$

let us determine states of equilibrium for fixed points and estimate the parameter values μ for which these points lose their stability. It is $c = -a$ for a point $O_0(0, 0, 0) \in D_0$. Therefore according to conditions (3.31) the area D_0 is dissipative area for the parameter values $\mu < 1/a$, and the fixed point O_0 is stable for $\mu < 0$. For the value $\mu = 0$ the characteristic equation (3.30) becomes

$$\lambda(\lambda^2 + \lambda + \beta) = 0,$$

and two its roots are complex conjugate with a negative real part. Hence, when parameter μ passes through the zero from left to right, the point O_0 changes its state of equilibrium from a stable focus to an unstable saddle–focus, having two-dimensional stable manifold W^s and one-dimensional unstable manifold W^u.

It is $c = b > 0$ for the points O_{-1} and O_1. Therefore areas D_{-1} and D_1 are dissipative areas for all positive parameter values μ. Another sign of the parameter c results according to the third inequality in (3.31) to that the bifurcation of change of an equilibrium state of points O_{-1} and O_1 for value $\mu = 0$ is opposite to that which we observed at the point O_0. Namely, at increasing of the parameter values μ a bifurcation of transition of the unstable saddle–focus with one-dimensional unstable manifold W^u and with two-dimensional stable manifold W^s into a stable focus takes place for the point $\mu = 0$. As it is easy to see from the second condition in (3.31), the points O_{-1} and O_1 remain stable up to the value

$$\mu_c = \frac{-1 + \sqrt{1 + 4b\beta/(1 - b)}}{2b},$$

for which an orbitally-stable limit cycles are born as a result of the Andronov–Hopf bifurcations.

Thus, all fixed points have equilibrium states of a saddle–focus type, that, apparently, defines a kind of irregular attractors, arising for the parameter values $\mu > \mu_c$.

Let us consider the scenario of occurrence of irregular attractors for the fixed values of parameters $a = 1/7$, $b = 2/7$, $\beta = 9$. For these values of parameters $\mu_c \approx 5.12$, and the limit cycle is stable in an interval $\mu \in (5.12, 5.895)$. For the values $\mu > 5.895$ there appears a double period cycle in system (3.29) and the cascade of Feigenbaum period doubling bifurcations is observed at the further increasing of parameter μ. So the cycle of period 4 is born for $\mu \approx 6.03$, the cycle of period 8 is born for $\mu \approx 6.058$, the cycle of period 16 is born for $\mu \approx 6.0616$, *etc.*

Some cycles of the Feigenbaum cascade and singular Feigenbaum attractors which are formed for $\mu = 6.07$ as a result of the period doubling bifurcations cascade are shown in Fig. 3.44. The cascade of period doubling bifurcations is followed by the subharmonic cascade of bifurcations. So the cycle of period 6 is observed for the value $\mu = 6.08$, the cycle of period 5 is observed for $\mu = 6.1098$ and the cycle of period 3 lies in the interval $\mu \in [6.138, 6.142]$. The Feigenbaum cascade of period doubling bifurcations occurs with the last cycle for the further increasing of the parameter μ. So the double cycle of the period 3 is observed for value $\mu = 6.144$. The Sharkovskii subharmonic cascade also comes to the end with formation of two irregular singular attractors. These attractors and some cycles of the subharmonic cascade are shown in Fig. 3.45.

With the further growth of the parameter μ, in system (3.29), cycles of

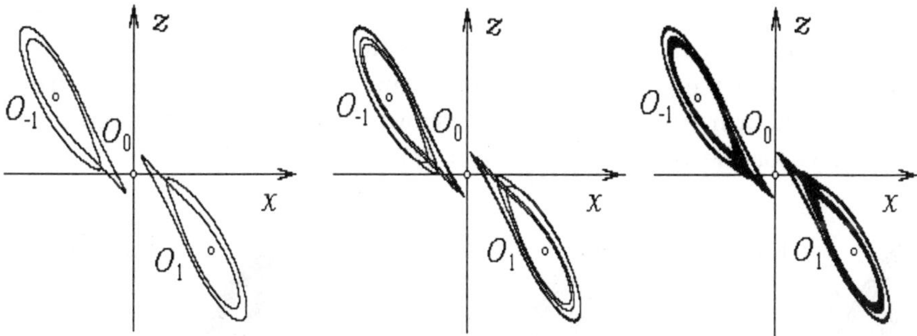

Fig. 3.44 Projections of period 2 cycle for $\mu = 5.9$, period 4 cycle for $\mu = 6.04$ and the Feigenbaum attractor for $\mu = 6.07$ to the plane (x, z) in the Chua system.

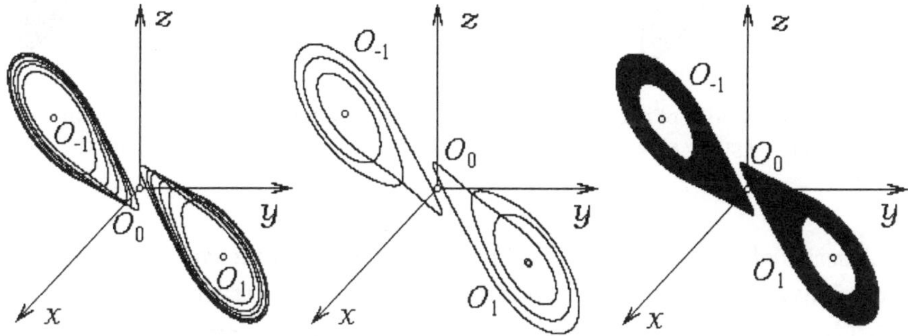

Fig. 3.45 Cycles of the Sharkovskii subharmonic cascade of period 6 for $\mu = 6.08$ and period 3 for $\mu = 6.14$ and irregular singular attractors for $\mu = 6.16$ in the Chua system.

C_n type of the homoclinic cascade of bifurcations can be observed. Such cycles have one turn extended aside of a point O_0, and n turns located in the domain of attraction of the point O_{-1} (or O_1). Obviously, the cycle of period 3 for $\mu \in [6.138, 6.142]$ gives the beginning for the homoclinic cascade. Further, in system (3.29) the cycle C_4 is observed for value $\mu = 6.176$, and the cycle C_6 is observed for $\mu = 6.212$ (Fig. 3.46). These cycles testify the presence in system (3.29) of homoclinic separatrix loops of saddle–focuses O_{-1} and O_1 which have two-dimensional stable and one-dimensional unstable manifolds. However it is not possible to find out a homoclinic contour for the pointed above fixed parameter values $a = 1/7$, $b = 2/7$, $\beta = 9$. Therefore there exist two incomplete homoclinic cascades of bifurcations in system (3.29) for the given above set of fixed parameters. With increasing of parameter μ the areas occupied by attractors, increase, being extended aside point O_0, i.e. towards each other. At some parameter value μ^* these areas so approach each other (in a neighbourhood of the point O_0), that a bifurcation of merging of attractors occurs (see Sec. 2.4.5) with formation of a uniform irregular attractor, shown in Fig. 3.46. We found that $\mu^* \approx 6.244$ for the values $a = 1/7$, $b = 2/7$, $\beta = 9$.

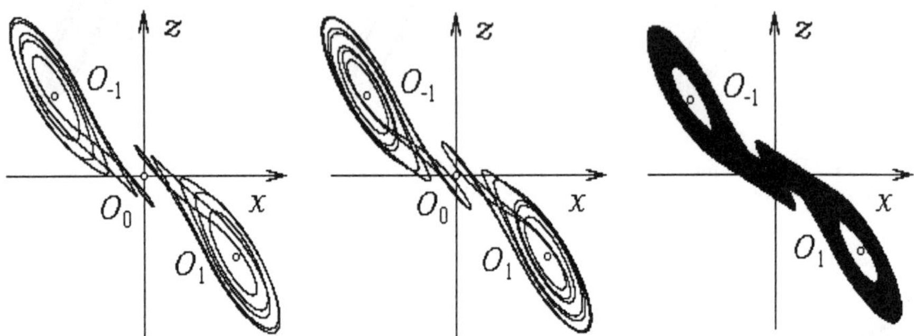

Fig. 3.46   Projections of cycles C_4 for $\mu = 6.176$ and C_6 for $\mu = 6.212$ of the homoclinic cascade and irregular singular attractor for $\mu = 6.245$ to a plane (x, z) in the Chua system.

A series of stable cycles C_{nn} of the double homoclinic cascade of bifurcations gives a representation of the structure of arisen attractor. So the cycles C_{11}, C_{33} and C_{77} of the homoclinic cascade of bifurcations exist accordingly for the values $\mu = 6.340$, $\mu = 6.390$ and $\mu = 6.522$ in system (3.29). Then the cycles C_{55}, C_{33}, C_{22} and C_{11} of the homoclinic cascade of bifurcations are observed accordingly for the values $\mu = 6.554$, $\mu = 6.660$, $\mu = 6.772$ and for $\mu \in [6.960, 6.985]$ (Fig. 3.47 and Fig. 3.48). Except for that the homoclinic cycle C_1 is found out in the Chua system for the parameter value $\mu = 7.208$ and for other fixed parameters specified above (Fig. 3.48).

For the values $\mu > \mu^*$ we have a double homoclinic singular attractor in the system (3.29). Finding the values of parameters, at which this attractor is complete, demands additional research. We shall notice, that it is accepted in the literature to name by one term "double scroll" various actually incomplete double homoclinic attractors, existing in the Chua system [Anishchenko *et al.* (1999)].

Thus, in system (3.29), the same mechanism of formation of irregular attractors is observed for $\mu > \mu^*$, as in the Lorenz system for the fixed values of parameters $\sigma = 10$, $b = 0.5$. Moreover, it is possible to say, that there are those and only those mechanisms of formation of irregular attractors in the

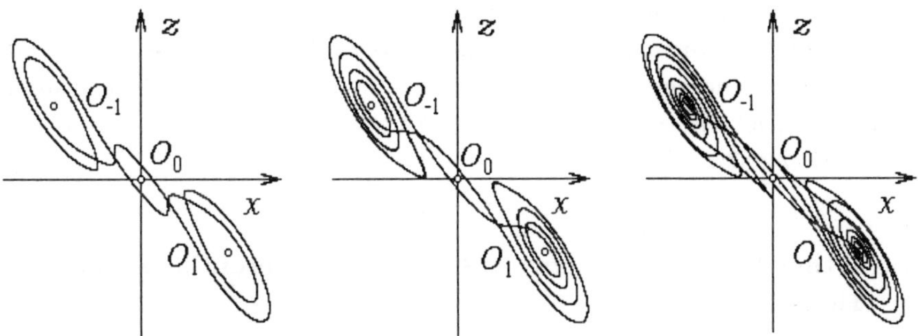

Fig. 3.47   Projections of homoclinic cycles C_{11} for $\mu = 6.340$, C_{33} for $\mu = 6.390$ and C_{77} for $\mu = 6.522$ to a plane (x, z) in the Chua system.

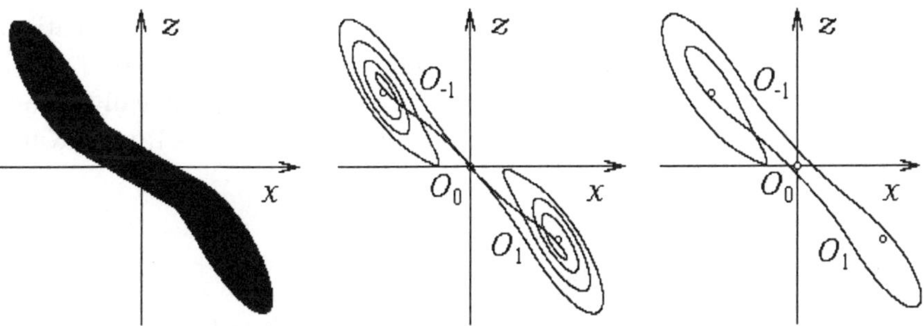

Fig. 3.48   Projections of complete double homoclinic singular attractor for $\mu = 6.544$ and homoclinic cycles C_{33} for $\mu = 6.660$ and C_1 for $\mu = 7.208$ to a plane (x, z) in the Chua system.

Chua system, as in the Lorenz and Rossler systems considered above. That is the Feigenbaum period doubling cascades, the Sharkovskii subharmonic cascades and the Magnitskii homoclinic cascades of bifurcations of stable cycles. At that the period doubling cascades and subharmonic cascades of bifurcations of stable cycles take place without dependence from, whether exist saddle–focuses separatrix loops or saddle–focuses of singular points in neighbourhoods of these cycles.

In conclusion of this section we may note, that similarity of homoclinic cascades in the Chua and Lorenz systems extends inclusive up to the cycle C_0 with which, as shown above in Secs. 3.1.1 and 3.1.2, the double homoclinic cascade of bifurcations begins in the Lorenz system. However, in the Chua system at the fixed values of parameters accepted by us $a = 1/7$, $b = 2/7$, and $\beta = 9$, this cycle lies in the domain of parameter values μ where solutions are unstable. This cycle is, thus, unstable and, hence, it is not detected. One can observe cycle C_0 in the Chua system, for example, for the values of parameters $a = 1/7$, $b = 2/7$, $\beta = 3$ and $\mu \in [2.854, 2.882]$.

3.5 Other Chaotic Systems of Ordinary Differential Equations

Results of investigations of scenarios of transition to dynamical chaos and formation of irregular attractors in the Vallis, Rikitaki and some other systems of nonlinear ordinary differential equations are presented in this section. We used the approach developed by us at studying the Lorenz attractor. It based on numerical research of cascades of soft bifurcations of stable cycles. Such approach is the most protected from errors of numerical calculations and allows to reproduce our results practically on any computer. The differential equations were integrated by the Runge–Kutta method of the fourth order.

3.5.1 *The Vallis systems*

In the year 1986 two mathematical models were offered by G. Vallis for description of fluctuations of temperature in western and eastern parts of near-equatorial area of the ocean. Such fluctuations of temperature render strong influence on the global climate of Earth. The first Vallis model does not consider the influence of trade winds and represents a system of three nonlinear ordinary differential equations [Vallis (1988)]

$$\begin{aligned}
\dot{x} &= \mu y - ax, \\
\dot{y} &= xz - y, \\
\dot{z} &= 1 - xy - z,
\end{aligned} \tag{3.32}$$

where x is speed of water motion on a surface of ocean,

$$y = (T_w - T_e)/2, \quad z = (T_w + T_e)/2,$$

T_w and T_e are temperatures accordingly in western and eastern parts of ocean, μ and a are some positive parameters. The system (3.32) has three singular (fixed) points

$$O_0(0,0,1), \quad O_i\left((-1)^i\sqrt{\frac{\mu-a}{a}}, \frac{ax_i}{\mu}, \frac{a}{\mu}\right), \quad i = 1,2.$$

It is convenient to shift the point O_0 to the point of origin by means of replacing $z \to z + 1$, and then the system (3.32) will take the form

$$\dot{x} = -ax + \mu y,$$
$$\dot{y} = x - y + xz, \tag{3.33}$$
$$\dot{z} = -xy - z.$$

Let us note, that the Vallis system in the form of (3.33) differs from the Lorenz system in essence only by signs of nonlinear members in the equations and, apparently, can have much similarity with it. Indeed, system (3.33) is dissipative everywhere for $a > 0$, since $\operatorname{div} F(x,y,z) = -a - 2$. Let us define the states of equilibrium or types of singular points

$$O_0(0,0,0), \quad O_i\left((-1)^i\sqrt{\frac{\mu-a}{a}}, \frac{ax_i}{\mu}, \frac{a}{\mu} - 1\right), \quad i = 1,2,$$

of system (3.33), using eigenvalues of the Jacobi matrix

$$J(x^*, y^*, z^*) = \begin{pmatrix} -a & \mu & 0 \\ 1 + z^* & -1 & x^* \\ -y^* & -x^* & -1 \end{pmatrix}.$$

Point O_0 of system (3.33) is a fixed point for any values of parameters μ, a. Its equilibrium state is defined by the characteristic equation

$$(\lambda + 1)(\lambda^2 + \lambda(a+1) + a - \mu) = 0.$$

Roots of the given equation have negative real parts for $\mu < a$ and therefore, point O_0 is stable. For the parameter values $\mu > a$ point O_0 is unstable, having an equilibrium state of a saddle–node type with one-dimensional unstable and two-dimensional stable manifolds. For $\mu > a$ two more singular points O_1 and O_2 are generated in the system (3.33). Their equilibrium states are defined by roots of the characteristic equation

$$\lambda^3 + \lambda^2(a+2) + \lambda(a + \mu/a) + 2(\mu - a) = 0. \tag{3.34}$$

Applying to the Eq. (3.34) the Routh–Hurwitz conditions for definition of stability of polynomials, it is easy to receive that for $a > 2$ points O_1 and O_2 are stable in the case of $a < \mu < \mu^* = (a^3 + 4a^2)/(a - 2)$. For $\mu > \mu^*$

the Eq. (3.34) has one real negative root and two complex conjugate roots with a positive real part, that is points O_1 and O_2 have an equilibrium state of a saddle–focus type. For analysis of the scenario of transition to chaos in system (3.33) we fix one of parameters $a = 5$ and consider parameter μ as a bifurcation parameter. We shall consider numerical solutions of system (3.33) in a range $\mu \in (a, +\infty)$. By analogy to the Lorenz system, the scenario of transition to chaos in the Vallis system is easier for understanding at reduction of parameter values μ from ∞ to a. So, unique attractor, a stable limit cycle C_0, exists in system (3.33) for great values of parameter $\mu \geq 180$. It surrounds both singular points O_1 and O_2. For the value $\mu \approx 180$ a bifurcation of birth of two stable cycles C_0^- and C_0^+ from the stable cycle C_0 is observed. Trajectories of these cycles are deformed accordingly to points O_1 and O_2 (Fig. 3.49).

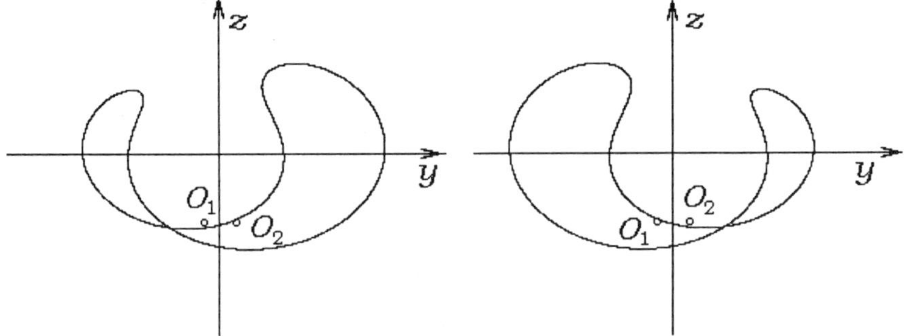

Fig. 3.49   Stable cycles C_0^- and C_0^+ in system (3.33) for $\mu = 170$.

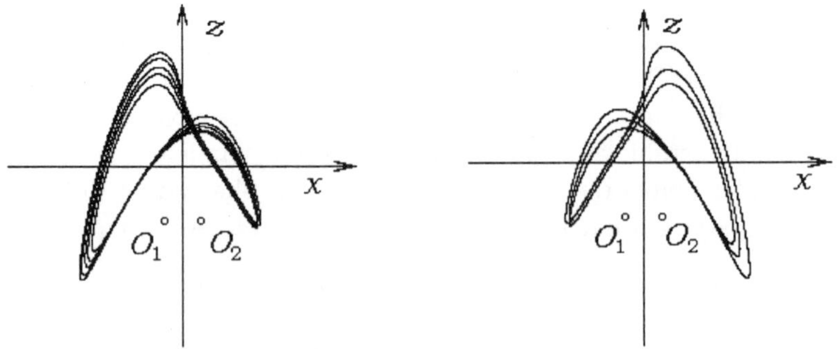

Fig. 3.50   Stable cycles C_0^+ of the period 5 and C_0^- of the period 3 accordingly for $\mu = 124.10$ and $\mu = 122.74$.

For the value $\mu \approx 134.25$ there occurs the period doubling bifurcation of these cycles which begins the Feigenbaum cascade of bifurcations. And an occurrence of cycles $C_0^{\pm}$ of the period 5 for $\mu \approx 124.1$ and the period 3 for $\mu \approx 122.74$ (Fig. 3.50) confirms the presence of subharmonic cascade of bifurcations of these cycles which comes to the end for the value $\mu \approx 122$ with formation of singular attractors in the form of two tapes. Each of these tapes is displaced aside by one of singular points O_1 or O_2 (Fig. 3.51). For parameter value $\mu \approx 119.06$ both tapes merge, forming a singular attractor, shown in Fig. 3.52.

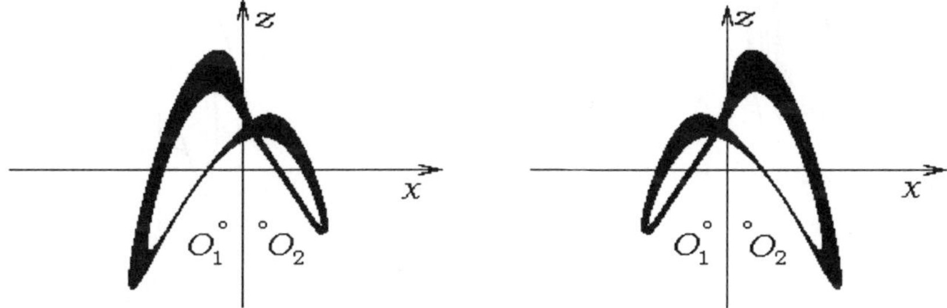

Fig. 3.51   Singular attractors in system (3.33) for the value $\mu = 122$.

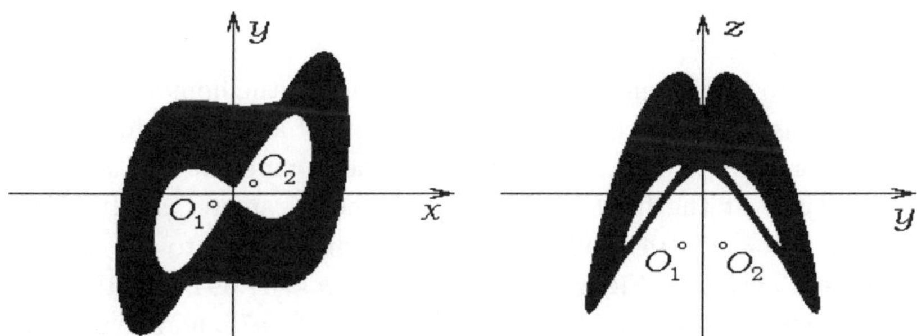

Fig. 3.52   Singular attractor in system (3.33) for the value $\mu = 119$.

Similarly to the Lorenz system, the cycles named by us as self-organizing cycles are observed in the Vallis system too. These cycles are found out after the formation of irregular attractor as a result of merging of two tapes which have arisen during the subharmonic cascade of bifurcations. The kind of given cycles becomes more and more simple when values of bifurcation parameter are moving away from the point of formation of the attractor. The cascade of self-organizing cycles comes to the end with generation of stable cycles of type C_{nn}, where n is the number of stage in the homoclinic cascade of bifurcations. Such self-organizing cycles in the Vallis system at first stage of the homoclinic cascade and a cycle C_{11} formed further are shown in Fig. 3.53 for $\mu = 117.6$, $\mu = 103.7$ and $\mu \approx 90$.

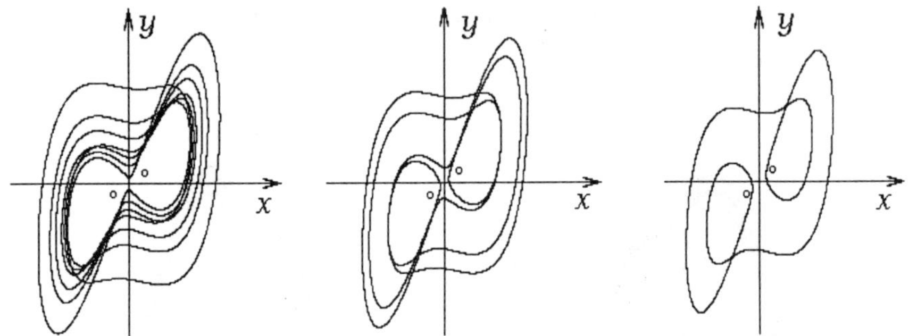

Fig. 3.53 Projections of stable self-organizing cycles in system (3.33) for the values $\mu = 117.6$ and $\mu = 103.7$ and the stable cycle C_{11} for $\mu = 90$.

The same bifurcations are typical for cycles C_{nn}, as for the cycle C_0. Therefore the formed cycle C_{11} generates then two stable cycles $C_{11}^{\pm}$ for the value $\mu \approx 88.3$. Each of these cycles generates its own subharmonic cascade of bifurcations.

Similarly to the Lorenz system the first stage of the homoclinic cascade of bifurcations comes to the end with formation of a cycle C_1 for the value $\mu \approx 59$ (Fig. 3.54). The second stage of the homoclinic cascade begins with this cycle at the further reduction of the parameter μ. It contains cycles C_{12} ($\mu = 47.814$) and C_{22} ($\mu = 53$) and comes to the end with generation of cycle C_2 (Fig. 3.54) for the value $\mu \approx 41.394$. For the chosen fixed parameter value $a = 5$ the critical value is $\mu^* = 75$, and consequently only incomplete homoclinic cascade of bifurcations can be realized in the system (3.33). The kind of projections of incomplete homoclinic singular attractor is shown in Fig. 3.55 for the value $\mu = 37$.

Thus, in the Vallis system similarly to the Lorenz system, we observe an irregular singular attractor, formed as a result of incomplete double homoclinic cascade of bifurcations. Depending on the values of other fixed parameter a here, apparently, can be born both complete and incomplete double homoclinic attractors as well as in the Lorenz system. Parameter space partition on areas with various kinds of singular attractors demands carrying out of special research.

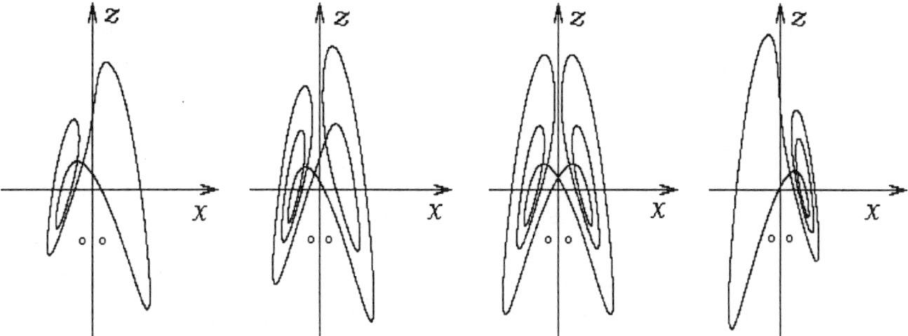

Fig. 3.54   Projections of stable cycles C_1^- ($\mu = 58$), C_{12} ($\mu = 47.814$), C_{22} ($\mu = 53$) and C_2^+ ($\mu = 41.354$) in system (3.33).

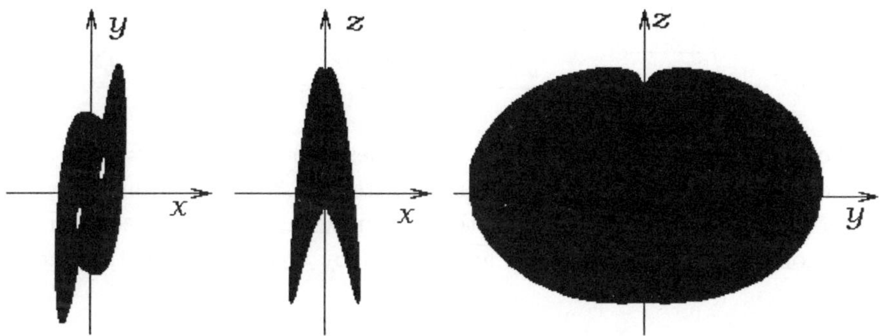

Fig. 3.55   Projections of incomplete double homoclinic attractor in system (3.33).

Another model, named as El-nino, has been offered by Vallis for description of nonlinear interactions of atmosphere, ocean and trade winds in equatorial area of Pacific ocean. This model is nonlinear and, in general, non-autonomous system of three differential equations [Vallis (1986); Vallis (1988)]

$$\dot{x} = \mu(y - z) - b(x - f(t)),$$
$$\dot{y} = xz - y + c, \qquad\qquad (3.35)$$
$$\dot{z} = -xy - z + c,$$

where $x(t)$ is the speed of surface ocean current, $y(t)$ and $z(t)$ are temperature of water accordingly on western and eastern bounds of water pool, $f(t) = a_0 + a_1 \sin 2\pi t$ is the periodic function considering influence of trade winds.

Let us consider the scenario of formation of irregular attractors for the case of $f(t) \equiv 0$, having resulted the system (3.35) to an autonomous kind

$$\dot{x} = \mu(y - z) - bx,$$
$$\dot{y} = xz - y + c, \qquad (3.36)$$
$$\dot{z} = -xy - z + c.$$

The system (3.36) is dissipative (div $F(x,y,z) = -2-b < 0$) in all phase space for the value $b > -2$. It is symmetric concerning the replacement of variables

$$x \to -x, \ y \to z, \ z \to y,$$

that finds reflection, in particular, in symmetry of singular (stationary, fixed) points: $O_0(0,c,c)$, $O_1(x^*, y^*, z^*)$ and $O_2(-x^*, z^*, y^*)$, where

$$x^* = \sqrt{\frac{2\mu c}{b} - 1}, y^* = \frac{b + \sqrt{2\mu bc - b^2}}{2\mu}, z^* = \frac{b - \sqrt{2\mu bc - b^2}}{2\mu}.$$

For further research we shall fix the values of parameters $b = 10$, $c = 12$ and consider behaviour of solutions of system (3.36) depending on parameter μ. We shall begin studying the system with studying of equilibrium states of singular points which are defined by eigenvalues of the Jacobi matrix

$$J(x^*, y^*, z^*) = \begin{pmatrix} -b & \mu & -\mu \\ z^* & -1 & x^* \\ -y^* & -x^* & -1 \end{pmatrix}.$$

In the El-nino system, also as well as in the Vallis system considered above, the point $O_0(0,c,c)$ is a singular point for any values of parameters μ, b, c. Characteristic equation for this point has the kind

$$(\lambda + 1)(\lambda^2 + \lambda(b + 1) + b - 2\mu c) = 0,$$

from which follows, that point O_0 has the equilibrium state of stable node type under the condition of $\mu < b/2c$. For $\mu > b/2c$ it loses stability, becoming a saddle–node, having two-dimensional stable and one-dimensional unstable manifolds. When the point O_0 losses its stability, two other points

O_1 and O_2 appear. Their equilibrium states are defined by roots of the characteristic equation

$$\lambda^3 + \lambda^2(b+2) + \lambda\left(b + \frac{2\mu c}{b}\right) + 4\mu c - 2b = 0. \qquad (3.37)$$

According to the Routh–Hurwitz criterion

$$b + 2 > 0, \quad (b+2)\left(b + \frac{2\mu c}{b}\right) > 4\mu c - 2b > 0$$

points O_1 and O_2 are stable up to the value $\mu < \mu^* = (b^3 + 4b^2)/2(b-2)$. For values $\mu > \mu^*$ the Eq. (3.37) has one negative real root and two complex conjugate roots with a positive real part and, hence, the points O_1 and O_2 are saddle–focuses. We shall note, that in the El-nino system the same situation, as in the Lorenz system is observed. At the moment of loss of stability of one of singular points there appear two new singular points which remain stable focuses up to some critical value μ^*. Apparently, it is necessary to expect, that the scenario of birth of irregular attractors in the El-nino system will be similar to the scenario of birth of the Lorenz attractor. Numerical experiments show that it is really so. And, depending on values of the fixed parameters, both complete and incomplete double homoclinic cascades of bifurcations can take place here, as well as in the Lorenz system. However, the detailed description of all scenarios of formation of irregular attractors in the El-nino system demands additional research, also as well as in the previous case. We shall be limited to discussion of those common mechanisms of generation of irregular singular attractors which have basic value for a large class of dissipative dynamical systems described by differential equations.

For the fixed values of parameters $b = 10$ and $c = 12$ indicated above there is an unique attractor (stable cycle C_0) in the El-nino system for the interval of parameter values $\mu \in (346, \infty)$, surrounding both points O_1 and O_2. For the value of $\mu \approx 346$ this cycle loses stability and two stable cycles C_0^+ and C_0^- appear as a result of bifurcation. At further reduction of the parameter μ, cycles $C_0^{\pm}$ generate subharmonic cascades of bifurcations, that is confirmed by generation of the period 3 cycles for the value $\mu = 231.5$. Then two irregular attractors are born in the form of two tapes (Fig. 3.56) for the interval of $\mu \in (224, 230)$. For $\mu \approx 223.9$ both tapes merge in one singular attractor. Cycle C_{11} is observed for values $\mu \in (138.7, 150.5)$. The first stage of homoclinic cascade comes to the end with the birth of the cycle C_1 for $\mu \approx 100.69$

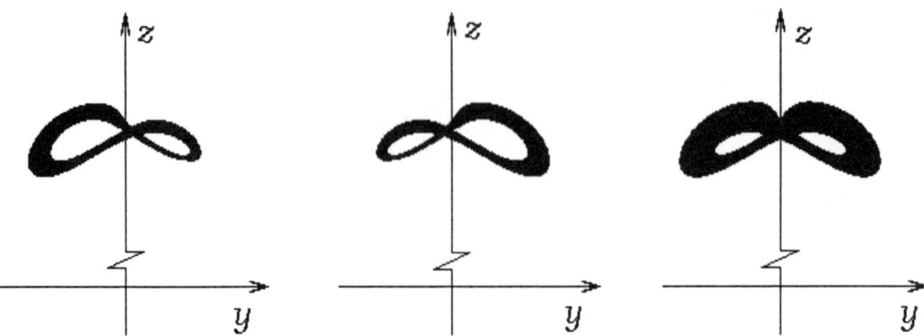

Fig. 3.56　Singular attractors in system (3.36) for the parameter values $\mu = 225$ and $\mu = 223.8$.

Let us note, that the subharmonic cascade of bifurcations of cycle C_1 for the fixed values of parameters $b = 10$, $c = 12$ indicated above comes to the end in the El-nino system for the value $\mu = 99.327$ with formation of the cycle C_1 of the period 3. Critical parameter value in this case is equal to $\mu^* = 87.5$. So, the first stage of the homoclinic cascade comes already to the end near a point of change of stability state at singular points O_1 and O_2. Therefore it is inconvenient to specify precisely the type of observable irregular attractor. But we can assert, that all irregular attractors in the El-nino system are formed as a result of a double homoclinic cascade of bifurcations. However, on the basis of the given research it is impossible to say whether it is complete or not, that is whether or not a point of accumulation of homoclinic cycles exists for the given values of the other fixed parameters. In conclusion of this section we shall specify one more feature of homoclinic cascades in the El-nino system. We have found out some cycles of homoclinic cascade for the interval $\mu \in (b/2c, \mu^*)$, that is in case of when the singular point O_0 remains unstable, and singular points O_1 and O_2 are stable. These cycles are: cycle C_{22} for $\mu \in (83.58,\ 83.62)$, cycle C_{12} for $\mu \in (77.345, 77.350)$ and cycle C_2 for $\mu \in (68.92,\ 68.97)$ (Fig. 3.57). It means that homoclinic cascades can exist outside of neighbourhoods of stable singular points. We have already met and shall meet in future with such systems. In particular similar features are observed in the Lorenz system for fixed values of parameters $\sigma = 10$ and $b = 0.5$ (see Sec. 3.1.3). Existence of noted features confirms that the leading role in formation of irregular attractors is not played by singular points, homoclinic and heteroclinic contours of saddle–nodes and saddle–focuses, but by some cycles which give start to all cascades of bifurcations and, first of all, to the period doubling cascade. The theory of such cycles named singular cycles and the mechanism of formation of various *singular attractors* generated by them is considered in Chapter 4.

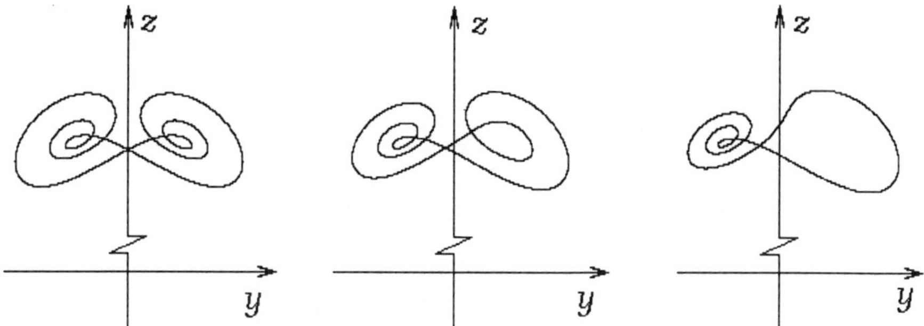

Fig. 3.57 Stable cycles of the homoclinic cascade C_{22}, C_{12}^- and C_2^- accordingly for the values $\mu = 83.60$, $\mu = 77.3475$ and $\mu = 68.95$.

Thus, transition to chaos in the Vallis systems of kinds (3.33) and (3.35) is carried out in the same way, as in the Lorenz, Rossler and Chua systems, namely: through the Feigenbaum period doubling cascade of bifurcations, the Sharkovskii subharmonic cascade of bifurcations and the Magnitskii homoclinic cascade of bifurcations.

3.5.2 *The Rikitaki system*

System of four nonlinear ordinary differential equations

$$
\begin{aligned}
\dot{x} &= -\mu x + yz, \\
\dot{y} &= -\mu y + xu, \\
\dot{z} &= 1 - xy - bz, \\
\dot{u} &= 1 - xy - cu
\end{aligned}
\tag{3.38}
$$

was offered by T. Rikitaki in the year 1958 for modelling of change in dynamics of magnetic poles of the Earth [Cook and Roberts (1970)]. It is obvious, that $\operatorname{div} F(x, y, z, u) = -(2\mu + b + c) < 0$ for positive values of parameters μ, b, c, and so the system (3.38) is dissipative. System (3.38) has three singular (stationary) points: $O_0(0, 0, 1/b, 1/c)$ and

$$
O_i\left((-1)^i\sqrt{\sqrt{\frac{c}{b}} - \mu c}, \ (-1)^i\sqrt{\sqrt{\frac{b}{c}} - \mu b}, \ \mu\sqrt{\frac{c}{b}}, \ \mu\sqrt{\frac{b}{c}}\right), \ i = 1, 2.
$$

Point O_0 is a singular point for any positive values of parameters, and points O_1 and O_2 exist only under the condition of $\mu^2 bc < 1$. Let us find eigenvalues of the Jacobi matrix for definition of conditions of stability of singular points

$$J(x^*, y^*, z^*, u^*) = \begin{pmatrix} -\mu & z^* & y^* & 0 \\ u^* & -\mu & 0 & x^* \\ -y^* & -x^* & -b & 0 \\ -y^* & -x^* & 0 & -c \end{pmatrix}.$$

At point O_0, they are defined by the characteristic equation

$$(\lambda + b)(\lambda + c)(\lambda^2 + 2\lambda\mu + \mu^2 - 1/bc) = 0,$$

from which follows, that the given point is a stable node for values $\mu^2 > 1/bc$, but for $\mu^2 < 1/bc$ it loses stability and becomes a saddle–node point.

Two other singular points O_1 and O_2 have the same equilibrium state, that is obvious from the characteristic equation

$$\begin{aligned}
\lambda^4 &+ \lambda^3(2\mu + b + c) + \lambda^2(2\mu(b + c) + bc + x^{*2} + y^{*2}) \\
&+ \lambda(x^*y^*(z^* + u^*) + x^{*2}(\mu + b) + y^{*2}(\mu + c) + 2\mu bc) \\
&+ \mu(bx^{*2} + cy^{*2}) + x^*y^*(bz^* + cu^*) = 0. \quad (3.39)
\end{aligned}$$

It is inconvenient to define this condition analytically, proceeding from a kind of given equation. We have numerically calculated eigenvalues of the Jacobian at points O_1 and O_2 depending on the parameter μ at the fixed values of other parameters $b = 0.004$ and $c = 0.002$. Numerical calculations show, that for all values of parameter $\mu \in (0, \sqrt{1/bc})$, two roots of the equation (3.39) are real and negative, and absolute value of one of them is proportional to the parameter μ. Two other roots are complex conjugate numbers with negative real parts. Thus, there are no conditions for the Andronov–Hopf bifurcation in the given system, and behaviour of solution of the system essentially depends on initial conditions.

If initial conditions are chosen near to singular points O_1 or O_2, the trajectory is attracted to a corresponding singular point. And the domain of stability of these points extends with growth of values of parameter μ. If initial conditions are chosen outside of the domain of attraction of these singular points, then both stable, and unstable periodic solutions are observed in the system (3.38). So, for example, a stable cycle of C_0 type is observed for values $\mu \in (2.3, \sqrt{1/bc})$. It surrounds by the eight both singular points O_1 and O_2. At value $\mu \approx 2.3$ a bifurcation of this cycle occurs and two stable cycles of kinds C_0^- and C_0^+ are formed as a result of this bifurcation. It occurs similarly to the Lorenz system (Fig. 3.58).

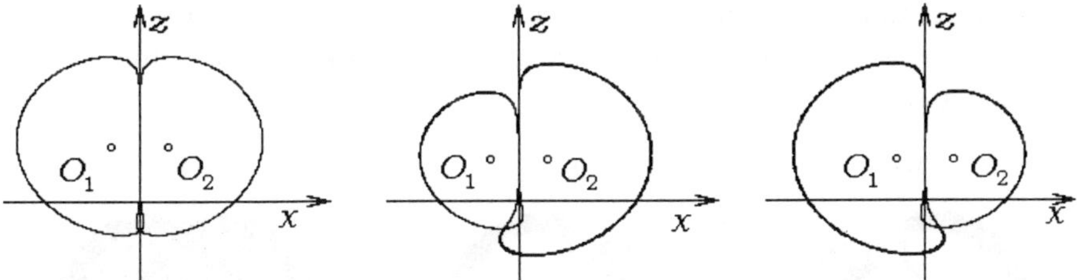

Fig. 3.58   Cycles C_0 and C_0^-, C_0^+ in the system (3.38) accordingly for values of the parameter $\mu = 3$ and $\mu = 2.23$.

However at given fixed parameter values $b = 0.004$ and $c = 0.002$ it is impossible to track a homoclinic cascade of bifurcations of stable cycles because they have extremely narrow domains of attraction. Fortunately the scenario of transition to chaos in the Rikitaki system is not limited by the above marked bifurcations. The same cycles C_0 and $C_0^\pm$ and other cycles of a homoclinic cascade are found out at small values of the parameter μ where they have large enough domains of attraction. It allows to consider the scenario of formation of irregular attractors, moving in the process of increasing the parameter μ. So a cycle of C_0 kind is observed for values $\mu \in (0.1, 0.448)$, and two stable cycles $C_0^\pm$ are formed as a result of its bifurcation at $\mu = 0.448$ (Fig. 3.59). These two cycles generate the Feigenbaum period doubling cascades of bifurcations at increasing the parameter μ.

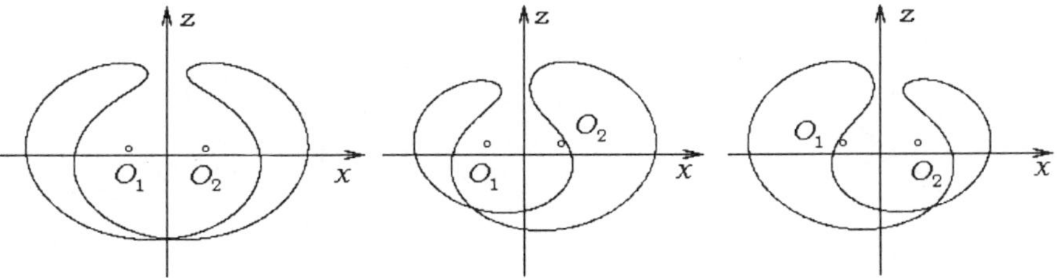

Fig. 3.59   Cycles C_0 and C_0^-, C_0^+ in system (3.38) accordingly for values of the parameter $\mu = 0.4$ and $\mu = 0.47$.

As an examples we present cycles $C_0^\pm$ of the period 2 at $\mu \in (0.489, 0.493)$, cycles of the period 4 at $\mu \in (0.4931, 0.4938)$ and cycles of the period 8 at $\mu \in (0.4938, 0.49454)$. Each of cycles C_0^- and C_0^+ generates an irregular singular attractor in the form of tape during its own subharmonic cascade of bifurcations, and then these tapes merge in one irregular attractor. Formation of attractors in a kind of two tapes comes to the end at value $\mu \approx 0.498$, and at value $\mu \approx 0.501$ there occurs the merging of these tapes with an appearance of an incomplete double subharmonic singular attractor (Fig. 3.60).

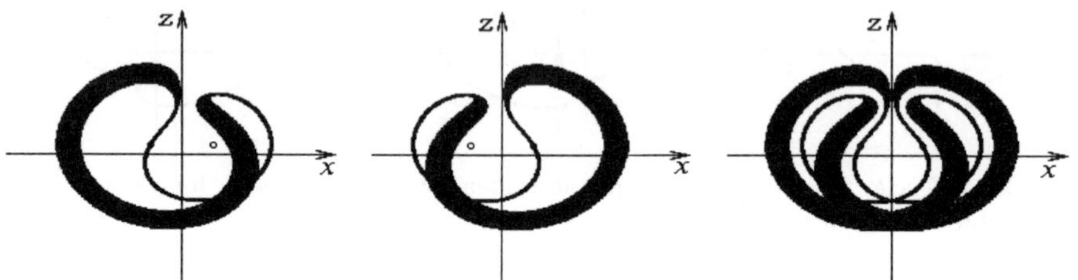

Fig. 3.60   Irregular singular attractors in the system (3.38) for the values of parameter $\mu = 0.498$ and $\mu = 0.505$.

Similarly to the scenario of a birth of a double homoclinic attractor in the Lorenz system, cycle C_{11} appears here at $\mu \in (0.523, 0.531)$ (Fig. 3.61), generating its own cycles $C_{11}^{\pm}$ and corresponding to them cascades of bifurcations. The first stage of the homoclinic cascade comes to the end here with the cycle C_1 ($\mu \approx 0.609$) too. The second stage of the homoclinic cascade comes to the end with formation of the cycle C_2 at value $\mu = 0.6786$, and the cycle C_3 lies in the region of values $\mu \in (0.72662, 0.72667)$. Projections of cycles C_1, C_2 and C_3 on the plane (x, z) are represented in Fig. 3.62.

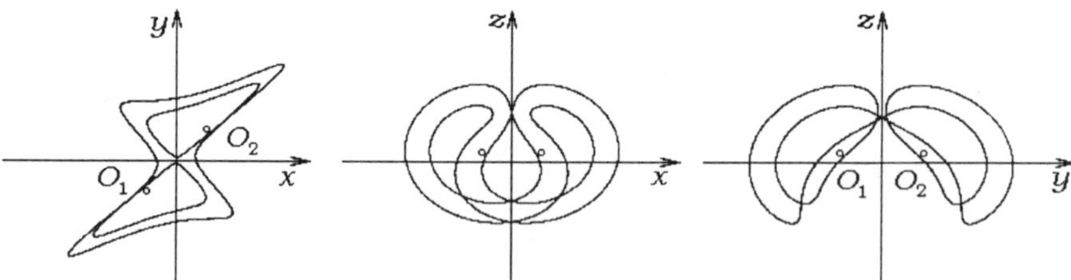

Fig. 3.61   Projections of the cycle C_{11} in system (3.38) for the value of parameter $\mu = 0.530$.

As it follows from the presented data, areas of stability of stable cycles are sharply reduced with each subsequent stage of the homoclinic cascade. Therefore it is difficult to establish a number of stages of the homoclinic

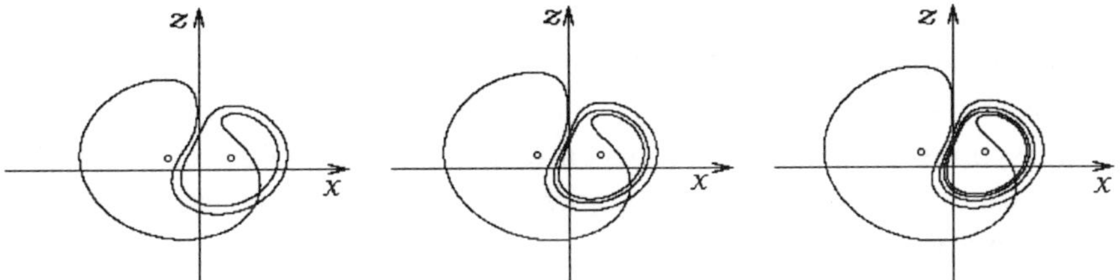

Fig. 3.62   Projections of cycles C_1, C_2 and C_3 of the homoclinic cascade in system (3.38) accordingly for values of parameter $\mu = 0.61$, $\mu = 0.6787$ and $\mu = 0.72665$.

cascade of bifurcations, participating in formation of irregular attractors. But as both singular points O_1 and O_2 are stable focuses in all area of existence of irregular attractors, then number of stages is finite, and all attractors are incomplete. Concerning the scenario of formation of irregular attractors it is possible to assert definitely that such attractors appear in the given system also as a result of a double homoclinic cascade of bifurcations.

Note, that the given cascade takes place as at reduction of parameter μ from the value $\mu \approx 2.3$, and at its increasing from the value $\mu \approx 0.1$. Projections of an incomplete double homoclinic singular attractor of the Rikitaki system are represented in Fig. 3.63 at $\mu = 1$.

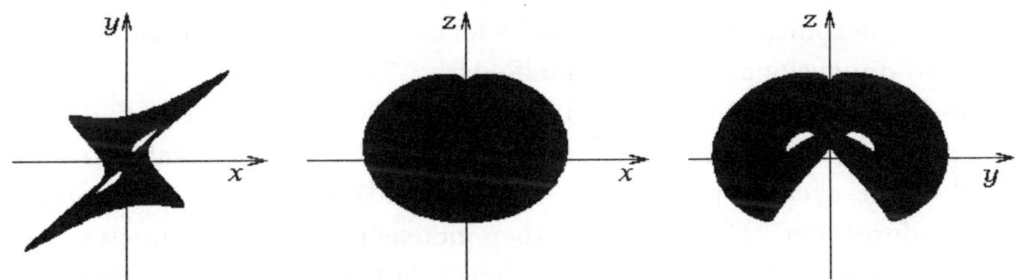

Fig. 3.63   Projections of an incomplete double homoclinic singular attractor of the system (3.38) for the parameter value $\mu = 1$.

3.5.3   *The "Simple" system*

System of equations [Novikov and Pavlov (2000)]

$$\dot{x} = 1 + \mu y z,$$
$$\dot{y} = x - y, \tag{3.40}$$
$$\dot{z} = 1 - xy$$

with quadratic nonlinearity in the right part and one parameter $\mu > 0$ models some dissipative self-oscillating process (div $F(x, y, z) = -1$). This system has two equilibrium states $O_1(-1, -1, 1/\mu)$ and $O_2(1, 1, -1/\mu)$, which are antisymmetric concerning signs on variables. Eigenvalues of a matrix of linearization are defined at these points by various characteristic equations: by the equation at point O_1

$$\lambda^3 + \lambda^2 + \lambda(\mu - 1) + 2\mu = 0 \tag{3.41}$$

and by the equation at point O_2

$$\lambda^3 + \lambda^2 + \lambda(\mu + 1) + 2\mu = 0. \tag{3.42}$$

It follows from the Routh–Hurwitz conditions that the point O_1 is unstable for any parameter values μ. For $\mu = -1$ and $\mu = 0$ the Eq. (3.41) can be solved precisely and it has accordingly following sets of solutions:

$$\lambda(-1) = \{-1, \pm\sqrt{2}\} \quad \text{and} \quad \lambda(0) = \{0, (-1 \pm \sqrt{5})/2\}.$$

Computer calculations show, that the point O_1 is a saddle–node with two-dimensional stable manifold in an interval of the parameter values $\mu \in (-1.146, 0)$, and in an interval $\mu \in (0, 0.08)$ it is a saddle–node with two-dimensional unstable and one-dimensional stable manifold. At $\mu > 0.08$ this point becomes an unstable saddle–focus and has one-dimensional stable and two-dimensional unstable manifolds.

It follows from the Routh–Hurwitz conditions for the Eq. (3.42) that the point O_2 is stable in an interval $\mu \in (0, 1)$. For the value $\mu = 1$ a stable limit cycle is born in the system (3.40) as a result of Andronov–Hopf bifurcation. During the further increasing of the parameter values the subharmonic cascade of bifurcations takes place, and a stable cycle of the period 3 is observed at $\mu = 2.036$ (Fig. 3.64).

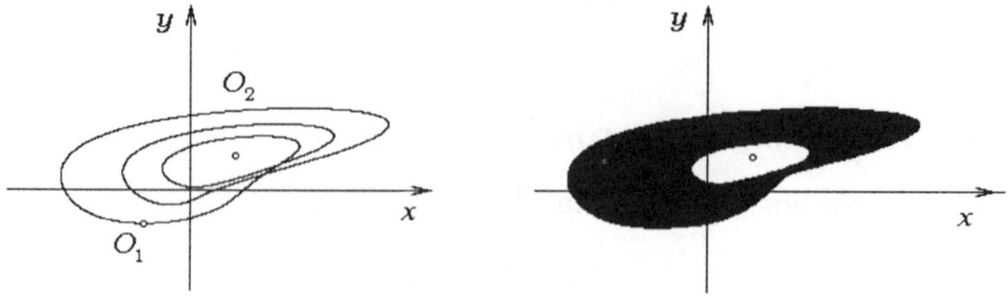

Fig. 3.64   Projections of a cycle of the period 3 for the value $\mu = 2.036$ and a singular attractor for the value $\mu = 2.1$.

Thus, a chaotic behaviour in the system for the parameter values $\mu > 2.036$ is caused by the subharmonic cascade of bifurcations of periodic solutions, generated as a result of loss of stability of the point O_2 (Fig. 3.65). However, for values $\mu \geq 2.3$ an influence of the point O_1 begins to be shown in formation of irregular attractors in the system (3.40).

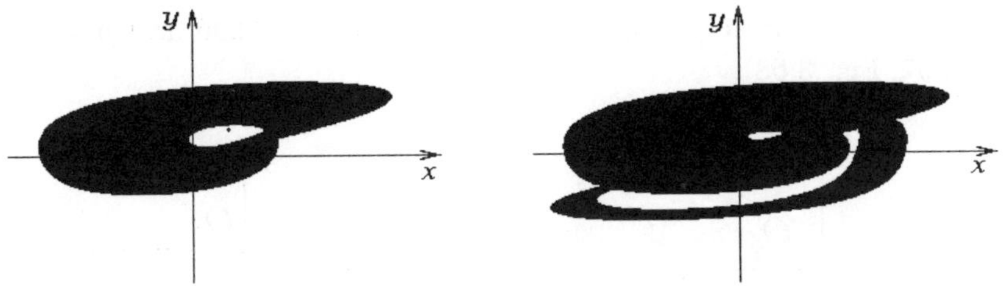

Fig. 3.65   Projections of singular attractors accordingly for values $\mu = 2.3$ and $\mu = 2.37$.

Let us consider how the scenario of transition to chaos is developed at reduction of parameter values μ from $+\infty$. At values $\mu \in (10.45, \infty)$ the system has a unique attractor, cycle C_0, surrounding both points O_1 and O_2. With reduction of the parameter values μ one can observe a subharmonic cascade of bifurcations of the cycle C_0. So one can observe the period 3 cycle at $\mu = 8.31$, and an irregular attractor, formed as a result of the subharmonic cascade of bifurcations of the cycle C_0 existing in the system (3.40) at $\mu \in (7.95, 8.14)$ (Fig. 3.66).

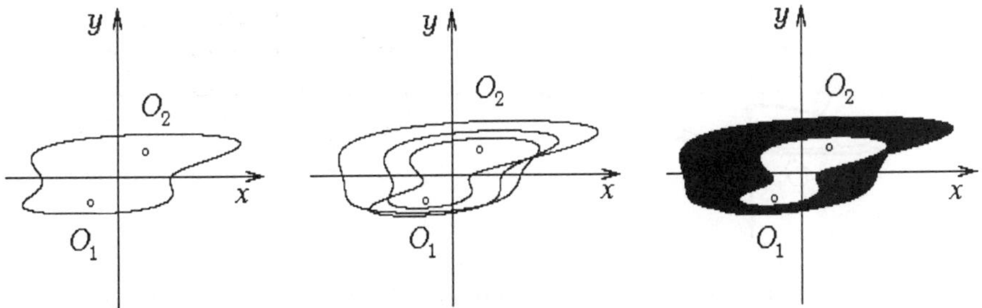

Fig. 3.66   Projections of the cycle C_0 ($\mu = 12$), cycle C_0 of the period 3 ($\mu = 8.31$) and irregular attractor ($\mu = 8$) of the system (3.40).

It is important to note, that cycle C_1^-, having one additional turn around the point O_1 is born in the system (3.40) for $\mu = 7.68$. Appearance of this cycle confirms the presence of the homoclinic cascade of bifurcations in the system, similarly to the cascade which takes place in the Lorenz system.

However, in the "Simple" system, unlike the Lorenz system, singular points O_1 and O_2 have different equilibrium states. Therefore cycles $C_k^{\pm}$, $k = 1, 2, \ldots$ are born here not simultaneously, but each of them appears at a corresponding parameter value. So the cycle C_1^+ is observed at value $\mu = 5.19$. Nevertheless each of these cycles generates the subharmonic cascade of bifurcations that is confirmed by the existence of cycles of the period 3 for cycles C_1^- and C_1^+ accordingly for values $\mu = 7.296$ and $\mu = 5.102$ (Fig. 3.67, Fig. 3.68).

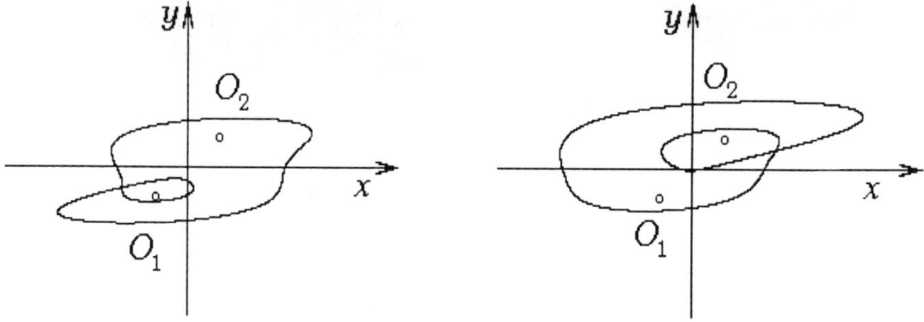

Fig. 3.67   Cycles C_1^- and C_1^+ accordingly for parameter values $\mu = 7.68$ and $\mu = 5.19$.

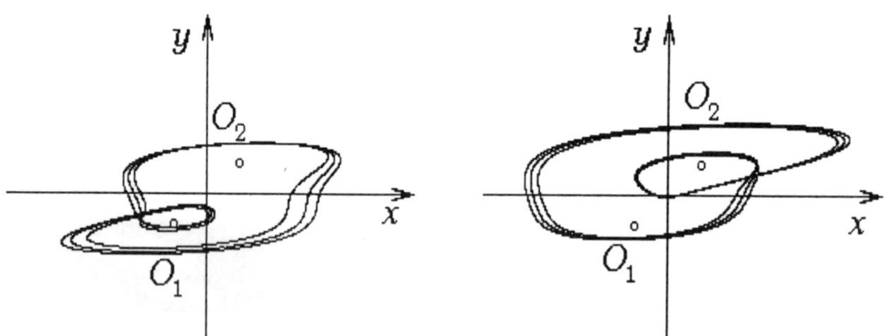

Fig. 3.68   Period 3 cycles C_1^- and C_1^+ accordingly for values $\mu = 7.296$ and $\mu = 5.102$.

Note that periodic solutions of cycles of self-organizing type are found in the "Simple" system also as well as in the Lorenz system. All these characteristic features including subharmonic and homoclinic cascades of bifurcations of stable cycles confirm that the "Simple" system has the same mechanisms of transition to dynamical chaos, as the systems considered above in previous sections.

3.5.4 *The Rabinovich–Fabrikant system*

In the system of differential equations

$$\dot{x} = y(z - 1 + x^2) + ax,$$
$$\dot{y} = x(3z + 1 - x^2) + ay, \tag{3.43}$$
$$\dot{z} = -2z(\mu + xy),$$

offered in the paper [Rabinovich and Fabrikant (1979)], chaotic behaviour
was revealed for values of parameters $\mu = 1.1$ and $a = 0.87$. We shall note,
that the given system is dissipative everywhere for $\mu > a$. Let us find its
singular points and define their equilibrium states. One of singular points
$O_0(0,0,0)$ coincides with the origin of coordinates and its equilibrium state
is defined by a characteristic equation

$$(\lambda + 2\mu)(\lambda^2 - 2a\lambda + a^2 + 1) = 0. \tag{3.44}$$

The Eq. (3.44) has the following roots $\lambda_1 = -2\mu$ and $\lambda_{2,3} = a \pm i$.
Hence, point O_0 is a saddle–focus, having one-dimensional stable and two-
dimensional unstable manifolds. Coordinates of other singular points are
defined by the system of equations

$$x^2(4\mu - 3a) + ay^2 = 4\mu, \ xy + \mu = 0, \ z = 1 - x^2\left(1 - \frac{a}{\mu}\right),$$

which solutions depend on values of parameters μ and a.

Transforming the last system to the equation of one variable,

$$ay^4 - 4\mu y^2 + \mu^2(4\mu - 3a) = 0,$$

we obtain, that this equation has real roots under the condition $4\mu - 3a \leq$
$4/a$. The last condition in common with the condition $\mu > a$ defines a set
of parameter values $a \in (0; 2]$ at which the other singular points exist. At
value $\mu = 1/a + 3a/4$ the system (3.43) besides the point O_0 has two more
singular points

$$O_1\left(-\sqrt{a\mu/2}, \sqrt{2\mu/a}, (4 + a^2)/8\right), \quad O_2\left(\sqrt{a\mu/2}, -\sqrt{2\mu/a}, (4 + a^2)/8\right).$$

Let us define the type of singular points O_1 and O_2 finding eigenvalues of
the Jacobi matrix

$$J(x^*, y^*, z^*) = \begin{pmatrix} a - 2\mu & -ax^*/y^* & y^* \\ -ay^*/x^* - 2x^{*2} & a & 3x^* \\ -2y^*z^* & -2x^*z^* & 0 \end{pmatrix}$$

at these points. These eigenvalues are defined by the characteristic equation

$$\lambda(\lambda^2 + 2\lambda(\mu - a) - 2a\mu\left(1 - \frac{a^2}{4}\right) + 2z^*(3x^{*2} + y^{*2})) = 0. \qquad (3.45)$$

Roots of the equation (3.45) are identical to both points O_1 and O_2. One of the roots is equal to zero, and other two roots have negative real parts at $\mu > a$. Having executed corresponding calculations, it is possible to show, that the singular points O_1 and O_2 of the system (3.43) are stable at value $\mu = 1/a + 3a/4$. However these points are not of great importance in the scenario of transition to chaos, as they are defined only at the unique point of the space of parameters. At other values $\mu \in (0, 1/a+3a/4)$ each of points O_1 and O_2 forms a pair of singular points accordingly $O_{11}(-x_1^*, -y_1^*, z_1^*)$, $O_{21}(x_1^*, y_1^*, z_1^*)$ and $O_{12}(-x_2^*, -y_2^*, z_2^*)$, $O_{22}(x_2^*, y_2^*, z_2^*)$, where

$$x_i^* = \sqrt{\frac{a\mu}{2 + (-1)^i\sqrt{4 - 4a\mu + 3a^2}}}, \; y_i^* = \frac{-\mu}{x_i^*}, \; z_i^*, \quad i = 1, 2.$$

Pairs of these points are symmetrical concerning the turn around an axis z on 180^0, that follows from a system of equations for definition of their coordinates. Eigenvalues of the Jacobi matrix are identical to each pair of singular points and, hence, each pair of points O_{11}, O_{21} and O_{12}, O_{22} has its own type of equilibrium state, identical to both points of the pair.

The analysis of equilibrium states of the considered pairs of singular points is inconvenient in general view. Therefore we have numerically defined eigenvalues of the Jacobi matrix accordingly for the pairs of points O_{11}, O_{21} and O_{12}, O_{22} at the fixed parameter value $a = 0.87$. It follows from numerical calculations that the points O_{11} and O_{21} are stable at value $\mu = \mu_0 \approx 1.7 < 1/a + 3a/4$. Then they lose stability at the value $\mu \approx 1.43$ as a result of a limit cycle birth bifurcation. This bifurcation begins the Feigenbaum cascade. So, there occurs a period doubling bifurcation of a limit cycle at the value $\mu \approx 1.182$. A cycle of the period 4 is born at $\mu \approx 1.159$, a cycle of the period 8 is born at $\mu \approx 1.155$, and a singular Feigenbaum attractor is born at $\mu \approx 1.1536$.

Thus, irregular attractors appear in the system (3.43) for the parameter value $a = 0.87$ as a result of the Feigenbaum cascade of bifurcations and the incomplete subharmonic cascade (Fig. 3.69). We shall note, that cycles of the period 3 appear in the system at smaller parameter values $a = 0.7$ and $a = 0.5$. It shows an opportunity of existence in the system (3.43) of transition to chaos through a complete subharmonic cascade of bifurcations and through a homoclinic cascade.

Projections to a plane (x, y) of cycles of period 4 and some singular attractors are shown in Fig. 3.69. Note that cascades of bifurcations develop in the system (3.43) independently around of each of points O_{11} and O_{21}, and there is no merging of tapes of attractors here as it is observed in a number of other systems.

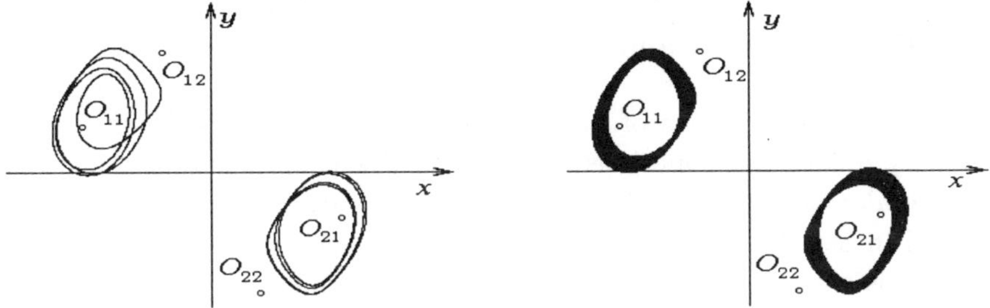

Fig. 3.69 Cycles of period 4 and singular attractors in system (3.43) accordingly for the values $\mu = 1.158$ and $\mu = 1.152$.

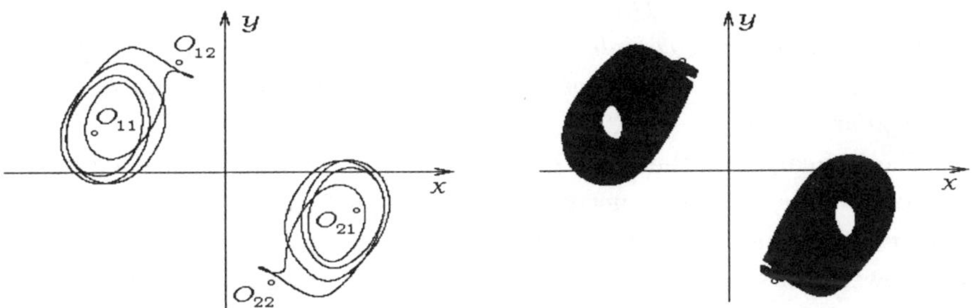

Fig. 3.70 Cycles of C_3 type at the value $\mu = 1.0978$ and singular subharmonic attractors at $\mu = 1.097$ in the system (3.43).

Another pair of singular points O_{12} and O_{22} loses stability at the values $\mu < 1/a + 3a/4$ and passes to an equilibrium state of a saddle–focus type having two-dimensional stable and one-dimensional unstable manifolds. At the further reduction of parameter values μ there appear some attributes of a homoclinic cascade of transition to chaotic regimes. It follows from appearance of the C_3 type cycle at $\mu = 1.0978$ (Fig. 3.70). Occurrence of cycles of homoclinic cascade is quite natural for this system in connection with that all singular points have saddle–focus type equilibrium states. However, the scenario of transition to chaos through a homoclinic cascade of bifurcations has no development in the given system.

Thus, in the system (3.43) there are the same mechanisms of formation of irregular (singular) attractors, as in the Lorenz, Rossler, Chua systems and in other systems. Those are the Feigenbaum period doubling cascade of bifurcations, the Sharkovskii subharmonic cascade of bifurcations and the Magnitskii homoclinic cascade of bifurcations.

3.6 Final Remarks and Conclusions

Results of investigations of nonlinear autonomous dissipative systems of ordinary differential equations produced in this chapter have shown, that *all irregular attractors of these systems are singular attractors* as they are defined in Sec. 1.4.3, and transition to chaotic behaviour in all considered systems is carried out with use of the same mechanisms.

Any scenario of transition to chaos always begins with the Feigenbaum period doubling cascade of bifurcations of some original stable cycle. Then it always continues with the Sharkovskii complete or incomplete subharmonic cascade of bifurcations of stable cycles of arbitrary period up to the cycle of period three. Then the further continuation of any scenario is always the Magnitskii complete or incomplete homoclinic cascade of bifurcations of stable cycles converging to homoclinic contours of singular points or singular cycles.

The indicated mechanisms generate an infinite variety of singular attractors among them it is necessary to distinguish *complete or incomplete, subharmonic and homoclinic singular attractors.* It depends on complexity and power of a set of cycles participating in their formation. The simplest singular attractor presenting in all systems is *the Feigenbaum attractor.*

Another important conclusion is that the presence of a saddle–node or a saddle–focus separatrix loop, or the presence of a saddle–node or a saddle–focus themselves are not necessary conditions for existence of chaotic dynamics in dissipative autonomous systems of nonlinear ordinary differential equations, and that *singular attractors of such systems are not structurally stable formations.* The theory of such singular attractors is considered in Chapter 4.

In Chapters 4–5 it will be shown that *the same mechanisms* lead to occurrence of chaotic dynamics in systems of non-autonomous nonlinear ordinary differential equations, in ordinary differential equations with delay argument and in partial differential equations.

Chapter 4

Principles of the Theory of Dynamical Chaos in Dissipative Systems of Ordinary Differential Equations

Results of numerous numerical experiments represented in Chapter 3 permitted the authors to come out with the proved suggestions that contrary to the generally accepted opinion there is one universal scenario of transition to chaos in all nonlinear dissipative systems of ordinary differential equations through the Feigenbaum period doubling cascade of bifurcations of stable cycles starting from some original singular cycle, then the Sharkovskii subharmonic cascade of bifurcations of stable cycles of arbitrary period up to the cycle of the period three and then the Magnitskii homoclinic cascade of bifurcations of stable cycles converging to homoclinic contour. Also the assumption was stated that any irregular attractor of three-dimensional systems is not some new stable in phase space and structurally stable in parameter space formation, but it is a singular attractor, i.e. it exists only at a separate point of accumulation of values of bifurcation parameter, being a closure of semi-stable nonperiodic trajectory. It follows that any irregular (singular) attractor lies on smooth submanifold of phase space (two-dimensional surfaces in a three-dimensional case) and it has no positive Lyapunov exponents. Thus, dimension of any irregular (singular) attractor of three-dimensional nonlinear dissipative system should not be more than two, and the dynamical chaos should be defined not by hyperbolicity of a system and not by exponential divergence of trajectories on an attractor, but by *a phase shift* of trajectories tending to the attractor. These assumptions also completely prove to be true by results of numerous numerical experiments with non-autonomous two-dimensional and autonomous many-dimensional dissipative systems of ordinary differential equations, with partial differential equations and differential equations with delay arguments represented in Chapters 4 and 5.

In the present chapter all assumptions formulated above are proved for a wide class of three-dimensional autonomous dissipative systems of nonlinear ordinary differential equations having originally a singular stable cycle. It was proved [Magnitskii (2004)], that all regular and singular attractors of such systems, arising after the loss of stability of a singular cycle at change of values of a bifurcation parameter during subharmonic, homoclinic and, probably, more complex cascades of bifurcations, belong to the closure of its smooth two-dimensional, at least two-sheeted, unstable invariant manifold. The mechanism was discovered by means of which a shift of phases of trajectories of a system at their rotation around an original singular cycle that enables to pass to some one-dimensional continuous nonmonotonic mapping of the segment into itself in some two-dimensional plane moving along a singular cycle. A singular point of a rotor type (see Sec. 4.4) of two-dimensional non-autonomous system of ordinary differential equations with periodic coefficients corresponds to an original singular cycle in this plain. Thus, a rotor type singular point, discovered in [Magnitskii (2004)], is a key element of the theory, it is a natural bridge between one-dimensional mappings and three-dimensional autonomous systems of ordinary differential equations. The established transition enables to explain the nature and principles of formation of singular attractors of autonomous three-dimensional systems on the basis of the theory of one-dimensional continuous nonmonotonic mappings foundation of which lay in works [Feigenbaum (1978); Feigenbaum (1980); Sharkovskii (1964); Li and Yorke (1975)]. It follows from this theory that transition to dynamical chaos in three-dimensional autonomous nonlinear systems of differential equations having singular cycles is carried out by that way which we observe in numerical experiments — through the Feigenbaum cascade of period doubling bifurcations of stable cycles, and then through the Sharkovskii subharmonic cascade of bifurcations. And as the further complication of behaviour of solutions of systems of differential equations goes through the Magnitskii homoclinic cascade of bifurcations, then this cascade should take place also in one-dimensional unimodal mappings. We shall notice at once, that it is necessary to develop the theory of multimodal one-dimensional mappings for description of subsequent more complex cascades of bifurcations of cycles in dissipative systems of differential equations.

The material of the present chapter is stated in the following order. The general theory of one-dimensional nonlinear mappings is stated in Sec. 4.1, the Feigenbaum theory of the cascade of period doubling bifurcations of cycles in one-dimensional unimodal mapping is stated in Sec. 4.2, the Sharkovskii theory of formation of cycles of arbitrary period in one-dimensional unimodal mapping according to the Sharkovskii order is stated in Sec. 4.3. *The Magnitskii theory of rotor type singular points* of two-dimensional non-autonomous nonlinear systems of ordinary differential equations is presented in Sec. 4.4. This theory is a bridge between one-dimensional unimodal mappings and two-dimensional non-autonomous nonlinear systems of ordinary differential equations. *The Magnitskii theory of singular cycles and singular attractors* of three-dimensional autonomous nonlinear systems of ordinary differential equations is presented in Sec. 4.5. This theory is a bridge between two-dimensional non-autonomous and three-dimensional autonomous nonlinear systems of ordinary differential equations. The analytical and numerical examples of two-dimensional non-autonomous and three-dimensional autonomous systems illustrating various aspects of the theory are presented in Sec. 4.4 and Sec. 4.5. As it was not really revealed yet any other scenarios of transition to chaos in three-dimensional autonomous dissipative systems of nonlinear ordinary differential equations, except for as through the cascade of period doubling bifurcations, subharmonic and then homoclinic cascades (see [Magnitskii (2004); Magnitskii (2005)] and [Magnitskii and Sidorov (2001)—(2005)], and Chapters 3–5), rather believable is the hypothesis on universality of the method of appearance of chaotic dynamics in three-dimensional nonlinear dissipative systems of ordinary differential equations described in the present chapter. In systems of greater dimension the scenarios of transition to chaos through the subharmonic cascade of bifurcations of two-dimensional tori can be realized (see Sec. 3.2 and Chapter 5), that also keeps within frameworks of the theory stated in this chapter.

4.1 Theory of One-Dimensional Smooth Mappings

Let us consider a dynamical system with the discrete time, given on an interval $I \subset \mathbb{R}$ and depending on a scalar parameter μ

$$x_{n+1} = f(x_n, \mu), \qquad x_n \in I \subset \mathbb{R}, \quad f \in C^1, \quad n = 0, 1, 2, \ldots . \qquad (4.1)$$

Continuously differentiable mapping $f : I \to I$ is one-dimensional and has evident geometrical interpretation, as it can be graphically presented in coordinates (x_n, x_{n+1}). In this case it is easy to find its *fixed points* by means of *the Lamerey diagram*. They lie in intersection of the graph $x_{n+1} = f(x_n, \mu)$ and the bisector $x_{n+1} = x_n$ (Fig. 4.1).

For one-dimensional mappings stability of a fixed point is defined by value of the module of a derivative of the mapping in this point. If the value $|f'(x^*, \mu)| < 1$ at point x^*, then the fixed point x^* is *stable* and it is an attractive point. If $|f'(x^*, \mu)| > 1$, then the fixed point is *unstable*. The given conditions are sufficient. A case, when $|f'(x^*, \mu)| = 1$ demands additional research. For example, a fixed point $x^* = b$ for a mapping $f(x) = x - a(x - b)^3$ in which $f'(b) = 1$, is stable. In this case the point $x = b$ is a flex point and the function $f(x)$ monotonously increases in the neighbourhood $\left(b - 1/\sqrt{3a}, \; b + 1/\sqrt{3a}\right)$ of this point, so as $f(x) > x$ at $x < b$ and $f(x) < x$ at $x > b$. Therefore the sequence $\{f^n(x)\}$ monotonously tends to the point b for all values $x \in \left[b - \sqrt{1/3a}, b + \sqrt{1/3a}\right]$.

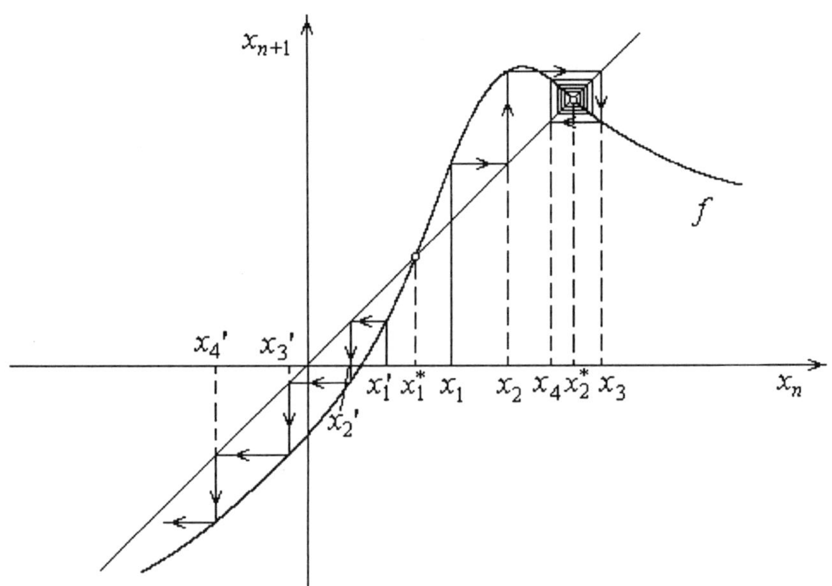

Fig. 4.1   The Lamerey diagram of some one-dimensional mapping f. The fixed point x_1^* is unstable, the fixed point x_2^* is stable.

The fixed point $x^* = 1 - 1/\mu = 2/3$ of the logistic map $f : \; x \mapsto \mu x(1 - x)$ is stable also for the value $\mu = 3$, though $f'(x^*, \mu) = -1$. Really, representing a mapping

$$f^2(x, \mu) = f \circ f(x, \mu) = \mu^2 x(1 - x)(1 - \mu x(1 - x))$$

in the form of a series on degrees $(x - x^*)$

$$f^2(x, \mu) = x^* + (\mu - 2)^2(x - x^*) - \mu(\mu - 2)(\mu - 3)(x - x^*)^2$$
$$- 2\mu^2(\mu - 2)(x - x^*)^3 - \mu^3(x - x^*)^4,$$

and substituting $\mu = 3$, i.e.

$$f^2(x, 3) = x - 18(x - x^*)^3 - 27(x - x^*)^4,$$

we see, that a coefficient at $(x - x^*)^3$ is negative and hence, here we have the same case, as in the previous example for the mapping f. If x is close to x^* and, for example, $x < x^*$, then $x < f^2(x, \mu) < f^4(x, \mu) < \cdots \to x^*$, and also $f(x, \mu) > f^3(x, \mu) > \cdots \to x^*$. Similarly we have $x > f^2(x, \mu) > f^4(x, \mu) > \cdots \to x^*$ and $f(x, \mu) < f^3(x, \mu) < \cdots \to x^*$ at $x > x^*$.

Let us notice, that a necessary and sufficient condition of stability of a fixed point of one-dimensional mapping $f \in C^0(I, I)$ is the inequality $f^2(x) > x$ at $x < x^*$ and $f^2(x) < x$ at $x > x^*$ [Sharkovskii *et al.* (1993)].

4.1.1 *Monotonic invertible mappings*

When a one-dimensional mapping $f : I \to I$ is strictly monotonic, then it is an invertible mapping, and a dynamical system on an interval I is arranged simply enough. At first, we shall consider a case when the function increases strictly monotonously, that is $f'(x) > 0$. Then $(f^m(x))' > 0$ for any $m \in \mathbb{N}$. Really, for $m = 2$ we have $x_{n+2} = f(x_{n+1}) = f \circ f(x_n) = f^2(x_n)$ at any point x_n. From here

$$[f^2(x_n)]' = f'(f(x_n)) \cdot f'(x_n) = f'(x_{n+1}) \cdot f'(x_n) > 0,$$

as each efficient is more than zero. Similarly, for any $m \in \mathbb{N}$ it is obvious that $f^m(x_n) = f \circ f \circ \cdots \circ f(x_n)$ at any point x_n and, hence,

$$[f^m(x_n)]' = f'(x_{n+m-1}) \cdot f'(x_{n+m-2}) \cdots f'(x_n) > 0.$$

Thus, each trajectory $x_0, x_1, x_2, \ldots, x_{n+1} = f(x_n)$ is monotonous (at $x_0 > x_1$ we have $x_0 > x_1 > x_2 > \cdots > x_n$, and at $x_0 < x_1$ accordingly we have $x_0 < x_1 < x_2 < \cdots < x_n$) and converges to one of fixed points. For the description of behaviour of a trajectory in case of monotonously increasing function $f(x)$ it is sufficiently to know a set $\text{Fix}\, f = \{x \in I : x = f(x)\}$ of fixed points of a mapping f and, besides, a $\text{sign}(f(x) - x)$ on each interval, additional to set $\text{Fix}\, f$: if $x_0 \in (a, b)$ and $a, b \in \text{Fix}\, f$, $(a, b) \subset I \setminus \text{Fix}\, f$, then $f^n(x_0) \xrightarrow[n \to \infty]{} a$, when $\text{sign}(f(x_0) - x_0) = -1$ and $f^n(x_0) \xrightarrow[n \to \infty]{} b$, when $\text{sign}(f(x_0) - x_0) = +1$. Thus, a splitting of the segment into stability domains of fixed points takes place. As stable and unstable fixed points alternate in a sequence x_1^*, x_2^*, ..., x_k^* of fixed points, then the interval (x_{i-1}^*, x_{i+1}^*) will be the attraction domain for each stable fixed point x_i^* (Fig. 4.2).

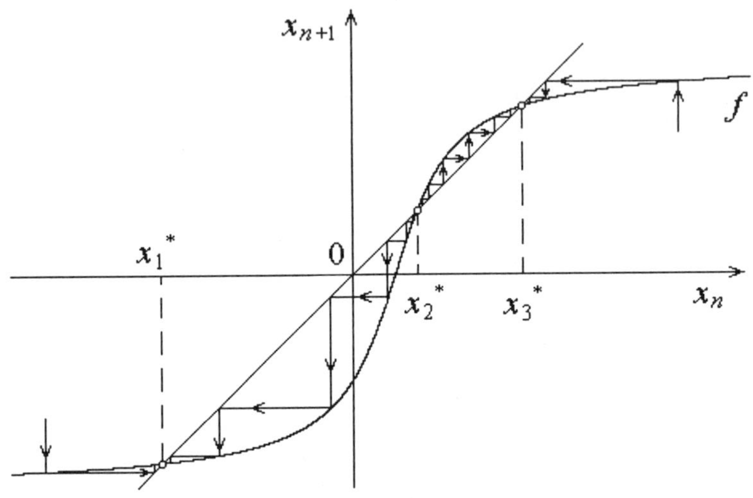

Fig. 4.2   Example of a mapping with monotonously increasing function $f(x)$. Fixed points x_1^* and x_3^* are stable, and a point x_2^* is unstable.

Bifurcations in dynamical system with monotonously increasing function are caused by a change of character of stability of fixed points, or their birth (disappearance). In this connection we shall note two cases, being analogue to bifurcations of cycles in the dynamical systems described by differential equations.

In the first case the graph of function $f(x)$ touches the bisector $x_{n+1} = x_n$ (curve 2 in Fig. 4.3a). This situation corresponds to a bifurcation as a result of which or merging and disappearance of stable and unstable fixed points occurs (if transition is carried out from curve 1 to curve 3), or, on the contrary, a birth of two new fixed points (stable and unstable) occurs (if transition is carried out from curve 3 to curve 1).

In the second case the fixed point of the function $f(x)$ coincides with a flex point of this function (curve 2 in Fig. 4.3b). In this case a change of stability of the fixed point and birth (or disappearance) of two other fixed points occurs as a result of such bifurcation. If, for example, originally stable fixed point x_0^* loses stability owing to this bifurcation at transition from curve 3 to curve 1, then the pair of new stable points x_1^* and x_2^* is born. On the contrary, transition from curve 1 to curve 3 corresponds to the bifurcation of disappearances of two stable fixed points and to change of an equilibrium state of the third fixed point x_0^* from unstable to stable.

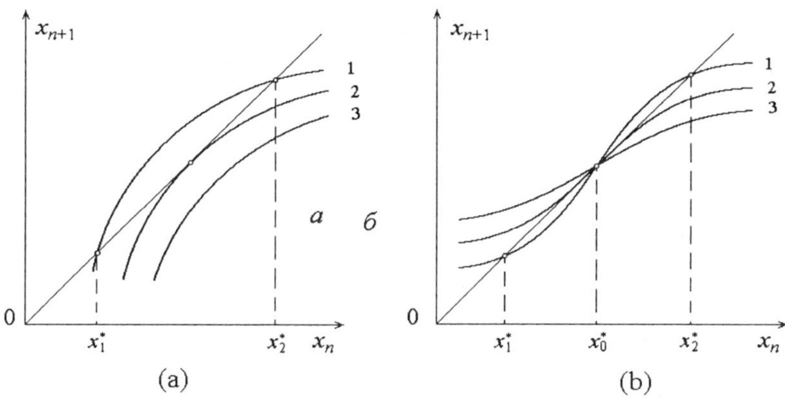

Fig. 4.3   Simultaneous birth (disappearance) of unstable x_1^* and stable x_2^* fixed points (a); change of stability of one fixed point x_0^* and a simultaneous birth (disappearance) of a pair of stable fixed points x_1^* and x_2^* (b).

If we consider a one-dimensional mapping as a Poincare mapping for some system of ordinary differential equations, then the first case corresponds to the saddle–node bifurcation of birth or disappearance of a pair (stable and unstable) limit cycles in a phase space of the system. The second case corresponds to the situation when a pair of new stable limit cycles is born in a system of differential equations as a result of the bifurcation of loss of stability of one limit cycle.

In the case when function $f(x)$ is strictly monotonously decreasing, that is $f'(x) < 0$, the second iteration $f^2(x) = f \circ f(x)$ has a positive derivative

$$f^{2'}(x) = (f \circ f(x))' = f'(f(x)) \cdot f'(x) > 0,$$

as both of the efficients in last expression are negative. Hence, f^2 is strictly monotonously increasing function. It is easy to see, that $\frac{df^{2k}(x)}{dx} > 0$, and $\frac{df^{2k+1}(x)}{dx} < 0$ and, hence, each trajectory of the mapping $x_{n+1} = f(x_n)$ is broken on two sequences $x_0, x_2, \ldots, x_{2k}$ and $x_1, x_3, \ldots, x_{2k+1}$, one of which increases, and another decreases.

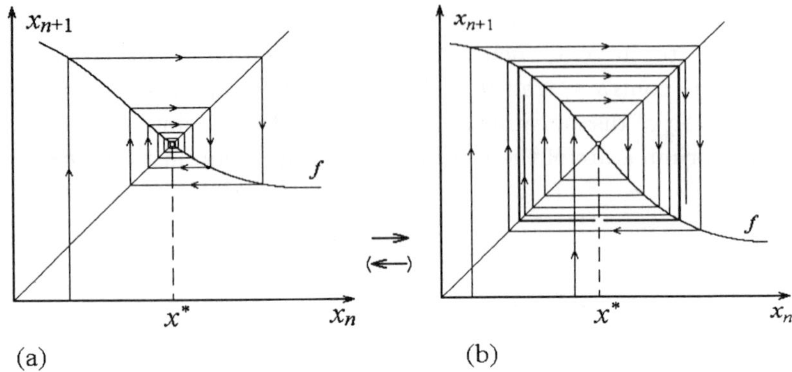

Fig. 4.4   Bifurcation of loss of stability of a fixed point x^* in case of a mapping with monotonously decreasing function $f(x)$ (a), and a birth of a cycle of period 2 (b) as a result of this bifurcation.

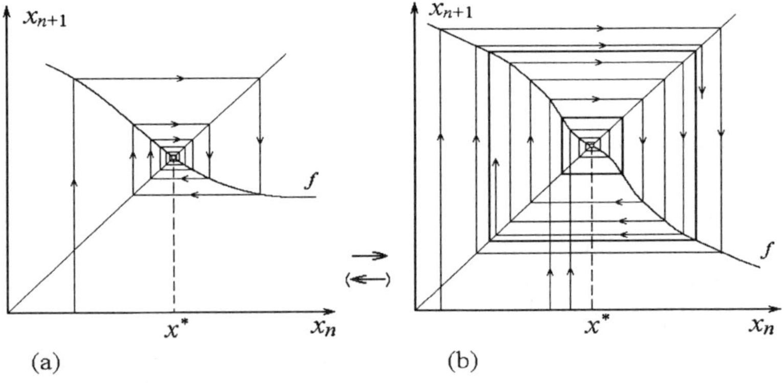

Fig. 4.5   Bifurcation of birth (disappearance) of a pair of period 2 cycles of a mapping with monotonously decreasing function $f(x)$ without change of the character of stability of the fixed point x^*.

Thus knowing a set Fix f^2 is sufficient for description of dynamical system in a case of monotonously decreasing function $f(x)$. This set consists of a fixed point x^* of the mapping $f(x)$, lying in intersection of the graph of the function $f(x)$ and the bisector $x_{n+1} = x_n$, and fixed points x_i^* of the mapping $f^2(x)$ to which the period 2 cycles of the mapping $f(x)$ correspond. Thus, all segment I is split into an attraction domain of the fixed point x^*

(if it is stable) and into attraction domains of the period 2 cycles. This splitting in essence defines a type of bifurcations in dynamical system with monotonously decreasing function: or change of a character of stability of the fixed point x^* owing to what the birth or disappearance of the period 2 cycle occurs (Fig. 4.4), or the birth (disappearance) of a pair of the period 2 cycles occurs without change of character of stability of the fixed point x^* (Fig. 4.5). In the first case the condition $f'(x^*) = -1$ obviously should be satisfied at loss of stability of the fixed point. Thus, this bifurcation of one-dimensional mapping is similar to the period doubling bifurcation of a stable cycle in nonlinear system of ordinary differential equations at transition of multiplier of the cycle through the point -1 of the unit circle.

4.1.2 *Nonmonotonic mappings*

Nonmonotony of a one-dimensional mapping $f : I \to I$ is the necessary condition for existence of complex behaviour in a dynamical system described by this mapping. It allows to carry out returning of some points into initial positions and, hence, to receive periodic points with any period, and not just with the periods 1 or 2, as in cases of monotonously increasing or monotonously decreasing function $f(x)$. The complex behaviour of dynamical system with nonmonotonic function is caused by many-valuedness of the inverse mapping. Really, following [Loskutov and Mikhailov (1990)], suppose, that inverse mapping $f^{-1} : x_{n+1} \mapsto x_n$ has p branches, i.e. it consists of a set of p invertible mappings f_i^{-1}, $i = 1, \ldots, p$. We shall write down f_i^{-1} in the form of

$$x_n = g_i(x_{n+1}), \quad i = 1, \ldots, p,$$

and consider the superposition of m invertible mappings

$$G(x) = g_{i_m} \circ g_{i_{m-1}} \circ \cdots \circ g_{i_2} \circ g_{i_1}(x), \quad m = 2, 3, \ldots,$$

where i_j is an arbitrary integer from the set $\{1, 2, \ldots, p\}$. The mapping $G(x)$ is monotonous and, hence, it has even one fixed point $x^*_{i_1 i_2 \ldots i_m}$. But this fixed point is also a fixed point of a mapping f^m. So, or this point is also a fixed point of the mapping f, or a cycle of the mapping f of the certain period corresponds to this fixed point of the mapping f^m. The first case is possible only for $i_1 = i_2 = \cdots = i_m$. And as it is possible to take arbitrary positive integer m and to choose numbers $i_1, i_2, \ldots, i_m$ from 1 up to p by any way, then nonmonotonic mapping can have infinite number of cycles of various periods and uncountable set of nonperiodic trajectories.

4.2 Feigenbaum Cascade of Period Doubling Bifurcations of Cycles of One-Dimensional Mappings

The analytical *theory of the Feigenbaum cascade* of period doubling bifurcations is complex enough, and its full and consecutive statement would demand a writing of the special monograph devoted to consideration both the theory, and numerous examples from nonlinear dynamics where sequences of the period doubling bifurcations take place. The purpose of our work is determined first of all by problems of nonlinear dynamics of dissipative systems described by the differential equations. Therefore we shall be limited here by consideration of the qualitative aspect of this theory and approximate estimations for Feigenbaum universal constants.

4.2.1 *Logistic mapping*

Studying the properties of logistic one-dimensional mapping

$$f(x,\mu) = \mu x(1-x), \quad x \in [0,1], \ \mu \in [1,4], \tag{4.2}$$

has resulted Feigenbaum to understanding the mechanism of the period doubling its cycles and to finding the equation for definition of parameter values μ at which this period doubling occurs. Subsequently it appeared, that the complex chaotic behaviour of one-dimensional dynamical system is not connected with the original features of the logistic mapping. It takes place in all difference equations of the first order of a kind $x_{n+1} = f(x_n, \mu)$ in which one-dimensional mapping $f : I \mapsto I$ is *unimodal* at a corresponding choice of scale, i.e. it has a unique extremum on the given interval I. Inverse mapping f^{-1} has thus two branches on the interval I.

Following to Feigenbaum, we shall consider the mapping (4.2) on the interval $x \in [0, 1]$. This mapping has a fixed point $x^* = 1 - 1/\mu$ which moves from the left to the right at increasing the parameter values μ and reaches the value $x^* = 1/2$ at $\mu = 2$. The fixed point loses stability at the value $\mu = 3$ and two new points appear in its neighbourhood forming a stable cycle of the period 2. These two points are stable fixed points of the mapping

$$f^2(x,\mu) = \mu^2 x(1-x)(1 - \mu x(1-x)).$$

Feigenbaum has established, that the main cause of the period doubling is the relation between derivatives of functions f and f^2. As it was already marked above, if $x_2 = f(x_1, \mu) = f^2(x_0, \mu)$, then

$$f^{2'}(x_0, \mu) = f'(x_1, \mu) \cdot f'(x_0, \mu) \tag{4.3}$$

by a rule of differentiation of composite function. Similarly

$$f^{n'}(x_0, \mu) = f'(x_{n-1}, \mu) \cdot f'(x_{n-2}, \mu) \cdots f'(x_0, \mu). \tag{4.4}$$

It follows from the relation (4.3) that

$$f^{2'}(x^*, \mu) = f'(x^*, \mu) \cdot f'(x^*, \mu) = [f'(x^*, \mu)]^2 \tag{4.5}$$

for a fixed point $x_0 = x^*$ of the mapping f taking into consideration that $x_1 = x_2 = x^*$. Similarly the relation (4.4) defines a value of a derivative of the mapping f^n at any point of a cycle of the period n

$$x_{k+1} = f(x_k, \mu), \quad k = 0, \ldots, n - 2, \quad x_0 = f(x_{n-1}, \mu)$$

in a kind of

$$f^{n'}(x_k, \mu) = f'(x_{n-1}, \mu) \cdot f'(x_{n-2}, \mu) \cdots f'(x_0, \mu). \tag{4.6}$$

Besides as $f'(1/2, \mu) = 0$ for the mapping (4.2), then for all $n \in \mathbb{N}$

$$f^{n'}(1/2, \mu) = 0.$$

It follows also from the last equation, that the mapping f^2 has extrema (maxima) in those points x_0 which are mapped to a point $x = 1/2$ by the function f, as in this case $x_1 = f(x_0, \mu) = 1/2$, and $f^{2'}(x_0, \mu) = f'(x_1, \mu) \cdot f'(x_0, \mu) = 0$.

The maximal value of the function f tends to $3/4$ at value $\mu \to \mu_1 = 3$, value $f'(x^*, \mu) \to -1$, and value $f^{2'}(x^*, \mu) \to 1$. Then we have $|f'(x^*, \mu)| > 1$, $|f^{2'}(x^*)| > 1$ for parameter values $\mu > 3$, and two new fixed points of the mapping $f^2 : I \to I$ appear besides the fixed point of the mapping f. That is the graph of the mapping f^2 intersects the bisector $x_{n+1} = x_n$ in two additional points (Fig. 4.6a and Fig. 4.6b accordingly for functions f and f^2 for $\mu = 3.14$). New fixed points x_1^* and x_2^* of the mapping f^2 are not fixed points of the mapping f which transforms one point to another, that is $x_2^* = f(x_1^*, \mu)$, and $x_1^* = f(x_2^*, \mu)$. This pair of points forms a cycle of the period 2 of the mapping f. We shall note, that the derivative of the mapping f^2 is more than 1 in fixed points of the mapping f (Fig. 4.6b). Hence, these points are unstable fixed points of the mapping f^2. On the contrary, a tangent of an angle of inclination of the graph of $f^2(x)$ to an abscissa axis is less 1 in two new fixed points, that is $f^{2'}(x_1^*, \mu) = f^{2'}(x_2^*, \mu) < 1$. Hence, these points are stable and each double iteration of the function f will be attracted either to x_1^*, or to x_2^*. At $n \to \infty$, sequence x_0, x_1, x_2, x_3, ... will approach to sequence x_1^*, x_2^*, x_1^*, x_2^*, ... which is a stable double cycle, or an attractor with the period 2 of the mapping f.

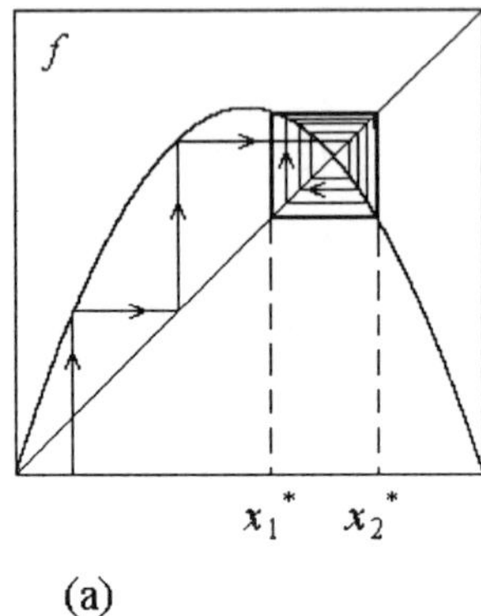

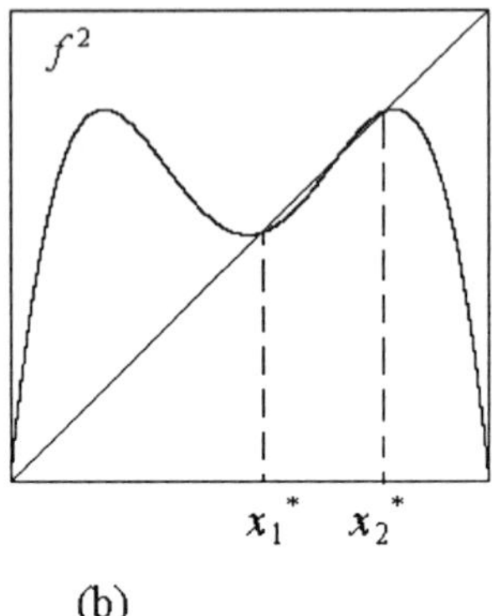

Fig. 4.6   The birth of the stable period 2 cycle of the logistic mapping (4.2).

As each point of cycle of the period n of the mapping f is also a fixed point of the mapping f^n, it follows from (4.6), that all points of the cycle lose stability simultaneously, at the same parameter value μ. It is a cause of infinite sequence of doubling of the period. So the minimum of the function f^2 at the value $x = 1/2$ goes down at the further increase of values μ, and values of its derivative grow at fixed points x_1^* and x_2^*. The fixed point x_1^* will accept the value $1/2$ at some parameter value $\mu = \mu_1^*$. Simultaneously with this the other fixed point x_2^* will accept the value corresponding to the right maximum of the function f^2 (Fig. 4.7) and thus the equality $f^{2'}(x_1^*, \mu) = f^{2'}(x_2^*, \mu) = 0$ will take place in both points of a double cycle (so-called *supercycle*).

At the further increase of the parameter values μ the derivative of the function f^2 becomes negative at points x_1^* and x_2^*, and it becomes equal to -1 at both fixed points x_1^* and x_2^* for $\mu = \mu_2 = 1 + \sqrt{6}$. That is the situation for the function $f^2(x, \mu)$ at the value $\mu = \mu_2$ is similar to the situation for the function $f(x, \mu)$ at the value $\mu = \mu_1 = 3$. Similarly to how a stable cycle of the period 2 appears from the fixed point of mapping $f(x, \mu)$ for $\mu = \mu_1$, and each fixed point of the mapping $f^2(x, \mu)$ forms its own cycle of the period 2 for $\mu = \mu_2$, that is a cycle of the period 4 for function f. To find points of a cycle of the period 4 of the mapping f, it is necessary to establish fixed points of the function f^4 which can be calculated from the function f^2 similarly to how the function f^2 has been calculated from the function f, that is $f^4 = f^2 \circ f^2$. Since this moment it is possible to forget about the function f and to consider f^2 to be the basic function.

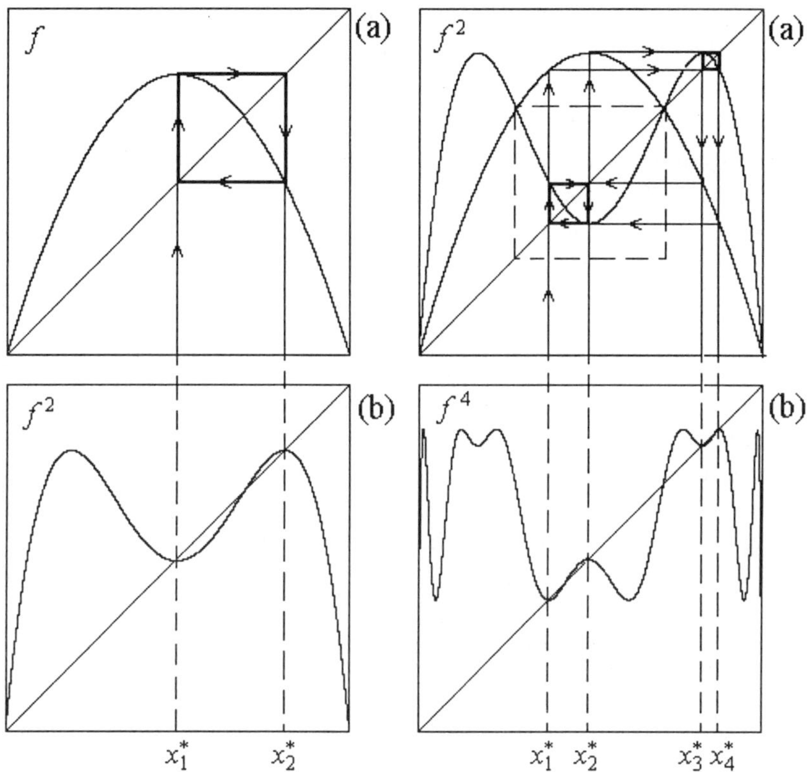

Fig. 4.7 The supercycle of period 2 of the mapping (4.2) for $\mu = \mu_1^*$.

Fig. 4.8 The supercycle of period 4 of the mapping (4.2) for $\mu = \mu_2^*$.

The fact, that the function f^2 is the second iteration of the mapping f, implies equality of values of its derivatives in fixed points. As it is correct for any number of iterations, it is enough to consider only the fixed point of the mapping f^4 nearest to the value $1/2$. And the behaviour of three other fixed points will be analogous. So, there appears recurrent procedure.

Again increasing μ up to a value μ_2^* at which one of fixed points of the mapping f^4 will accept the value $1/2$, and derivatives in fixed points will be equal to zero (Fig. 4.8), we shall obtain the most stable cycle (supercycle) of the period 4. At the further increasing of the parameter values μ the derivative of the function f^4 will become again negative at fixed points, and then it will accept the value -1 for $\mu = \mu_3$. And again the period doubling bifurcation will occur, owing to which a stable cycle of the period 8 of the mapping f will appear. And again we shall obtain $f^8 = f^4 \circ f^4$.

Thus, the equation always takes place

$$f^{2^{n+1}} = f^{2^n} \circ f^{2^n}, \tag{4.7}$$

that is the same mechanism leads to doubling of the period of any iteration f^{2^n}. Function $f^{2^{n+1}}$ is formed from the function f^{2^n} under the formula (4.7). Similarly function $f^{2^{n+2}}$ is formed from the function $f^{2^{n+1}}$. It follows from here, that there is a certain operator, the result of action of which on function f^{2^n} at the parameter value $\mu = \mu_n$ defines the function $f^{2^{n+1}}$ at the value $\mu = \mu_{n+1}$. As, besides we consider the function f^{2^n} only in some interval containing a fixed point with the value close to $1/2$, and the size of this interval constantly decreases with increasing of the parameter values μ, then the mapping, forming this interval, also is compressed into very small domain of a curve near to a point $x = 1/2$. The behaviour of the mapping f far from the point $x = 1/2$ is insignificant for property of the period doubling, and the nature of a maximum of the function f is important only in a limit at $n \to \infty$. It is possible to conclude from here that behaviour of all functions with quadratic extremum is the same in a limit of infinite number of the period doubling. Hence, the operator, acting on mapping, has a stable fixed point in a space of functions. And this point (function) is the common universal limit for repeated iterations of any concrete function.

4.2.2 *Period doubling operator*

The idea of construction of such operator consists in the following. We shall mark off by dotted line a square in Fig. 4.8a containing a part of the function f^2. We shall again use this function as an example for construction of the period doubling operator. Then we shall invert this square concerning the point $(1/2,\ 1/2)$ and stretch it so that the square formed by a cycle of the period 2 of the mapping f^2 and shown in Fig. 4.8a by a continuous line, has coincided with the similar square formed by a cycle of the period 2 of the mapping f and represented in Fig. 4.7a. In both squares, there will be curves with an identical type of a maximum at the point $x = 1/2$ and being equal to zero in the right bottom corner. Since the function f defines a corresponding part of the function f^2 at increasing μ from the value μ_1 up to the value μ_2, and the function f^2 defines a corresponding part of the function f^4 after a necessary stretching and inversion. Feigenbaum has calculated the first five such functions turned out as a result of such procedure. Difference between values of three last functions has been so a little, that they have practically coincided [Feigenbaum (1980)].

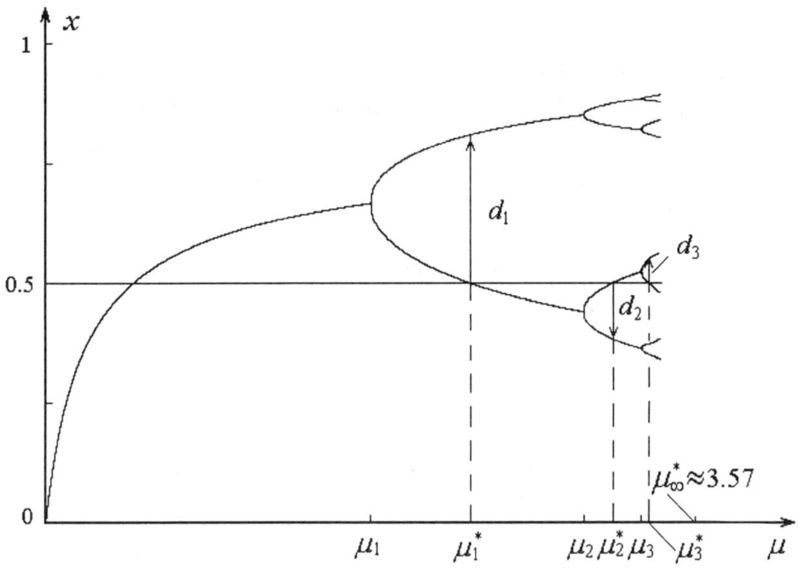

Fig. 4.9   Bifurcation diagram of the Feigenbaum cascade for the logistic mapping (4.2).

We shall note, that change of scale of the function f^4, defined at similar transformation of the function f^2, is based only on properties of composition of functions. Therefore, if the described curves for functions f^{2^n} and $f^{2^{n+1}}$ converge to one limit at $n \to \infty$, then the scale of changing values (scaling) in the period doubling cascade of bifurcations at transition from one level to another also will converge to the certain value. Let us pay attention, that the side of each square, formed by a cycle at each following period of doubling iteration, is equal to distance d_n between the point $x = 1/2$ and the fixed point x^* of the mapping f^{2^n}, nearest to the point $x = 1/2$ and satisfying to the condition $f^{2^{n'}}(x^*) = 0$ (Fig. 4.9). The last condition defines, how it was marked above, the so-called supercycles, i.e.

cycles with the greatest stability domains. Hence, the distance between these neighboring elements of attractor (stable cycle) decreases asymptotically in a constant number of times during the period doubling cascade. Besides, the nearest to the point $x = 1/2$ element of attractor moves from one hand of the point $x = 1/2$ to another at each subsequent doubling of the period. Thus, transition to a cycle of the period 2^{n+1} at $\mu = \mu_{n+1}$ is carried out by a compression of scale approximately in α times, that is

$$\frac{d_n}{d_{n+1}} \approx -\alpha. \tag{4.8}$$

It follows also from Fig. 4.9, that value d_n is the distance between the point $x = 1/2$ and an element of a cycle of the period 2^n nearest to this point for $\mu = \mu_n^*$, where μ_n^* is the parameter value μ, corresponding to a supercycle of the period 2^n. This nearest element is 2^{n-1} iteration of the point $x = 1/2$ and consequently

$$d_n = f^{2^{n-1}}(1/2, \mu_n^*) - 1/2. \tag{4.9}$$

For the further statement we shall make a shift transformation of $x + 1/2 \to x$. Then the expression (4.9) will become

$$d_n = f^{2^{n-1}}(0, \mu_n^*).$$

Using the representation (4.8), we obtain, that the limit takes place

$$\lim_{n \to \infty} (-\alpha)^n d_{n+1} = d. \tag{4.10}$$

In view of (4.9) we obtain, that the sequence of the scaled iterations $f^{2^n}(0, \mu_{n+1}^*)$ converges, that is

$$\lim_{n \to \infty} (-\alpha)^n f^{2^n}(0, \mu_{n+1}^*) = d. \tag{4.11}$$

Represented above Figs. 4.7–4.8 illustrate a more strong assertion, namely, the function f^{2^n} stretched in $(-\alpha)^n$ times converges to some quite certain function, and the expression (4.11) is a limit of this function at $x = 0$. Hence, the expression (4.11) can be generalized for all interval, and then the limiting function designated as $g_1(x)$, will look like

$$g_1(x) = \lim_{n \to \infty} (-\alpha)^n f^{2^n}\left(\frac{x}{(-\alpha)^n}, \mu_{n+1}^*\right). \tag{4.12}$$

At repeated iterations ($n \to \infty$) lesser parts of a curve f near to a maximum remain essential (see, for example, Fig. 4.8), and, hence, function $g_1(x)$ is defined by behaviour of the function $f^{2^n}\left(\dfrac{x}{(-\alpha)^n}, \mu^*_{n+1}\right)$ only near to the point $x = 0$. Therefore function $g_1(x)$ should be universal for all functions f with a quadratic extremum.

Generalizing an expression (4.12), we shall define a family of universal functions

$$g_i(x) = \lim_{n\to\infty} (-\alpha)^n f^{2^n}\left(\frac{x}{(-\alpha)^n}, \mu^*_{n+i}\right), \quad i = 0, 1, \dots . \tag{4.13}$$

It is easy to see, that

$$g_{i-1}(x) = \lim_{n\to\infty} (-\alpha)^n f^{2^n}\left(\frac{x}{(-\alpha)^n}, \mu^*_{n+i-1}\right)$$

$$= \lim_{n\to\infty} (-\alpha)(-\alpha)^{n-1} f^{2^{n-1}+1}\left(-\frac{1}{\alpha}\frac{x}{(-\alpha)^{n-1}}, \mu^*_{n+i-1}\right)$$

$$= \lim_{m\to\infty}\left[(-\alpha)(-\alpha)^m f^{2^m}\left(\frac{1}{(-\alpha)^m}(-\alpha)^m f^{2^m}\left(-\frac{1}{\alpha}\frac{x}{(-\alpha)^m}, \mu^*_{m+i}\right), \mu^*_{m+i}\right)\right]$$

$$= -\alpha g_i\left(g_i\left(-\frac{x}{\alpha}\right)\right) = T g_i(x), \tag{4.14}$$

where T is the period doubling operator. Thus, all functions of the family (4.13) are connected among themselves by transformation of doubling

$$g_{i-1}(x) = -\alpha g_i\left(g_i\left(-\frac{x}{\alpha}\right)\right) \equiv T g_i(x).$$

We shall designate $\quad g(x) = \lim_{i\to\infty} g_i(x) \quad$ and obtain, that

$$g(x) = T g(x) = -\alpha g\left(g\left(-\frac{x}{\alpha}\right)\right). \tag{4.15}$$

Thus the function $g(x)$ is a fixed point of the period doubling operator T.

4.2.3 *Feigenbaum universality*

The equation (4.15) basically allows to find a universal constant α. For example, for $x = 0$ we shall obtain the equation

$$g(0) = -\alpha g(g(0)). \tag{4.16}$$

However, there are, at least, two essential features in solution of the functional Eq. (4.15). First of them is to the effect that the Eq. (4.15) is invariant concerning scaling of the function $g(x)$. Indeed, having substituted the function $\lambda g(x/\lambda)$ in the Eq. (4.15) instead of the function $g(x)$, we obtain an equation

$$\lambda g\left(\frac{x}{\lambda}\right) = -\alpha\lambda g\left(\frac{1}{\lambda}\lambda g\left(-\frac{x}{\alpha\lambda}\right)\right),$$

or

$$g(u) = -\alpha g\left(g\left(-\frac{u}{\alpha}\right)\right),$$

where $u = x/\lambda$. It means that the function $\lambda g(x/\lambda)$ is the solution of the Eq. (4.15) with the same value α for arbitrary $\lambda \neq 0$. Since the theory allows to choose an arbitrary scale factor λ, then we can choose it so that, for example, $g(0) = 1$.

Other feature of the functional Eq. (4.15) consists in absence of methods for exact solution of this equation. For approximate solution of this equation we shall choose function $g(x)$ from a class of smooth functions with a quadratic maximum at the zero point. With this purpose we shall represent a solution in a form of series on even powers of the variable x

$$g(x) = 1 + a_1 x^2 + a_2 x^4 + \cdots + a_n x^{2n} + \cdots \tag{4.17}$$

and we shall use two members of this series for approximate estimation of both the function $g(x)$ and the universal constant α. Then the Eq. (4.15) for a fixed point of the operator of doubling will become

$$1 + a_1 x^2 = -\alpha(1 + a_1) - \left(\frac{2a_1^2}{\alpha}\right)x^2 + O(x^4).$$

Discarding the remainder and solving the system of the equations

$$\alpha + 2a_1 = 0,$$
$$1 + \alpha(1 + a_1) = 0,$$

concerning unknown coefficients α and a_1, we shall find that $a_1 = (-1 \pm \sqrt{3})/2$. As function $g(x)$ has a maximum at the zero point, we should choose $a_1 = (-1 - \sqrt{3})/2 \approx -1.366$, and then $\alpha = 1 + \sqrt{3} \approx 2.732$. These approximate estimations differ about 10% from the numerical results obtained by Feigenbaum at using four members of the equation (4.17)

$$g(x) = 1 - 1.52763x^2 + 0.104815x^4 - 0.0267057x^6.$$

In Feigenbaum's research $\alpha = 2.502907875$.

The obtained universal constant α defines scaling of the variable x in the mapping $f(x,\mu)$. We shall consider now, whether the other variable,

μ refers to scaling. We shall notice, that the parameter values of $\mu = \mu_n^*$ are defined from a condition of existence of supercycles and corresponds to such values of μ, at which an equation $\frac{d}{dx} f^{2^n}(x, \mu_n^*)\big|_{x_0^*} = 0$ takes place. Supercycles contain the fixed point $x_0^* = 1/2$ as the element. Therefore, we have the equation $f^{2^n}(1/2, \mu_n^*) = 1/2$ which after the shift $x + 1/2 \to x$ transforms to the equation

$$f^{2^n}(0, \mu_n^*) = 0 \tag{4.18}$$

for finding the values μ_n^*.

The Eq. (4.18) has an infinite set of solutions, as supercycles, taking place in the windows of chaotic regimes, also satisfy to it. To find the values of μ_n, concerning to the period doubling cascade, that means the sequence

$$\mu_1 < \mu_1^* < \mu_2 < \mu_2^* < \mu_3 < \cdots < \mu_n < \mu_n^* < \cdots, \tag{4.19}$$

the Eq. (4.18) should be solved, since $n = 0$ and ordering μ_n according to (4.19).

Obviously, values μ_n define a speed of approach to the value μ_∞, which corresponds to the end of the period doubling cascade and to appearance of the Feigenbaum attractor. We shall show, that a scaling

$$\mu_n^* - \mu_\infty \sim \delta^{-n} \tag{4.20}$$

also takes place on a variable μ, where δ is another universal Feigenbaum constant. Then, in view of (4.19), it will be true also that

$$\mu_n - \mu_\infty \sim \delta^{-n}.$$

Let us expand the function $f(x, \mu)$ in a series in a neighbourhood of the point μ_∞

$$f(x, \mu) \approx f(x, \mu_\infty) + \frac{\partial f(x, \mu)}{\partial \mu}\bigg|_{\mu_\infty} (\mu - \mu_\infty)$$

$$= f_0(x) + (\mu - \mu_\infty) f_1(x) \tag{4.21}$$

and apply the period doubling operator T to (4.21) (it is possible to use a sign of equality in (4.21) in the case of the logistic mapping)

$$Tf(x, \mu) = T(f_0(x) + (\mu - \mu_\infty) f_1(x))$$

$$= Tf_0(x) + (\mu - \mu_\infty) \delta T_{f_0} f_1(x). \tag{4.22}$$

In view of the period doubling operator appearance (4.15), we find

$$
\begin{aligned}
\delta T_{f_0} u(x) &= T(f_0(x) + u(x)) - T f_0(x) \\
&= -\alpha(f_0 + u)\left(f_0\left(-\frac{x}{\alpha}\right) + u\left(-\frac{x}{\alpha}\right)\right) + \alpha f_0\left(f_0\left(-\frac{x}{\alpha}\right)\right) \\
&= -\alpha\left[(f_0 + u)\left(f_0\left(-\frac{x}{\alpha}\right) + u\left(-\frac{x}{\alpha}\right)\right) - f_0\left(f_0\left(-\frac{x}{\alpha}\right) + u\left(-\frac{x}{\alpha}\right)\right)\right. \\
&\qquad\left. + f_0\left(f_0\left(-\frac{x}{\alpha}\right) + u\left(-\frac{x}{\alpha}\right)\right) - f_0\left(f_0\left(-\frac{x}{\alpha}\right)\right)\right] \\
&= -\alpha\left[f_0'\left(f_0\left(-\frac{x}{\alpha}\right)\right) \cdot u\left(-\frac{x}{\alpha}\right) + u\left(f_0\left(-\frac{x}{\alpha}\right)\right)\right] + o(u) \\
&= L_{f_0} u + o(u).
\end{aligned}
\tag{4.23}
$$

Operator L is a linear nonself-adjoint operator of a kind

$$
L_f u = -\alpha\left[f'\left(f\left(-\frac{x}{\alpha}\right)\right) \cdot u\left(-\frac{x}{\alpha}\right) + u\left(f\left(-\frac{x}{\alpha}\right)\right)\right].
$$

It is defined in relation to function f. Some small values of the higher order in comparison with u are designated in (4.23) as $o(u)$. After n-fold application of the period doubling operator T to function $f(x, \mu)$, we obtain

$$
T^n f(x, \mu) = T^n(f_0) + (\mu - \mu_\infty) L_{T^{n-1} f_0(x)} \cdots L_{f_0(x)} f_1
$$
$$
+ o((\mu - \mu_\infty) f_1(x)).
\tag{4.24}
$$

As it was marked above, values of the operator $T^n f_0(x)$ converge to the fixed point of the Eq. (4.15), that is

$$
\lim_{n \to \infty} T^n f(x, \mu_\infty) = \lim_{n \to \infty} (-\alpha)^n f^{2^n}\left(\frac{x}{(-\alpha)^n}, \mu_\infty\right) = \lim_{i \to \infty} g_i(x) = g(x).
$$

Then the (4.24) becomes

$$
T^n f(x, \mu) \approx g(x) + (\mu - \mu_\infty) L_g^n f_1(x).
\tag{4.25}
$$

Let us expand function $f_1(x)$

$$
f_1(x) = \sum_k c_k \varphi_k(x), \quad k = 0, 1, 2, \ldots
\tag{4.26}
$$

in a series on eigenfunctions $\varphi_k(x)$ of the operator L_g

$$
L_g \varphi_k(x) = \lambda_k \varphi_k(x), \quad k = 0, 1, 2, \ldots
\tag{4.27}
$$

and obtain that

$$L_g^n f_1(x) = \sum_k c_k \lambda_k^n \varphi_k(x), \quad k = 0, 1, 2, \ldots \qquad (4.28)$$

Feigenbaum has established, that only one real eigenvalue of the operator L_g is more than unit, that is $\lambda_0 > 1$, and $\lambda_k < 1$, $k \neq 0$. Later it has been proved analytically in [Collet *et al.* (1980)]. Not stopping here on the analytical proof of this fact, we shall consider the given statement from the qualitative side. The one-parametrical family of functions $f(x, \mu)$ is a line in space of functions. For each of functions in this space there is an isolated parameter value μ_∞, at which the iterations of this function converge to the function $g(x)$ at multiple application of the period doubling operator T. We shall fill a space of functions with the lines corresponding to functions $f(x, \mu)$. The set of points in space of the functions corresponding to various functions $f(x, \mu_\infty)$, defines a surface, repeated application to which points of the operator T gives the functions converging to $g(x)$. It is a stable manifold g of the operator T. But only one line parametrized by values of one parameter μ passes through each point of this surface. Therefore, there is only one direction in a space of functions defined by eigenvector with eigenvalue, greater than the unit. In other words, there is one eigenfunction φ_0 of operator L_g with corresponding eigenvalue $\lambda_0 > 1$. Therefore, we shall obtain from the expression (4.27), being limited to the contribution only from λ_0, that

$$L_g^n f_1(x) \approx c_0 \lambda_0^n \varphi_0(x) \qquad (4.29)$$

at $n \to \infty$. Then (4.25) will become

$$T^n f(x, \mu) \approx g(x) + (\mu - \mu_\infty) \delta^n c_0 \varphi_0(x), \qquad (4.30)$$

where the designation $\delta = \lambda_0$ is entered. Indeed, it is easy to show, that eigenvalue λ_0 coincides with the universal Feigenbaum constant δ. We shall consider a case when $\mu = \mu_n^*$ and $x = 0$. Then it follows from (4.30) that

$$T^n f(0, \mu_n^*) \approx g(0) + (\mu_n^* - \mu_\infty) \delta^n c_0 \varphi_0(0). \qquad (4.31)$$

Taking into account that $g(0) = 1$ and that, according to the condition (4.18),

$$T^n f(0, \mu_n^*) = (-\alpha)^n f^{2^n}(0, \mu_n^*) = 0, \qquad (4.32)$$

we come to the required result

$$\lim_{n\to\infty}(\mu_n^* - \mu_\infty)\delta^n = -\frac{1}{c_0\varphi_0(0)} = \text{const.} \qquad (4.33)$$

Numerical value of the universal constant δ can be obtained from the equation on eigenvalues of the operator L_g

$$L_g\varphi_0(x) = \delta\varphi_0(x),$$

that is

$$-\alpha\left[g'\left(g\left(-\frac{x}{\alpha}\right)\right)\cdot\varphi_0\left(-\frac{x}{\alpha}\right) + \varphi_0\left(g\left(-\frac{x}{\alpha}\right)\right)\right] = \delta\varphi_0(x). \qquad (4.34)$$

We shall use only the first member $\varphi_0(0)$ from the power expansion of function $\varphi_0(x)$ for an approximate solution of the Eq. (4.34). In this case we shall obtain the algebraic expression for an estimation of the constant δ

$$\delta = -\alpha[g'(1) + 1]. \qquad (4.35)$$

It is possible to calculate the value $g'(1)$ for a function with a quadratic extremum $(g'(0) = 0,\ g''(0) \neq 0)$, twice differentiating the Eq. (4.15) for a fixed point of the period doubling operator

$$g''(x) = -\frac{1}{\alpha}\left[g''\left(g\left(-\frac{x}{\alpha}\right)\right)\left(g'\left(-\frac{x}{\alpha}\right)\right)^2 + g'\left(g\left(-\frac{x}{\alpha}\right)\right)g''\left(-\frac{x}{\alpha}\right)\right].$$

Substituting $x = 0$ in the last expression, we shall obtain that

$$g'(1) = -\alpha. \qquad (4.36)$$

Thus, (4.35) becomes

$$\delta = \alpha^2 - \alpha. \qquad (4.37)$$

Using the value $\alpha \approx 2.732$ obtained above, we find, that $\delta \approx 4.732$. The obtained value of δ differs from the Feigenbaum's value calculated with greater accuracy $(\delta \approx 4.6692016)$ approximately to 1%.

4.2.4   *Dimension of the Feigenbaum attractor*

We shall consider the important question on dimension of irregular attractor of a one-dimensional mapping which is born at the end of the Feigenbaum cascade of bifurcations. This irregular attractor is named the Feigenbaum attractor. Arbitrary neighbourhood of any its point contains points belonging to some unstable cycle. Therefore the Feigenbaum attractor is, obviously, nondense set of points, and use of the concept of geometrical dimension for finding of its dimension is unsuitable. The use of fractal dimension is most acceptable, probably, in this case (see Sec. 1.4.4). The upper estimate of a fractal dimension d_F of a supercycle of the period 2^n at $n \to \infty$ can be calculated as follows. It required $N(\varepsilon) = 2^n$ segments to cover all points of a supercycle of the period 2^n with segments of length ε. According to estimations [Schuster (1984)], the average minimal length of segments ε_n for covering all points of a supercycle of the period 2^n at $n \to \infty$ is equal to

$$\varepsilon_n = \left(\frac{\alpha+1}{2\alpha^2}\right)^n. \tag{4.38}$$

Substituting these values in the formula (1.9), we obtain that

$$d_F = \frac{\ln 2}{\ln \frac{2\alpha^2}{\alpha+1}} \approx 0.543. \tag{4.39}$$

Thus, the Feigenbaum attractor of a unimodal one-dimensional mapping is, possibly, a fractal, and it has a fractional dimension with value smaller than unit. At the same time the value of dimension presented in (4.39) is only upper estimate. Therefore, as it seems to us, the given question demands an additional studies.

4.3 Sharkovskii Subharmonic Cascade of Bifurcations of Cycles of One-Dimensional Mappings

The Sharkovskii order established in the paper [Sharkovskii (1964)], is a discovery which is not less important, than the Feigenbaum universality, being the evidence of complexity of structure of one-dimensional dynamical systems.

4.3.1 *The Sharkovskii's theorem*

It follows from *the Sharkovskii's theorem*, rediscovered in [Li and Yorke (1975)], that complication of structure of cycles of iterations of one-dimensional unimodal mappings can not come to the end with the Feigenbaum cascade of bifurcations. It can proceed some other, more complex cascade of bifurcations according to the order established in the theorem.

Definition 4.1 Ordering in the set of natural numbers, looking like

$$1 \lhd 2 \lhd 2^2 \lhd 2^3 \lhd \cdots \lhd 2^2 \cdot 7 \lhd 2^2 \cdot 5 \lhd 2^2 \cdot 3 \lhd \cdots$$

$$\cdots \lhd 2 \cdot 7 \lhd 2 \cdot 5 \lhd 2 \cdot 3 \lhd \cdots \lhd 9 \lhd 7 \lhd 5 \lhd 3 \quad (4.40)$$

is called *the Sharkovskii order*.

Theorem 4.1 (The Sharkovskii's theorem). *If a continuous mapping $I \to I$ has a cycle of the period n, then it has also cycles of each period n' such, that $n' \lhd n$ in sense of the Sharkovskii order.*

Corollary 4.1 *If mapping f has a periodic point of the period 3, then it has periodic points of all periods.*

The proof of the Sharkovskii's theorem is based on some statements considered below. In the beginning we shall notice, that if mapping $f \in C^0(I, I)$ has a cycle of the period $n > 1$, than it has also a fixed point by virtue of a continuity. Moreover, if I is a segment, then any continues mapping has always a fixed point in I. Now we shall consistently consider some statements according to which the sequence order of cycles is defined in one-dimensional mappings.

Lemma 4.1 *If mapping $I \to I$ is continuous, then at least one point of some cycle of the period $n' < n$ lies between any two points of a cycle of the period $n > 1$.*

Proof. Let $a < b$ be points of a cycle of the period n, and n_a, n_b be quantities of points of this cycle lying to the left of points a and b accordingly. Obviously, $0 \leq n_a < n_b < n$. There exists n_b various integers s_i, $i = 1, \ldots, n_b$, smaller n and such, that $f^{s_i}(b) < b$, where s_i is a duration of transition in iterations from the point b into one of points of the cycle, located to the left of b. As $n_a < n_b$ there will be a positive integer $s_{i'}$, $1 \leq i' \leq n_b$, such, that $f^{s_{i'}}(b) < b$, $f^{s_{i'}}(a) > a$. Hence, there exists point $x_0 \in (a, b)$, for which $f^{s_{i'}}(x_0) = x_0$ takes place, i.e. x_0 is a point of cycle of the period $n' \leq s_{i'} < n$. $\square$

Lemma 4.2 *If a continuous mapping has a cycle of the period $n > 2$, then this mapping has also a cycle of the period 2.*

Following to Sharkovskii, we shall prove more common statement from which Lemma 4.2 follows, namely: if continuous mapping has a cycle of the period $n > 2$, then this mapping has also a cycle of the smaller period $2 \leq n' < n$.

Proof. If there are no fixed points between any two points of a cycle of the period n, then the given statement follows directly from Lemma 4.1. Now we shall assume, that there are fixed points between two points of a cycle. We shall take any point c, belonging to a cycle and distinct from the least point a and from the greatest point b. We shall assume $f(c) > c$ for definiteness. Let k be the least number of iterations for which the point c passes into the point a, i.e. $f^k(c) = a$, $2 \le k < n$. Under the assumption there are fixed points between points a and c. We shall choose one of them, the nearest to the point c, and we shall designate it as d. As $f(d) = d$ and $f(c) > c$, then the inequality $f(x) > x$ is fair for $\forall x \in (d, c)$. Hence, $f^k(d) = d$ for function $y = f^k(x)$, and, moreover, there exists a neighbourhood of a point d, where $f^k(x) > x$ for $x > d$. But as $f^k(c) = a < c$, then there is point x^* in the interval (d, c) for which $f^k(x^*) = x^*$. Thus, as there are no fixed points in the interval (d, c), then x^* is a periodic point of mapping f of the period $n' \ge 2$, and $n' \le k < n$. $\qquad\square$

Corollary 4.2 *If mapping f has a cycle of the period 2^l, $l \ge 1$, then this mapping has all cycles of periods 2^i, $i = 0, 1, \ldots, l - 1$.*

Corollary 4.3 *If mapping f has a cycle of the period, which is unequal to 2^i, $i = 0, 1, 2, \ldots$, then this mapping has all cycles of periods 2^i, $i = 0, 1, 2, \ldots$.*

We shall apply Lemma 4.2 to mapping $g = f^{2^{k-1}}$ to prove that the given mapping f has a cycle of the period 2^k. Then the periodic point of the mapping f of the period $2^l m$ with odd number $m > 1$, is a periodic point of mapping g of the period $2^{l-k+1} m$ if $k \le l$ and of the period m otherwise. According to Lemma 4.2 the mapping g has a periodic point of the period 2 which is, obviously, a periodic point of the period 2^k for the mapping f.

Thus, the Corollary 4.2 already contains a part of statement of the theorem, concerning cycles of periods 2^i, $i = 0, 1, 2, \ldots$. Corollary 4.3 also contains a part of statement of the theorem. It concludes that if mapping has a cycle, the period of which is not equal to one of the periods 2^i, $i = 0, 1, 2, \ldots$, for example, the period is equal to $n = 3$, then this mapping has at least a countable set of cycles $2^i \lhd n$, $i \ge 0$ among which there are cycles of arbitrary large periods.

For conclusion of the proof of the theorem we shall involve some reasons connected with symbolic dynamics.

Let I' and I'' be two intervals. We shall say, that the interval I' f-covers the interval I'', if $f(I') \supset I''$, i.e. the image of the interval I' contains the interval I''. In that case there will be a subinterval $K \subset I'$, for which $f(K) = I''$ and $f(\partial K) = \partial I''$, i.e. the image of subinterval K precisely coincides with the interval I'', including their borders. The subinterval K is named h-interval if there is no other subinterval $K' \subset K$ with the same properties. It is said, that the interval I' f-covers the interval I'' n times, if there exist n not intersected in pairs h-intervals $K_1, K_2, \ldots, K_n$, $K_j \subset I'$, such, that $f(K_j) = I''$.

Cyclic substitution π, oriented graph and (or) matrix of transition P can be associated with any cycle. If points of cycle B of the period n are ordered in a kind of $x_1^* < x_2^* < \cdots < x_n^*$, $f(x_i^*) = x_{s_i}^*$, where $1 \le s_i \le n$, $i = 1, \ldots, n$, then the cyclic substitution looks like

$$\pi_n = \begin{pmatrix} 1 & 2 & \ldots & n \\ s_1 & s_2 & \ldots & s_n \end{pmatrix}.$$

The given points create a splitting of segment I into intervals $I_s = [x_{s-1}^*, x_s^*]$. Oriented graph of transitions is given by nodes I_1, $I_2, \ldots,$ I_{n-1} and oriented ribs connecting the nodes I_i and I_j, if $f(I_i) \supset I_j$. Thus it is accepted to write $I_i \to I_j$ when an interval I_i f-covers an interval I_j. The number of arrows from I_i to I_j is equal to number of h-intervals $I_{ik} \subset I_i$, for which $f(I_{ik}) = I_j$.

We shall name the graph of transitions as B-graph of a cycle, following to Sharkovskii. The matrix of admissible transitions (points of intervals I_i) is defined as follows

$$p_{ij} = \begin{cases} 0, & \text{if} \quad f(I_i) \not\supset I_j, \\ 1, & \text{if} \quad f(I_i) \supset I_j. \end{cases}$$

For example, the mapping presented in Fig. 4.10a, has a cycle of the period 3, formed by points x_1^*, x_2^*, x_3^*. For this cycle $\pi_3 = \begin{pmatrix} 1 & 2 & 3 \\ 2 & 3 & 1 \end{pmatrix}$, $f(I_1) \supset I_2$, $f(I_2) \supset I_1 \cup I_2$, a matrix of transitions $P = \begin{pmatrix} 0 & 1 \\ 1 & 1 \end{pmatrix}$, and B-graph of the cycle looks like it shown in Fig. 4.10b.

Now we shall consider, how the intervals covering each other and forming the closed path, and the periodic points of a cycle are connected.

Lemma 4.3   *If B-graph contains the closed path $I_{s_0} \to I_{s_1} \to I_{s_2} \to \cdots \to I_{s_{k-1}} \to I_{s_0}$ $(1 \le s_i \le n - 1)$, then there is a periodic point x^* such, that*

$$f^i(x^*) \in I_{s_i}, \quad i = 0, 1, \ldots, k-1, \quad f^k(x^*) = x^*.$$

Moreover, it is possible to take the periodic point x^* of the period k which is the least period of sequence s_0, s_1, ..., s_{k-1}, s_0.

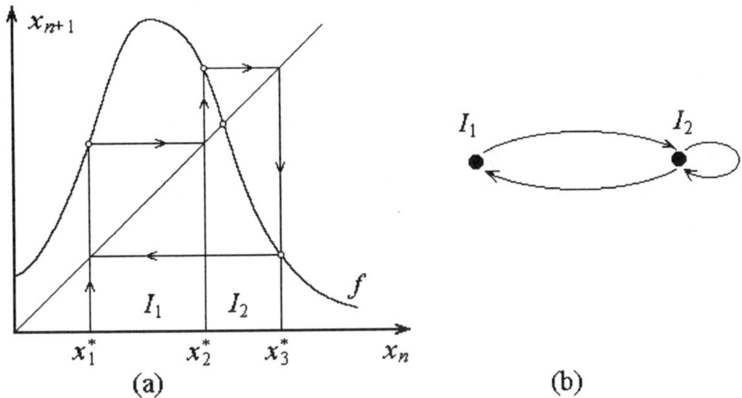

(a)   (b)

Fig. 4.10   The mapping having a cycle of period 3 (a) and its graph of transitions (b).

Proof.   As f^i defines a continuous mapping, then there exists a h-interval, such that $I' \subset I_{s_0}$, $f^i(I') \subset I_{s_i}$, $i = 1, \ldots, k-1$ and $f^k(I') = I_{s_0}$. Hence, there exists a point $x^* \in I'$, for which $f^k(x^*) = x^*$. $\square$

Let us consider now a lemma concerning a mapping having a cycle of the period 3.

Lemma 4.4   *If a continuous mapping has a cycle of the period 3, then it has cycles of all periods.*

Proof. B-graph of cycle $B = \{x_1^*, x_2^*, x_3^*\}$ of the period 3 looks like it is presented in Fig. 4.10. It contains only two intervals. For any n it is always possible to make a periodic sequence in accordance with the B-graph of this cycle

$$I_1 \to \underbrace{I_2 \to \ldots \to I_2}_{n-1} \to I_1 \to \underbrace{I_2 \to \ldots}_{n-1}.$$

According to Lemma 4.3, a cycle of the period n corresponds to this periodic sequence, which points pass intervals I_1, I_2 in the indicated order. $\square$

Let us notice, that one of important properties of the B-graph, having a cycle of the period $n > 2$ is the existence of such natural number as $1 \leq s^* \leq n-1$, that $I_{s^*} \to I_{s^*}$ takes place, i.e. the interval I_{s^*} f-covers itself. Indeed, we shall denote $s^* = \max\{i : f(x_i^*) > x_i^*\}$. As $f(x_n^*) < x_n^*$, then $s^* < n$, and as $f(x_{s^*+1}^*) < x_{s^*+1}^*$, then $f(I_{s^*}) \supset I_{s^*}$ or $I_{s^*} \to I_{s^*}$. Moreover, as each point of the B-graph has an image and a prototype, then for any $1 \le i \le n-1$ there exist i_1 and i_2 ($1 \le i_1, i_2 < n$), such that $I_{i_1} \to I_i \to I_{i_2}$. Hence, the B-graph of a cycle of the period $n > 2$ contains a subgraph $I_{s^*} \to \ldots \to I_s$ for any $1 \le s \le n-1$.

Lemma 4.5 *If a continuous mapping f has a cycle of the odd period $n = 2k + 1$, $k > 1$, and if it has no other cycles of the period $2k - 1$, then B-graph of a cycle of the period n contains a subgraph represented in Fig. 4.11, or obtained from it by replacement I_i on I_{n-i}.*

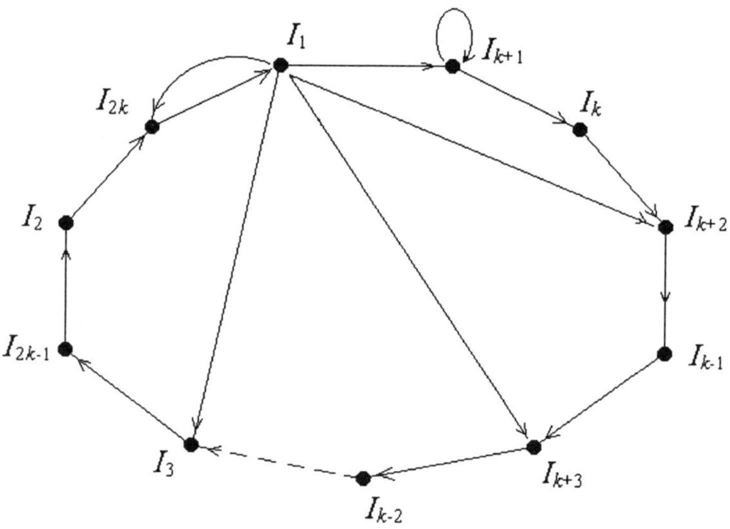

Fig. 4.11   Subgraph of a cycle of the odd period $n = 2k + 1$.

Proof.   The value $s^* = \max\{i : f(x_i^*) > x_i^*\}$ divides all set B of points of a cycle by two subsets: $B^- = \{x^* \in B : x^* \le x_{s^*}^*\}$ and $B^+ = \{x^* \in B : x^* > x_{s^*}^*\}$. As n is an odd number, then B^- and B^+ contain a various number of points of a cycle. Let, for definiteness, the number of points in B^- be more. Then there are points $x^* \in B^-$, for which $f(x^*) \in B^-$. Let $s = \max\{i < s^* : f(x_i^*) \le x_{s^*}^*\}$. As $x_{s+1}^* \in B^-$, then $f(x_{s+1}^*) \ge x_{s^*+1}^*$. Hence, $f(I_s) \supset I_{s^*}$ or $I_s \to I_{s^*}$. On the other hand, as it was marked above, there always exists a subgraph $I_{s^*} \to \ldots \to I_s$. Thus, we obtain a closed graph $I_{s^*} \to \ldots \to I_s \to I_{s^*}$.

It is possible to consider this path as the shortest path. Then its length is equal to n. Otherwise we can apply Lemma 4.3 to one of the closed paths $I_{s^*} \to \ldots \to I_s \to I_{s^*}$ or $I_{s^*} \to \ldots \to I_s \to I_{s^*} \to I_{s^*}$. Then we shall obtain a periodic point of the odd period n', $1 < n' < n$ in an interval I_{s^*} that contradicts to conditions of the present Lemma.

If the path $I_{s^*} \to \ldots \to I_s \to I_{s^*}$ is the shortest path, then each of intervals I_i, $i = 1, \ldots, n-1$, except for I_{s^*}, can be found in it only once, as the length of the path is equal to n. Having designated $s^* = s_1$, and indexes of the intervals following the interval I_{s^*}, through s_2, s_3, $\ldots$, s_{n-1}, we shall obtain a path $I_{s_1} \to I_{s_2} \to \ldots \to I_{s_{n-1}} \to I_{s_1}$. As this path is the shortest path, then B-graph of a cycle does not contain the edges going from interval I_{s_i} into interval I_{s_j} for $j > i+1$, $i = 1, 2, \ldots, n-2$. Let us show now, that up to orientation elements of the B-graph are situated in segment I by the way of $I_{s_{n-1}}, I_{s_{n-3}}, \ldots, I_{s_4}, I_{s_2}, I_{s_1}, I_{s_3}, I_{s_{n-4}}, I_{s_{n-2}}$.

For $n = 3$ we have only two intervals, and the statement about arrangement of intervals in the segment is obvious. We shall assume that $n > 3$ and consider an interval $I_{s_1} = [a, b]$. Then $f(a) \geq b$, $f(b) \leq a$. Both equalities cannot be carried out simultaneously, as in this case the point a will be either a fixed point, or a periodic point of the period 2. But one of them is executed always. Let $f(a) = b$, $f(b) = a_2 < a$ and $I_{s_2} = [a_2, a]$. It is clear, that $f(I_{s_1}) \supset I_{s_1} \cup I_{s_2}$. If $f(a_2) < a$, then $f(I_{s_2}) \supset I_{s_1}$, and we have a subgraph shown in Fig. 4.10b from which existence of a cycle of the period 3 follows. Then existence of cycles of any odd period $n' < n$ follows from Lemma 4.4, that contradicts to conditions of the present Lemma.

Thus, point $f(a_2)$ lies to the right of the point b and $f(I_{s_2}) = I_{s_3}$ is an interval which adjoins to the interval I_{s_1} from the right, unlike interval I_{s_2}, adjoining to I_{s_1} from the left. This corresponds to that for any point $x^* \in B^-$, except for $x^*_{s_{n-1}}$ its image $f(x^*) \in B^+$, and, on the contrary, for any point $x^* \in B^+$, its image $f(x^*) \in B^-$. It means, that the interval I_{s_2} will be adjacent with the interval I_{s_1}, the interval I_{s_3} will be adjacent with the interval $I_{s_1} \cup I_{s_2}$ and will adjoin to the interval I_{s_1}, the interval I_{s_4} will be adjacent with the interval $I_{s_2} \cup I_{s_1} \cup I_{s_3}$ and will adjoin to the interval I_{s_2}, etc. Thus, two arrangements of intervals on the segment I are possible only: $I_{s_{n-1}}, I_{s_{n-3}}, \ldots I_{s_4}, I_{s_2}, I_{s_1}, I_{s_3}, I_{s_{n-4}}, I_{s_{n-2}}$, or in reverse order. In the first case we have $s_{n-1} = 1$, $s_{n-3} = 2, \ldots$, $s_1 = k+1, \ldots$, $s_{n-2} = 2k$, (Fig. 4.12 for $n = 5$), and in the second case $s_{n-1} = 2k$, $s_{n-3} = 2k-1, \ldots$. In both cases $I_1 \to I_{k+1}$ and $I_2 \to I_{2k}$. Hence $I_1 \to I_{k+i}$, $i = 1, \ldots, k$. The proof of the Lemma is completed. $\qquad\square$

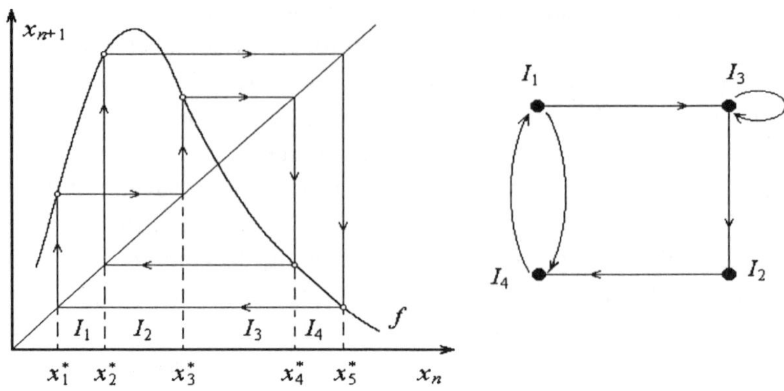

Fig. 4.12   Mapping with a cycle of period 5 and its graph of transitions.

In conclusion of the proof of the theorem it is necessary to prove the following statement.

Lemma 4.6   *If a continuous mapping f has a cycle of the odd period $n > 1$, then it has cycles of any odd period, greater than n, and it also has cycles of any even period.*

Proof.   Let $n = 2k + 1$ be the least odd period, greater than 1. According to Lemma 4.5, B-graph of a cycle of the period n contains a closed path $I_1 \to I_{2k+1-n'/2} \to I_{n'/2} \to \ldots \to I_2 \to I_{2k} \to I_1$ for any even $n' < n$. Applying Lemma 4.3 to this path, we obtain the proof of the statement for a case $n' < n$. If $n' > n$, then Lemma 4.3 should be applied to the closed path $I_{k+1} \to I_k \to I_{k+2} \ldots \to I_1 \to I_{k+1} \ldots \to I_{k+1}$. $\square$

Using Lemma 4.6 we shall prove now the last statement of the theorem.

Lemma 4.7   *If a continuous mapping has a cycle of the period $2^l(2k + 1)$, $k \geq 1$, then it has also cycles of the periods $2^l(2r + 1)$ and $2^{l+1}s$, $r > k$, $s \geq 1$.*

Proof.   Indeed, if a mapping f has a cycle of the period $2^l(2k + 1)$, then a mapping f^{2^l} has a cycle of the period $2k + 1$. It follows from Lemma 4.6 that a mapping f^{2^l} contains a cycle of the period $2s$ for arbitrary $s \geq 1$. Thus, a mapping f has a cycle of the period $2^{l+1}s$. Besides it follows from Lemma 4.6 also, that a mapping f^{2^l} has a cycle of the period $2r + 1$ which points are periodic points of the period $2^l(2r + 1)$ for the mapping f. $\square$

The proof of the Sharkovskii's theorem is completed.

Let us note, that the closedness condition of the B-graph in Lemma 4.3 is essential. Consider, for example, a cycle of the period 4. If to use the Sharkovskii's theorem on existence of cycles, it is possible to affirm only that there are cycles of the periods 2 and 1. However, as it is easy to see, two various cyclic substitutions can be put in conformity to the cycle of the period 4

$$\pi_4^{(1)} = \begin{pmatrix} 1\ 2\ 3\ 4 \\ 3\ 4\ 2\ 1 \end{pmatrix} \quad \text{and} \quad \pi_4^{(2)} = \begin{pmatrix} 1\ 2\ 3\ 4 \\ 2\ 3\ 4\ 1 \end{pmatrix}.$$

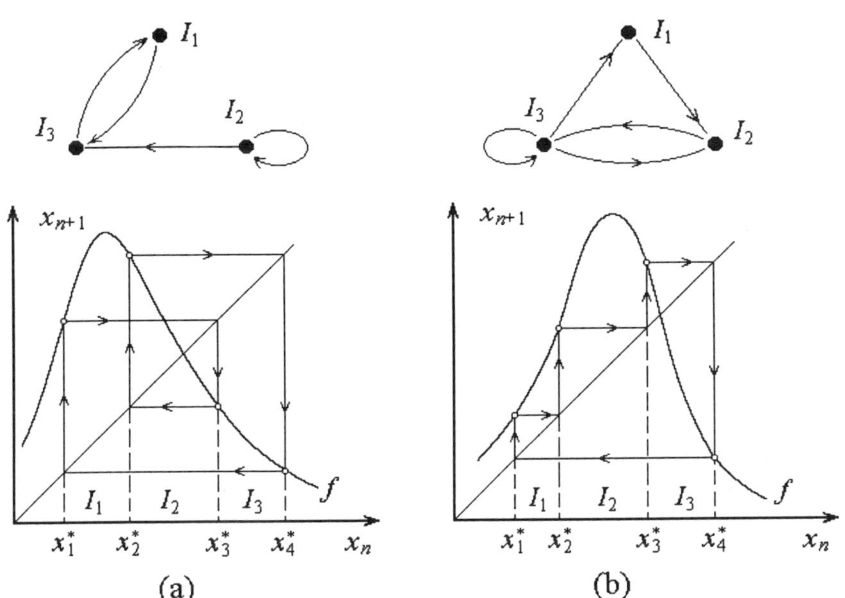

Fig. 4.13   Two mappings with cycles of period 4 and their graphs of transitions.

Substitution $\pi_4^{(1)}$ corresponds to such mapping f, when $f(I_1) \supset I_3$, $f(I_3) \supset I_1$, $f(I_2) \supset I_2 \cup I_3$. So, B-graph of the cycle of the period 4 has no closed path of length, greater than two (Fig. 4.13a). In this case, only cycles of the periods 2 and 1 are really possible. If a mapping f has a cycle of the period 4, corresponding to the cyclic substitution $\pi_4^{(2)}$, then $f(I_1) \supset I_2$, $f(I_2) \supset I_3$, $f(I_3) \supset I_3 \cup I_2 \cup I_1$, and B-graph of the cycle has the closed path of the length 3 (Fig. 4.13b). In this case, according to Lemma 4.3, the mapping f has a cycle of the period 3 and, hence, it has cycles of arbitrary periods.

4.3.2   *Behind the Feigenbaum cascade*

In the Sharkovskii's theorem nothing is spoken about stability of cycles simultaneously existing in one-dimensional continuous mappings in accordance with the order (4.40). However, numerous examples including the logistic mapping (4.2), show, that the birth of various stable cycles of one-dimensional unimodal mappings $f(x, \mu)$ occurs exactly in the sequence set by order (4.40) at change of values of some bifurcation parameter μ. First the period doubling cascade of bifurcations of the original stable simple cycle occurs at $\mu_0 < \mu < \mu_\infty$. Thus mapping f has a unique stable cycle of the period 2^n and a family of unstable cycles of all periods 2^i, $i = 0, 1, \ldots, n-1$ when μ belongs to an interval $\mu_n < \mu < \mu_{n+1}$. At the value $\mu = \mu_\infty$ a mapping $f(x, \mu)$ has a nonperiodic semistable trajectory, the singular Feigenbaum attractor. Points of a countable set of unstable cycles of all periods 2^n, $n = 0, 1, \ldots$, lie in any neighbourhood of arbitrary point of this trajectory. Thus, the Feigenbaum cascade of bifurcations occurs according to the Sharkovskii order, and it is an initial stage of the Sharkovskii subharmonic cascade of bifurcations described by this order.

All interval of change of parameter values μ at $\mu > \mu_\infty$ consists of infinite number of subintervals (windows) of periodicity $\mu_\alpha < \mu < \mu_{\alpha+1}$, separated by isolated parameter values μ, at which a mapping has singular irregular attractors, that is semistable nonperiodic trajectories. In each window of periodicity a basic cycle of the period k from the Sharkovskii order is stable or one of cycles of the period $2^l \cdot k$, $l = 1, 2, \ldots$, is stable from the cascade of period doubling bifurcations of original cycle. At that a mapping has also unstable cycles of all periods m, satisfying the condition $m \vartriangleleft 2^l k$. In the case of the logistic mapping (4.2), a cycle of the period 3, born at the value $\mu \approx 3.828$, has the largest window of periodicity (Fig. 4.14).

Complexity of irregular nonperiodic attractors increases with growth of parameter values for $\mu > \mu_\infty$, because a quantity of unstable cycles presented in system increases. Moreover, as numerical experiments show, a bifurcation of coherence occurs after each cascade of the period doubling bifurcations of some cycle, i.e. a joining of various parts of a previous irregular attractor takes place. There are bases to assume, that only the set of points of the most elementary singular attractor, the Feigenbaum attractor is a non-dense set in a segment. Points of all more complex singular subharmonic attractors are presumably dense in some even as much as small intervals. Cumulative length of these intervals grows with growth of parameter values μ, covering finally the whole segment. Let us emphasize, however, that questions of power, dimension and measure of sets of points belonging to various subharmonic singular attractors of one-dimensional mappings are at present investigated insufficiently. Some results connected with these problems can be found, for example, in [Sinai (1995); Jakobson (1981)] where it is shown, in particular, that dynamics of logistic mapping (4.2) possesses the properties of ergodicity and mixing for $\mu = 4$.

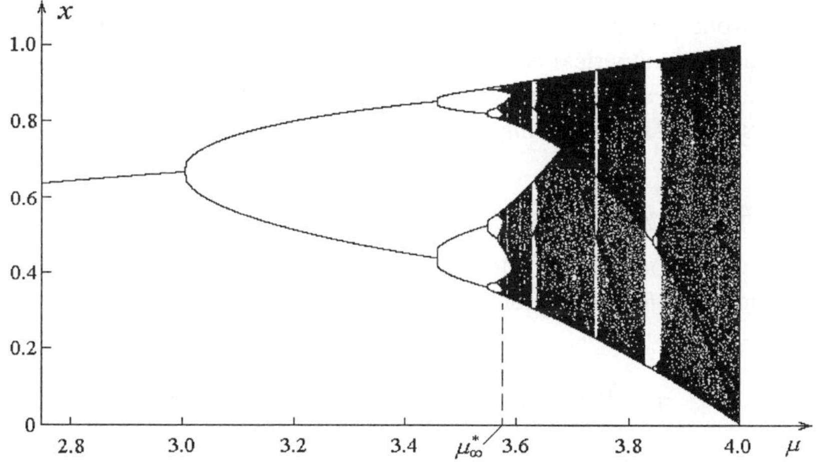

Fig. 4.14   The bifurcation diagram of the logistic mapping for $\mu \leq 4$.

4.4 Dynamical Chaos in Two-Dimensional Non-Autonomous Systems of Differential Equations

It was shown in previous sections of Chapter 4 that transition to chaos under the variation of a system parameter in one-dimensional continuous unimodal mappings occurs in accordance with a unique universal scenario. This scenario begins with a cascade of the Feigenbaum period doubling bifurcations of stable cycles and then continues by a subharmonic cascade of Sharkovskii bifurcations of a generation of stable cycles of an arbitrary period up to a cycle of the period 3.

In the present section, we consider two-dimensional non-autonomous systems of ordinary differential equations and present the Magnitskii's theory of rotor type singular points for such systems. This theory establishes a bridge between one-dimensional mappings and differential equations with rotor type singular points. It permits one to apply Feigenbaum–Sharkovskii theory to description of dynamical chaos in differential equations. We show that two-dimensional non-autonomous systems of differential equations with rotor type singular points have the same above-mentioned universal scenario of transition to chaos. Moreover, we show that in the case of differential equations this universal scenario continues by a homoclinic cascade of Magnitskii bifurcations of stable cycles converging to a homoclinic contour and possibly more complex cascades. All analytic results are justified by numerical calculations for corresponding examples of both model and classical two-dimensional non-autonomous systems of differential equations, such as the Duffing–Holmes and the Matie equations.

4.4.1   *Rotor type singular points*

Consider a smooth family (depending on a scalar parameter μ) of two-dimensional real nonlinear non-autonomous systems of ordinary differential equations

$$\dot{u} = D(t,\mu)u(t) + H(u,t,\mu), \ \ H(0,t,\mu) \equiv 0, \tag{4.41}$$

with a singular point $O(0,0)$, which have the $T(\mu)$-periodic matrix $D(t,\mu)$ of the leading linear part at the singular point and the nonlinear part $H(u,t,\mu)$ whose expansion in the components of the vector $u(t)$ at the singular point starts from second-order terms and has bounded coefficients. It follows from the Floquet theory that the fundamental matrix solution of the linear part

$$\dot{u} = D(t,\mu)u(t) \tag{4.42}$$

of system (4.41) can be represented in the form $U(t,\mu) = P(t,\mu)e^{B(\mu)t}$, where $P(t,\mu)$ is some T-periodic complex matrix and $B(\mu)$ is some constant complex matrix whose eigenvalues are the Floquet exponents of the linear system (4.42) with T-periodic coefficients. It is important that system (4.42) can have various complex but not complex-conjugate Floquet exponents $\alpha_1(\mu)$ and $\alpha_2(\mu)$. Indeed, it also follows from the Floquet theory that the fundamental matrix solution of the linear system (4.42) can be represented in the form of the product

$$U(t,\mu) = R(t,\mu)e^{E(\mu)t} = R(t,\mu)\,\mathrm{diag}(e^{\beta_1(\mu)t},\ e^{\beta_2(\mu)t}) \tag{4.43}$$

of a real $2T$-periodic matrix $R(t, \mu)$ and a diagonal real matrix $e^{E(\mu)t}$. Then, obviously, $\Im\{\alpha_1(\mu)\} = \Im\{\alpha_2(\mu)\}$, $\Re\{\alpha_1(\mu)\} = \beta_1(\mu)$, $\Re\{\alpha_2(\mu)\} = \beta_2(\mu)$, and $R(t, \mu) = P(t, \mu)e^{i\Im\{\alpha_1(\mu)\}t}$, i.e. the real parts of Floquet exponents of the linear system (4.42) are different and the imaginary parts coincide.

Definition 4.2 [Magnitskii (2004)] The singular point of a two-dimensional non-autonomous real system (4.41) with a periodic leading linear part that has complex Floquet exponents with equal imaginary parts and different real parts is referred to as *a rotor*.

The above-considered singular point has no analogs among singular points of two-dimensional autonomous real systems. Its name reflects the permanent rotation of trajectories of the system around the singular point; however, the mechanism of that rotation differs from the rotation mechanism of trajectories of an autonomous system around a focus-type singular point. In the latter case, a singular point has either a two-dimensional stable manifold or a two-dimensional unstable manifold. The rotor can have a one-dimensional stable manifold and a one-dimensional unstable manifold combining some features of focus and saddle.

The canonical form of a rotor type singular point was found in [Magnitskii and Sidorov (2004c)]. That is a real system

$$\dot{u}_1 = \frac{\beta_1 + \beta_2 + (\beta_1 - \beta_2)\cos\omega t}{2}u_1 + \frac{(\beta_1 - \beta_2)\sin\omega t - \omega}{2}u_2,$$
$$\dot{u}_2 = \frac{(\beta_1 - \beta_2)\sin\omega t + \omega}{2}u_1 + \frac{\beta_1 + \beta_2 - (\beta_1 - \beta_2)\cos\omega t}{2}u_2, \tag{4.44}$$

with $2\pi/\omega$-periodic coefficients. Here β_1 and β_2 are arbitrary real constants. We perform the change of variables

$$u(t) = R(t)v(t) = \begin{pmatrix} \cos\dfrac{\omega t}{2} & -\sin\dfrac{\omega t}{2} \\ \sin\dfrac{\omega t}{2} & \cos\dfrac{\omega t}{2} \end{pmatrix} v(t). \tag{4.45}$$

Then $\dot{v}(t) = Ev(t)$, where the matrix

$$E = R^{-1}(t) \begin{pmatrix} \dfrac{\beta_1 + \beta_2 + (\beta_1 - \beta_2)\cos\omega t}{2} & \dfrac{(\beta_1 - \beta_2)\cos\omega t - \omega}{2} \\ \dfrac{(\beta_1 - \beta_2)\cos\omega t + \omega}{2} & \dfrac{\beta_1 + \beta_2 - (\beta_1 - \beta_2)\cos\omega t}{2} \end{pmatrix} R(t) - R^{-1}(t)\dot{R}(t)$$

is a constant diagonal real matrix: $E = \mathrm{diag}(\beta_1, \beta_2)$. Therefore, the fundamental matrix $U(t)$ of system (4.44) admits a real representation of the form (4.43):

$$U(t) = R(t)e^{Et} = \begin{pmatrix} \cos\dfrac{\omega t}{2} & -\sin\dfrac{\omega t}{2} \\ \sin\dfrac{\omega t}{2} & \cos\dfrac{\omega t}{2} \end{pmatrix} \begin{pmatrix} e^{\beta_1(\mu)t} & 0 \\ 0 & e^{\beta_2(\mu)t} \end{pmatrix} \qquad (4.46)$$

which is a representation of solution of the linear system (4.44) with a $2\pi/\omega$-periodic matrix in the form of the product of a $4\pi/\omega$-periodic real matrix of double period by the real diagonal matrix $exp(Et)$. To pass to the Floquet representation, we rewrite the matrix (4.46) in the form of the product of a $2\pi/\omega$-periodic complex matrix by the complex matrix e^{Bt}:

$$U(t) = \begin{pmatrix} \dfrac{1 + \exp(-i\omega t)}{2} & -\dfrac{1 - \exp(-i\omega t)}{2i} \\ \dfrac{1 - \exp(-i\omega t)}{2i} & \dfrac{1 + \exp(-i\omega t)}{2} \end{pmatrix} \begin{pmatrix} e^{(\beta_1 + i\omega/2)t} & 0 \\ 0 & e^{(\beta_2 + i\omega/2)t} \end{pmatrix}.$$

The diagonal complex entries $\alpha_1 = \beta_1 + i\omega/2$ and $\alpha_2 = \beta_2 + i\omega/2$ of the matrix B are the Floquet exponents of the original linear non-autonomous system (4.44), and their imaginary parts coincide. Consequently, the zero singular point of system (4.44) is a rotor; moreover, since β_1 and β_2 are arbitrary quantities, it follows that system (4.44) can be treated as a canonical form of a rotor type singular point. The multipliers of the rotor, that is, the real multipliers of system (4.44) corresponding to the Floquet exponents, are equal to

$$\lambda_j = \exp\left(\left(\beta_j + \dfrac{i\omega}{2}\right)\dfrac{2\pi}{\omega}\right) = \exp\left(\dfrac{2\pi\beta_j}{\omega} + i\pi\right) = -\exp\left(\dfrac{2\pi\beta_j}{\omega}\right), \quad j = 1, 2.$$

Let us return to the nonlinear system (4.41). Since the change of variables $u(t) = R(t, \mu)v(t)$ reduces system (4.41) to a system with a constant diagonal matrix $B(\mu)$ of the leading linear part, it follows from the Lyapunov theorem on stability in the first approximation that if the real parts of the Floquet exponents of the linear system (4.42) are negative, then the rotor is stable (asymptotically stable); otherwise (if one of the real parts is positive), the rotor is unstable and has a one-dimensional stable manifold and a one-dimensional unstable manifold. Multipliers of a stable rotor obviously lie on a negative part of the real axis inside the unit disk.

Note that if a rotor type singular point of system (4.41) loses stability, then a $2T$-periodic stable solution can appear around the rotor as a result of a bifurcation related to the crossing of an imaginary axis by one of Floquet exponents from left to right. That is equivalent to crossing of the unit circle at the point -1 by some of rotor's multipliers lying inside the unit disk before the bifurcation. In this case, the second Floquet exponent remains in the left half-plane.

4.4.2 *Scenario of transition to chaos*

Let us show that two-dimensional nonlinear non-autonomous dissipative systems of ordinary differential equations with rotor type singular points can have arbitrarily complicated chaotic dynamics on the plane. And the mechanism of such dynamics is the same as in one-dimensional continuous mappings. Suppose that a rotor type zero singular point $O(0,0)$ of the family of two-dimensional non-autonomous systems (4.41) is asymptotically stable for all $\mu < 0$. Let for $\mu = 0$ one of its multipliers crosses the unit circle at the point -1, and the bifurcation of generation of $2T$-periodic stable solution without self-intersections occurs in accordance with the representation (4.43). Since, in the plane (u_1, u_2), the trajectory of the solution $u(t)$ of system (4.41) rotates around the rotor O, we can define a monotone decreasing continuous mapping f of a segment of the one-dimensional line (for example, the segment $c \leq u_1 \leq d$ of the line $u_2 = 0$ such that $c < 0 < d$) into itself for a half-turn around the rotor O. Obviously, the one-dimensional mapping $f(u_1)$ has the unstable fixed point $u_1 = 0$ and the stable cycle (c, d) corresponding to the stable cycle of double period of system (4.41), which is generated in the plane (u_1, u_2) and has no self-intersections.

Suppose that the length of the interval (c, d) grows with the parameter $\mu > 0$, and, starting with some value of the parameter μ, trajectories of the two-dimensional non-autonomous system of differential equations (4.41) become self-intersecting and twist around its stable cycle of double period. In terms of the one-dimensional mapping $f(u_1)$, this corresponds to the appearance of a maximum point on its graph in the domain of $u_1 < 0$, which leads to the appearance of a double-valued inverse mapping $f^{-1}(u_1)$. Then the following assertion is valid.

Theorem 4.2 *First stages of a scenario of transition to chaos in two-dimensional non-autonomous systems of ordinary differential equations with rotors coincide with the stages of scenario of transition to chaos for iterations of a continuous self-mapping of the unit segment with a double-*

valued inverse mapping. A cascade of Feigenbaum period doubling bifurcations of the original $2T$-periodic stable limit cycle is first realized, and then a subharmonic cascade of bifurcations of stable limit cycles with arbitrary period takes place in accordance with the Sharkovskii order

$$1 \lhd 2 \lhd 2^2 \lhd 2^3 \cdots \lhd 2^2 \cdot 7 \lhd 2^2 \cdot 5 \lhd 2^2 \cdot 3 \cdots$$
$$\cdots \lhd 2 \cdot 7 \lhd 2 \cdot 5 \lhd 2 \cdot 3 \cdots \lhd 9 \lhd 7 \lhd 5 \lhd 3. \quad (4.47)$$

Proof. Validity of the desired assertion follows from the above-represented constructions of the continuous one-dimensional mapping $f(u_1)$ of a segment into itself, which has a double-valued inverse mapping $f^{-1}(u_1)$, and from the results of Feigenbaum and Sharkovskii on iterations of such mappings (see Sec. 4.2 and Sec. 4.3).

In this case, a periodic or nonperiodic trajectory of the one-dimensional mapping $f(u_1)$ uniquely corresponds to a periodic or nonperiodic trajectory of the system (4.41) lying in the two-dimensional plane of variables (u_1, u_2). All stable cycles are induced either as a result of bifurcations of double period for previous stable cycles of the cascade or as a result of bifurcations of singular attractors and then undergo a cascade of bifurcations of double period and become unstable. Unstable cycles do not vanish but remain in the system. The ordering sign $\lhd$ occurring in (4.47) implies that existence of a cycle of period k results in existence of all cycles of the period n as $n \lhd k$. Moreover, there can be several such cycles. So, if the system (4.41) has a stable limit cycle of period 3, then it has also all unstable cycles of all periods in accordance with the Sharkovskii order (4.47). $\qquad\square$

It follows from the Theorem 4.2 that there exist infinitely many intervals of values of the parameter μ for which the family of systems (4.41) has regular attractors (asymptotically orbitally stable periodic trajectories, even of large period). Any irregular attractor of the family of systems (4.41) with rotor type singular point is a singular attractor, as it is defined in Sec. 1.4.3, i.e. it is a closure of nonperiodic semistable trajectory. The family of systems (4.41) has singular attractors at infinitely many accumulation points of various infinite subcascades of period doubling bifurcations of various cycles. Obviously, the simplest singular attractor is the Feigenbaum attractor, the first nonperiodic attractor existing in the family of systems (4.41) for $\mu = \mu_\infty$, where the value of μ_∞ corresponds to the limit of the sequence of values μ, for which the period doubling bifurcations of the original cycle take place.

Remark 4.1 *Obviously, the above-defined one-dimensional mapping of the segment $c \le u_1 \le d$ of the line $u_2 = 0$ into itself is not the unique mapping that can be defined on the basis of trajectories of the system (4.41) rotating around the rotor O. The role of the line $u_2 = 0$ can be played by any line passing through the rotor. In this case, inverse mapping, as well as direct mapping, can be double-valued or even multivalued.*

Examples show that the Sharkovskii subharmonic bifurcation cascade does not exhaust the entire complexity of transition to chaos in two-dimensional non-autonomous systems of ordinary differential equations. It can be continued at least by the Magnitskii homoclinic cascade of bifurcations of stable cycles which tend to a rotor homoclinic separatrix loop. After generation of the period 3 cycle and a cascade of its period doubling bifurcations that terminates the subharmonic cascade, for a further growth of the parameter μ, turns of regular and singular attractors approach the rotor O, which, in terms of mapping $f(u_1)$, implies that $f(c)$ tends to zero. But $f(u_1)$ cannot intersect the line $u_2 = 0$ at any point other than $\mu = 0$. Therefore, there exists a value of parameter μ_c such that a third branch appears in the mapping $f^{-1}(u_1)$ for $\mu > \mu_c$. This point corresponds to the rotor homoclinic separatrix loop. Further bifurcations should be described by the theory of one-dimensional mappings with multivalued inverse mappings. It means that *the Feigenbaum–Sharkovskii–Magnitskii theory* should be supplemented and expanded by the theory of one-dimensional continuous mappings with multivalued inverse mappings.

As an example, we consider the simplest two-dimensional non-autonomous system of the form (4.41) with periodic leading linear part of the form (4.44) with $\beta_1(\mu) = 2\mu$ and $\beta_2(\mu) = 2\mu - 4$:

$$\begin{aligned}
\dot{u}_1 &= 2(\mu - 1 + \cos \omega t)u_1 + (2\sin \omega t - \omega/2)u_2 - u_2^2, \\
\dot{u}_2 &= (2\sin \omega t + \omega/2)u_1 + 2(\mu - 1 - \cos \omega t)u_2.
\end{aligned} \tag{4.48}$$

Let us point out some solutions from the cascade of period doubling bifurcations, which are observed in the numerical integration of system (4.48) in the course of variation of the parameter μ, for $\omega = 4$. The original simple cycle preserves the stability up to the value of $\mu \approx 0.0972$. A cycle of period 2 is stable in the range $\mu \in (0.0972, 0.1105)$, a cycle of period 4 is observed for $\mu = 0.112$, a cycle of period 8 is observed for $\mu = 0.11363$, a cycle of period 16 is observed for $\mu = 0.11405$, and so on. The cascade of period doubling bifurcations is completed by generation of the Feigenbaum attractor for $\mu \approx 0.1143$.

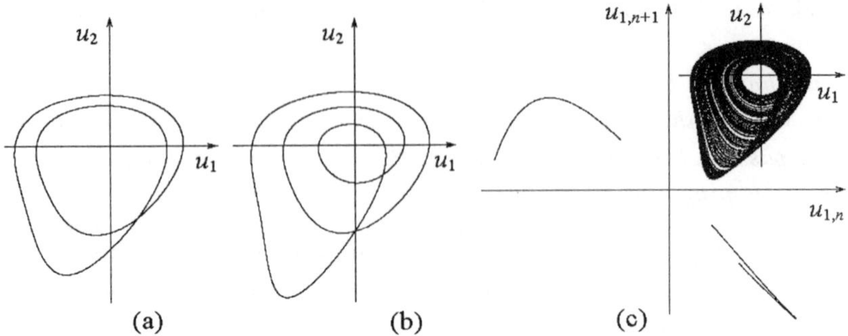

Fig. 4.15   The cycle of period 2 (a) for $\mu = 0.0975$; the cycle of period 3 (b) for $\mu = 0.1225$; the mapping $f(u_1)$ and the Sharkovskii attractor (c) of the non-autonomous two-dimensional system (4.48) for $\mu = 0.1245$.

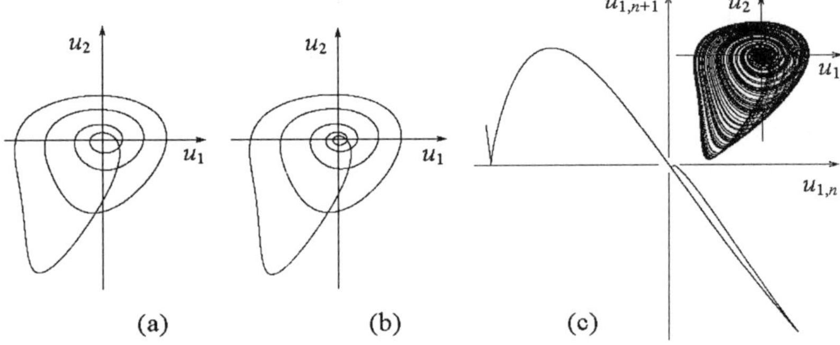

Fig. 4.16   The cycle C_4 (a) for $\mu = 0.12776$; the cycle C_5 (b) for $\mu = 0.12976$; the mapping $f(u_1)$ and the singular attractor (c) of the non-autonomous two-dimensional system (4.48) for $\mu = 0.136$.

Next, stable periodic solutions arranged in accordance with the Sharkovskii ordering are generated in the system for further growth of the parameter μ. Thus we have a cycle of period $12 = 3 \cdot 2^2$ for $\mu = 0.11451$, a cycle of period $6 = 3 \cdot 2$ for $\mu = 0.1159$, a cycle of period 7 for $\mu = 0.11794$, a cycle of period 5 for $\mu = 0.11908$, and a cycle of period 3 for $\mu \approx 0.1222$. The period of this cycle doubles for $\mu = 0.12282$, and further growth of the parameter μ, leads to appearance of a Feigenbaum attractor on a cycle of period 3. A Sharkovskii attractor is formed for the value $\mu \approx 0.1245$ (see Fig. 4.15). The further growth of the parameter μ leads to appearance of periodic solutions of a homoclinic cascade: one observes the cycle C_4 for $\mu = 0.127795$, the cycle C_5 for $\mu = 0.12976$, the cycle C_6 for

$\mu = 0.13055$, and so on up to the appearance of a homoclinic contour for the value $\mu \approx 0.13115$. For large values of the parameter μ, the mapping $f^{-1}(u_1)$ becomes triple-valued (see Fig. 4.16).

4.4.3 Dynamical chaos in some classical two-dimensional non-autonomous systems

In the present section we shall consider some classical nonlinear non-autonomous differential equations such as Duffing–Holmes equation, Matie equation and others. We shall show, that the uniform universal scenario of transition to dynamical chaos through the cascade of Feigenbaum–Sharkovskii–Magnitskii bifurcations is realized in all these equations. The theory of this universal scenario is stated above in Chapter 4.

4.4.3.1 *The Duffing–Holmes equation*

The ordinary differential equation of the second order

$$\ddot{x} + k\dot{x} - x + \mu x^3 = f_0 \cos \Omega t, \tag{4.49}$$

with some constants $k > 0$, μ, f_0, and Ω is called as the Duffing–Holmes equation [Guckenheimer and Holmes (1983)]. It is obtained at modification of the Duffing equation

$$\ddot{x} + k\dot{x} + \omega^2 x + \mu x^3 = f_0 \cos \Omega t, \tag{4.50}$$

in which the restoring force $f(x)$ is a monotonous nonlinear function $f(x) = -\omega^2 x - \mu x^3$. Cubic nonlinearity in the Duffing equation essentially expands a spectrum of its periodic solutions in comparison with the corresponding linear equation. In particular, harmonic fluctuations $x = A(\omega) \cos \Omega t$ with an amplitude depending on the frequency are possible in the Eq. (4.50). Some kinds of fluctuations of various amplitudes take place in the Eq. (4.50) for some values of frequency, and also subharmonic fluctuations with frequencies Ω/n can be realized in the Duffing equation for $k = 0$, where n is a natural number. However, there are no solutions with chaotic fluctuations in the classical Duffing equation.

Change of function of a restoring force $f(x)$ from monotonic on non-monotonic $f(x) = x - \mu x^3$ leads to appearance of chaotic fluctuations. We shall consider the scenario of transition to chaos for one of periodic solutions of the Eq. (4.49). We shall rewrite the Eq. (4.49) in the form of two-dimensional non-autonomous system

$$\dot{x} = y,$$
$$\dot{y} = x - ky - \mu x^3 + f_0 \cos \Omega t. \tag{4.51}$$

At the fixed values of parameters $k = 0.1$, $f_0 = 2$, $\Omega = 1$ a periodic solution, a singular limit cycle exists in the parameter values domain $\mu \in (2.455, 4.235)$. A kind of this cycle in the phase space is shown in Fig. 4.17. At the value $\mu \approx 4.235$ the period doubling bifurcation of the original singular cycle occurs (see Fig. 4.17). It gives rise to the Feigenbaum cascade of the period doubling bifurcations of stable cycles.

So, a cycle of period 4 is observed in system (4.51) for the parameter values $\mu \in [5.59, 5.759]$, a cycle of period 8 for $\mu = 5.78$, a cycle of period 16 for $\mu = 5.801$ *etc.* The singular Feigenbaum attractor is generated in the system for $\mu = \mu_\infty \approx 5.808$ (Fig. 4.17c). Further numerical research shows, that other stable cycles appear in the system (4.51) for $\mu > \mu_\infty$ in accordance with the Sharkovskii order. In particular one can observe a cycle of period $22 = 2 \cdot 11$ for $\mu = 5.813$, a cycle of period 12 for $\mu = 5.828$, a cycle of period $6 = 2 \cdot 3$ (Fig. 4.17d) for $\mu \in [5.9, 5.933]$. Computer calculations show that the subharmonic cascade of bifurcations is incomplete in the system (4.51) and, therefore, an irregular attractor represented in (Fig. 4.17e) is incomplete subharmonic singular attractor.

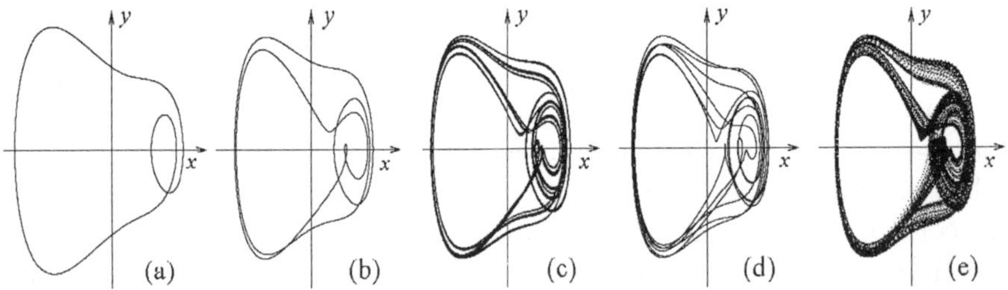

Fig. 4.17   Original singular cycle (a), a cycle of period 2(b) for $\mu = 5.0$, Feigenbaum attractor (c) for $\mu = 5.808$, a cycle of the period 6 of the subharmonic cascade (d) for $\mu = 5.95$ and a singular subharmonic attractor (e) for $\mu = 6.053$ in non-autonomous two-dimensional system (4.51).

In conclusion we shall notice, that the system (4.51) does not vary at change of signs of variables x and y. Hence, alongside with solutions shown in Fig. 4.17, there exist other solutions of the system (4.51) which are symmetric to solutions shown in Fig. 4.17 relatively to the origin.

4.4.3.2 *The Matie equation*

Unlike the Duffing equation, the Matie equation

$$\ddot{x} + (\delta + \varepsilon \cos \omega t)x = 0 \tag{4.52}$$

has periodic coefficients. Modified Matie equation

$$\ddot{x} + \mu\dot{x} + (\delta + \varepsilon \cos \omega t)x + \alpha x^3 = 0 \tag{4.53}$$

differs from the equation (4.52) by presence of a friction factor ($\mu > 0$) and nonlinearity. In the oscillation theory, solutions of the equation (4.53) are investigated at small values of parameters μ, ε, α. It has been shown that presence of nonlinearity leads to restriction of amplitude of oscillations, and a viscous friction stabilizes the system in the sense that at increasing of the parameter values μ the area of space of parameters in which an equilibrium point is asymptotically stable also increases. At the same time the area of existence of oscillatory mode decreases, and oscillations are absent in general for $\mu \geq 0.5$.

We investigated numerically the behavior of solutions of the Eq. (4.53) presented in the form of non-autonomous two-dimensional system

$$\begin{aligned} \dot{x} &= y, \\ \dot{y} &= -(\delta + \varepsilon \cos \omega t)x - \mu y - \alpha x^3. \end{aligned} \tag{4.54}$$

We found that for the fixed values of parameters $\delta = 5$, $\varepsilon = 14$, $\omega = 2$, $\alpha = 1$, the system (4.54) has asymptotically stable zero solution for values of the parameter $\mu > 2.46$. Then a stable singular cycle with the frequency of $\omega/2$ is generated in the system (4.54) at $\mu \approx 2.46$. This singular cycle generates the Feigenbaum period doubling cascade of bifurcations. In particular, one can observe a cycle of period 2 for $\mu \in [1.1425, 1.3375]$, a cycle of period 4 for $\mu = 1.130$, a cycle of period 8 for $\mu = 1.1184$, and a cycle of period 16 for $\mu = 1.1156$ *etc.* The period doubling cascade comes to the end with a formation of Feigenbaum attractor at parameter value $\mu \approx 1.140$ (Fig. 4.18).

At further reduction of a friction factor μ, the stable periodic solutions appear in the system (4.54) in accordance with the Sharkovskii order. For example, a stable cycle of period 9 exists for the value $\mu = 1.08395$, a stable cycle of period 7 exists for $\mu = 1.08005$ and a stable cycle of period 5 exists for $\mu = 1.067$. Existence of a stable cycle of period 3 for the value $\mu = 1.0588$ finishes the Sharkovskii complete subharmonic cascade of bifurcations. We shall note, that the cycle of period 3 generates in this system its own complete subharmonic cascade of bifurcations that is proved by the existence of a cycle of period 2 for this cycle at $\mu = 1.055$ and a cycle of period 3 for this cycle at $\mu = 1.051$. A complete subharmonic singular attractor takes place in the system (4.54) for $\mu \approx 1.050$ (Fig. 4.18).

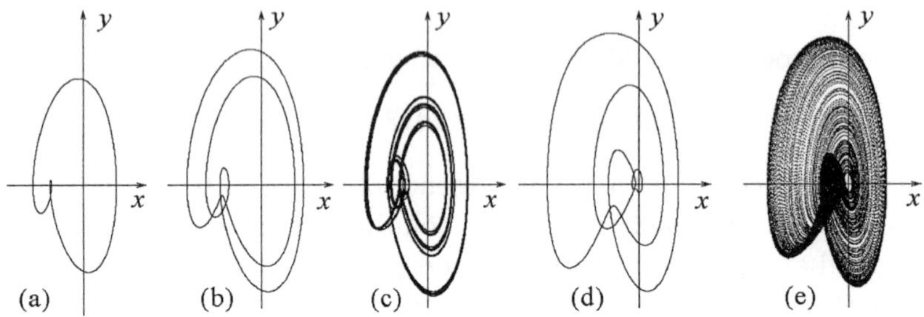

Fig. 4.18   Singular cycle (a) for $\mu = 2$, a cycle of period 2(b) for $\mu = 1.2$, Feigenbaum attractor (c) for $\mu = 1.140$, a cycle of period 3 for $\mu = 1.0588$ and a complete subharmonic attractor for $\mu = 1.050$ in non-autonomous two-dimensional Matie system (4.54).

A kind of the system (4.54) does not change with changing of signs of variables x and y. Therefore there are some other solutions of the system (4.54) which are symmetric to the considered above solutions relatively to the origin.

4.4.3.3   *The Croquette equation*

Non-autonomous two-dimensional system

$$\dot{x} = y,$$
$$\dot{y} = -\mu y - \alpha \sin x - \beta \sin(x - \omega t),$$

$$(4.55)$$

presented in [Berger *et al.* (1984)], models rotatory oscillations of a magnet in an external magnetic field at presence of friction. The parameter μ defines a value of friction at movement in an environment. The system (4.55) has been intended for research of dependence of behavior of a nonlinear parametrical pendulum on the value of friction. Therefore we shall also numerically investigate dependence of solutions of the system (4.55) on the parameter values μ at the fixed values of other parameters $\alpha = 1.15$, $\beta = 1$, $\omega = 1$.

The system (4.55) is dissipative for $\mu > 0$ and besides there are parametrical oscillations with a frequency ω in this system for $\mu \gg 1$. The frequency of such oscillations is equal to the frequency of driving force. At the indicated above fixed values of the parameters α, β and ω, the period of these oscillations is constant with reduction of values of the parameter μ up to the value $\mu \approx 0.625$ at which the period doubling bifurcation of oscillations occurs. Then the Feigenbaum cascade of the period doubling bifurcations of stable cycles takes place for smaller values of μ. So, a cycle of period 4 is observed for the value $\mu = 0.5622$, a cycle of period 8 is observed for $\mu = 0.5548$, and a cycle of period 16 is observed for $\mu = 0.5527$. The cascade of the period doubling bifurcations comes to the end with formation of the Feigenbaum singular attractor for the value $\mu \approx 0.5518$ (Fig. 4.19).

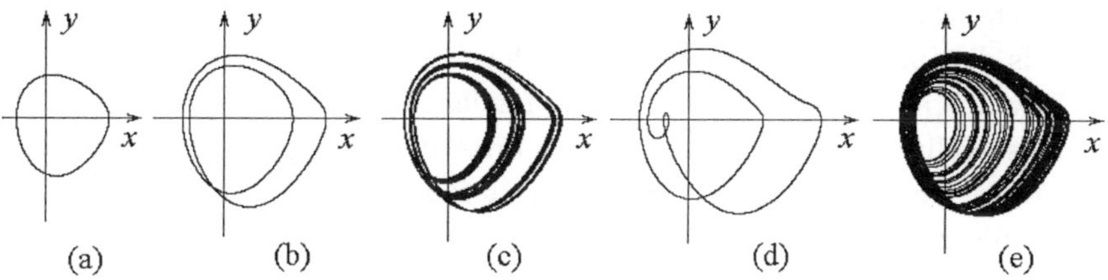

Fig. 4.19 A singular cycle (a) for $\mu = 1$, a cycle of the double period (b) for $\mu = 0.6$, the Feigenbaum attractor (c) for $\mu = 0.5518$, a cycle of period 3 for $\mu = 0.515$ (d) and a complete subharmonic singular attractor for $\mu = 0.510$ (e) in the system (4.55).

The cascade of the period doubling bifurcations is followed by the subharmonic cascade of bifurcations of stable cycles of arbitrary period according to the Sharkovskii order. One can observe a cycle of period $6 = 2 \cdot 3$ for $\mu = 0.54525$, a cycle of period 9 for $\mu = 0.53855$, a cycle of period 5 for $\mu = 0.5334$ and a cycle of period 3 for $\mu = 0.515$ (Fig. 4.19).

4.4.3.4 *The Krasnoschekov equation*

We shall consider now the following nonlinear non-autonomous equation of the second order

$$\frac{d}{dt}\left[(1 + \alpha_1 \cos(\omega_1 t + \varphi_1))^2 \dot{x}\right] + \mu \dot{x}$$
$$= (1 + \alpha_1 \cos(\omega_1 t + \varphi_1))(1 - \alpha_2 \omega_2^2 \cos(\omega_2 t + \varphi_2)) \sin x \quad (4.56)$$

with periodic coefficients and with the limited nonlinear function $\sin x$.

Let us rewrite this equation in the form of a two-dimensional non-autonomous system

$$\dot{x} = y,$$
$$\dot{y} = \frac{g_2(t)}{g_1(t)} \sin x + \frac{2g_1(t)\dot{g}_1(t) - \mu}{g_1^2(t)} y, \tag{4.57}$$

where $g_1(t) = 1 + \alpha_1 \cos(\omega_1 t + \varphi_1)$, $g_2(t) = 1 - \alpha_2 \omega_2^2 \cos(\omega_2 t + \varphi_2)$. It is easy to see, that the condition

$$2g_1(t)\dot{g}_1(t) - \mu < 0 \tag{4.58}$$

defines the values of parameter $\mu : \mu > 2|1 - \alpha_1||\alpha_1\omega_1|$ for which the system (4.57) is dissipative. The system (4.57) has a set of fixed points $O_k(\pi k, 0)$, $k \in \mathbb{Z}$ which are asymptotically stable for large values of parameter μ. At loss of stability of these points, stable limit cycles appear in their neighbourhoods which are under influence practically of all parameters of the Eq. (4.56). Numerical investigations show, that all these cycles are singular cycles.

For example we shall consider evolution of solutions of the system (4.57) in the neighbourhood of point $(0,0)$ under fixed values of the parameters: $\alpha_1 = 0.8$, $\alpha_2 = 3$, $\omega_1 = \omega_2 = 2$, $\varphi_1 = 0$, $\varphi_2 = 1$ and at reduction of the parameter μ. Point $(0,0)$ loses stability for the parameter value $\mu \approx 5.74$ and a stable singular cycle is born. This cycle generates the Feigenbaum cascade of the period doubling bifurcations. One can observe a cycle of period 2 for the value $\mu = 4.17$, a cycle of period 4 for $\mu = 4.155$, a cycle of period 8 for $\mu = 4.1532$, a cycle of period 16 for $\mu = 4.1526$ *etc.* This cascade is terminated by formation of the Feigenbaum singular attractor for the parameter value $\mu \approx 4.1524$ (Fig. 4.20).

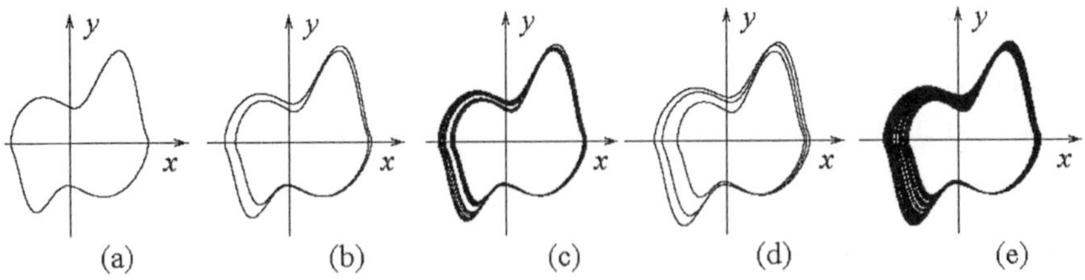

Fig. 4.20   A singular cycle (a) for $\mu = 4.2$, a cycle of period 2(b) for $\mu = 4.17$, the Feigenbaum attractor (c) for $\mu = 4.1524$, a cycle of period 3 (d) for $\mu = 4.05$ and a complete subharmonic singular attractor (e) for $\mu = 4.04$ in the system (4.57).

Further reduction of the parameter values μ leads to realization of the

Sharkovskii complete subharmonic cascade of bifurcations of stable cycles in accordance with the Sharkovskii order. It is proved by existence of a stable cycle of period 6 for the parameter value $\mu = 4.148$, a stable cycle of period 5 for $\mu = 4.0895$ and a stable cycle of period 3 for $\mu = 4.05$ in the system (4.57).

4.5 Dynamical Chaos in Three-Dimensional Autonomous Systems of Differential Equations

As it was shown in Chapter 3, the transition to chaos under variation of a system parameter in a wide class of three-dimensional autonomous nonlinear dissipative systems of ordinary differential equations, including all classical chaotic systems such as the Lorenz hydrodynamic system, the Rossler chemical system, the Chua electrotechnical system, the Magnitskii macroeconomic system *etc.*, occurs in accordance with a unique universal scenario of transition to chaos. This scenario begins with the Feigenbaum cascade of period doubling bifurcations of stable cycles and then continues by the Sharkovskii subharmonic cascade of bifurcations of a generation of stable cycles of the arbitrary period and by the Magnitskii homoclinic cascade of bifurcations of stable cycles converging to a homoclinic contour. Therefore, neither the presence of a positive Lyapunov exponent in the system nor the presence of a saddle–node or saddle–focus separatrix loop nor the presence of a saddle–node or a saddle–focus itself is necessary for the existence of chaotic dynamics in a nonlinear dissipative autonomous system of ordinary differential equations. The above-mentioned universal scenario of transition to chaos was theoretically justified for three-dimensional nonlinear dissipative autonomous systems of ordinary differential equations by N. Magnitskii in [Magnitskii (2004); Magnitskii and Sidorov (2004b); Magnitskii and Sidorov (2004c)]. In this section, we present the Magnitskii's theory of this universal scenario for such systems. We show that this scenario is based on a bifurcation, after which an originally stable singular limit cycle with complex but not complex-conjugate Floquet exponents becomes a singular saddle cycle and generates a stable cycle of a double period. In this case, the original singular cycle of a three-dimensional autonomous system corresponds to the zero rotor-type singular point of a two-dimensional non-autonomous system, whose coordinates rotate together with the trajectory of the original cycle transversally to it. This permits one to apply the theory of dynamical chaos in two-dimensional non-autonomous systems (presented in Sec. 4.4) to the substantiation of the above-mentioned universal scenario of transition to chaos in three-dimensional autonomous systems. Such a passage is impossible in the usually considered Poincare section. Thus, we show that any irregular attractor of a three-dimensional autonomous dissipative system is a singular attractor, i.e. it is a closure of some semistable nonperiodic trajectory and exists only at accumulation points of values of a system parameter. The simplest singular attractor is the Feigenbaum attractor. We prove that each of the appearing attractors of a three-dimensional autonomous dissipative system (either periodic or singular,) lies on a two-dimensional (many-sheeted in general) surface that is the closure of a two-dimensional invariant unstable manifold (a separatrix surface) of a singular saddle cycle. We prove also that any singular attractor has no positive Lyapunov exponents and chaotic dynamics in an autonomous three-dimensional dissipative system is due to the phase shift between trajectories forming a separatrix surface of a singular saddle cycle.

All analytic results are justified by corresponding examples of three-dimensional autonomous systems of differential equations presented in this section and by numerous examples of classical chaotic three-dimensional autonomous systems presented in Chapter 3.

4.5.1   *Singular cycles of three-dimensional autonomous systems*

Consider a smooth family of nonlinear autonomous systems of ordinary differential equations

$$\dot{x} = F(x, \mu), \quad x \in M \subset \mathbb{R}^3, \ \mu \in I \subset \mathbb{R}, \ F \in C^\infty, \tag{4.59}$$

defined in the three-dimensional phase space M by smooth vector fields F that depend on the scalar system parameter μ lying in a segment I of the real axis $\mathbb{R}$. Let the limit cycle $x_0(t, \mu)$ of period $T = T(\mu)$ be a solution of the family of systems (4.59) for all $\mu \in I$. By linearizing the family (4.59) in a neighborhood of the cycle, we obtain a parameter-dependent system of non-autonomous ordinary differential equations with a periodic matrix of the linear part:

$$\dot{y} = A(t, \mu)y + O(|y|^2), \ A(t, \mu) = \left\{ \frac{\partial F_i}{\partial x_j}(x_0(t, \mu), \mu) \right\}, \ i, j = 1, 2, 3, \tag{4.60}$$

where $y(t) = x(t) - x_0(t, \mu)$, and $A(t + T, \mu) = A(t, \mu)$. In this case, the vector $y = 0$ is a solution of system (4.60) for all $\mu \in I$.

It follows from the Floquet theory that each fundamental matrix solution of the linear system with periodic coefficients:

$$\dot{y} = A(t,\mu)y, \qquad A(t+T,\mu) = A(t,\mu) \tag{4.61}$$

can be represented in the form of $Y(t,\mu) = P(t,\mu)V(t,\mu)$, where $P(t,\mu)$ is some, in general, complex T-periodic matrix and the matrix $V(t,\mu) = e^{B(\mu)t}$ is a fundamental matrix of solutions of the linear system of equations with constant, in general, complex coefficients.

Therefore, the transformation $P(t,\mu)$ reduces the linear system (4.61) with periodic coefficients to a linear system with constant coefficients. In this case, stability of a periodic solution is determined by eigenvalues of the matrix B, that is, the Floquet exponents of the original cycle, or, which is equivalent, by the eigenvalues of the real matrix $C = e^{BT}$, which are the multipliers of the cycle. One Floquet exponent corresponding to the motion along the cycle always vanishes. If the cycle is stable for $\mu < 0$, then the remaining two exponents have negative real parts (one simple multiplier has the value $+1$, and two other multipliers have absolute values less than 1, i.e. lie inside the unit disk on the plane of the complex variable).

Suppose that, for $\mu = 0$, the cycle loses stability after a bifurcation related to crossing of the imaginary axis by one of the Floquet exponents from left to right or, which is equivalent, to intersection of the unit circle by one of multipliers lying inside the unit disk for $\mu < 0$. By using the change of variables $y(t) = Q(t,\mu)z(t)$, where $Q(t,\mu)$ is a T-periodic matrix, we simplify analysis of the system (4.60) by transition to a coordinate system attached to the cycle. In such a system, one of coordinate vectors is given by the vector $\dot{x}_0(t)$ tangent to the cycle, and two other vectors lie in the plane S transverse to the cycle. The cycle corresponds to the point $(0,0,0)$. Since multiplier of the cycle corresponding to the vector $\dot{x}_0(t)$ is always equal to unity and the Floquet exponent vanishes and is not bifurcational, it follows that coordinates of the normal form of bifurcation necessarily lie in the plane S defined by the last two coordinates of the vector $z(t)$. Therefore, without loss of generality, we assume that the first component of the vector $z(t)$ identically vanishes. Then the analysis of possibly bifurcation of the cycle is reduced to the analysis (in the plane S) of solutions of the two-dimensional system of differential equations

$$\dot{u}(t) = D(t,\mu)u(t) + H(u,Q(t,\mu),\mu) \tag{4.62}$$

with respect to the second and third components of the vector $z(t)$, where the matrix $D(t,\mu)$ is obtained from the matrix $L(t,\mu) = Q^{-1}(t,\mu)A(t,\mu)Q(t,\mu) - Q^{-1}(t,\mu)\dot{Q}(t,\mu)$ by deleting its first row and the first zero column. The expansion of the function H in components of the vector $u(t)$ at the singular point $O(0,0)$ of system (4.62) corresponding to the cycle starts from second-order terms. The linear part of system (4.62) has the same Floquet exponents (except for the zero one) as the linearized system (4.60).

The following two essentially different cases are possible: the case of a constant matrix $D(t,\mu) = D(\mu)$ and the case of a variable T-periodic matrix $D(t,\mu)$. The first case implies that $Q(t,\mu) = P(t,\mu)$ and $L(t,\mu) = B(\mu)$ are real matrices, and the Floquet exponents other than zero are given by the eigenvalues of the matrix $D(\mu)$. In other words, the passage to a coordinate system attached to the cycle with the use of the transformation $Q(t,\mu)$ already performs the reduction of system (4.61) with periodic coefficients to a system with constant real coefficients. In this case, the cycle bifurcation corresponds to either the passage of one real eigenvalue or two complex conjugate eigenvalues of the matrix $D(\mu)$ through the imaginary axis from left to right (the passage of one multiplier through the point $+1$ of the unit circle or the passage of two complex conjugate multipliers through the unit circle). As a result of such bifurcation either a pair of new stable limit cycles is generated, or cycles exchange the stability, or the cycle vanishes together with a similar unstable cycle, or a stable two-dimensional torus is generated from the cycle. The zero singular point of the two-dimensional autonomous system (4.62) corresponding to the cycle in the plane S transverse to the cycle is a stable node or focus for $\mu < 0$ and is an unstable focus or saddle with a one-dimensional stable manifold and a one-dimensional unstable manifold for $\mu > 0$. Such a cycle is referred to as *a regular stable cycle*, and a saddle cycle generated from it as a result of some of the above-described bifurcations is referred to as *a regular saddle cycle* [Magnitskii (2004)]. In any case, bifurcations of such a cycle do not lead to the appearance of a chaotic dynamics in three-dimensional systems of ordinary differential equations.

We have a much more complicated situation in the case of a variable T-periodic matrix $D(t,\mu)$. This case implies that the transition to the co-ordinate system attached to the cycle with the use of the transformation $Q(t,\mu)$ reduces system (4.60) with periodic coefficients to system (4.62) of smaller dimension, which also have T-periodic real coefficients. In this case, the zero singular point $O(0,0)$ of the two-dimensional non-autonomous system (4.62) is a rotor, and its Floquet exponents coinciding with the Floquet exponents of the original cycle of the autonomous three-dimensional system are complex but not complex conjugate numbers with equal imaginary parts. A stable cycle, which has such Floquet exponents, is referred to as *a singular stable cycle* [Magnitskii (2004)]. The bifurcation of the singular cycle $x_0(t, \mu)$ corresponds to the passage of one of its complex Floquet exponents $\alpha(\mu)$ through the imaginary axis for $\mu = 0$ from left to right. The appearing saddle cycle is referred to as *a singular saddle cycle* [Magnitskii (2004)]. In this case, the corresponding real multiplier of the cycle for $\mu = 0$ has the form $\lambda = e^{\alpha_1(0)T} = e^{i\pi} = -1$.

As the parameter μ passes through the bifurcation value $\mu = 0$, the original singular stable cycle $x_0(t, \mu)$ becomes a singular saddle cycle, and the zero singular point O of system (4.62) corresponding to the original cycle becomes an unstable rotor. In this case, naturally, the singular cycle has a two-dimensional invariant unstable manifold W^u. The structure of this manifold is locally sufficiently simple but can be very complicated globally. Since the transformation (4.45) reduces the system (4.62) to a system whose singular point is locally a saddle with a one-dimensional invariant unstable manifold, it follows that the latter manifold locally rotates in the plane S around a rotor O with period $2T$. It simultaneously rotates together with the plane S in the three-dimensional space M with period T and obviously, circumscribes *a Mobius band*. The phases of these two rotations (along the cycle and around the cycle) are uniquely related. Therefore, generation of a simple stable cycle of double period without self-intersections in the two-dimensional plane S around the unstable rotor O implies generation of a stable cycle of double period from the original singular cycle $x_0(t)$ of the autonomous three-dimensional system. This new stable cycle is an outer boundary of the Mobius band, the two-dimensional surface G that is the closure of the unstable invariant manifold W^u of the original singular cycle. Separatrices of the original singular saddle cycle tend to the generated stable cycle of the double period untwisting along the Mobius band.

As numerous examples show (see Chapter 3), it is exactly the period doubling bifurcation of a singular stable cycle, that is the beginning of the Feigenbaum cascade of period doubling bifurcations of stable cycles as well as the beginning of the Sharkovskii subharmonic and Magnitskii homoclinic cascades of bifurcations, which imply the appearance of chaotic dynamics in autonomous systems of ordinary differential equations. It will be shown in Chapter 5 that also exactly this bifurcation begins the subharmonic cascade of bifurcations of two-dimensional tori of transition to diffusion chaos in nonlinear partial differential equations.

4.5.2 *Singular attractors of three-dimensional autonomous systems*

Recall that singular attractors are attractors generated at all stages of all above-mentioned bifurcation cascades at accumulation points of the bifurcation parameter. By definition, in any neighborhood of any singular attractor, there is an unstable limit cycle. Let us show that the phase shift plays a basic role in all bifurcation cascades of the original singular cycle and in formation of all singular attractors. More precisely, the phase shift permits the trajectories of an autonomous system of ordinary differential equations to lie on a two-dimensional many-sheeted surface, which corresponds to the existence of not one-to-one and even multi-modal one-dimensional mappings in the above-defined two-dimensional plane S. This means that three-dimensional autonomous systems can have an arbitrary complex dynamics. Note that in any Poincare section transversal to the original cycle, such a situation is impossible, since it would contradict the uniqueness theorem for an autonomous system. The reason is that the Poincare mapping neglects the phase of periodic solutions; therefore, one would apply it to the analysis of solutions of systems of differential equations with great care.

Thus let the singular limit cycle $x_0(t, \mu)$ of period T be a stable solution of the set of systems (4.59) for all $\mu < 0$, and let the above-described period doubling bifurcation takes place for it for $\mu = 0$. In fact, the limit cycle is described not by a single trajectory $x_0(t, \mu)$ but by an infinite family of such trajectories $x_0(t + \varphi, \mu)$ depending on the phase φ. By using the above-represented method, we pass to the analysis of solutions of the family of two-dimensional non-autonomous systems (4.62) with the T-periodic matrix $D(t + \varphi, \mu)$ in the plane S rotating transversally to the original cycle. In this case, the matrix Q occurring in (4.62) also depends on the argument $t + \varphi$. Let the singular point $O(0,0)$ of the non-autonomous system (4.62) be an unstable rotor for all μ, $0 < \mu < \mu^*$.

Lemma 4.8 *For any μ, $0 < \mu < \mu^*$, the trajectories*

$$x(t + \varphi, \mu) = x_0(t + \varphi, \mu) + Q(t + \varphi, \mu)(0, u_1(t), u_2(t))^T$$

of system (4.59) and only these trajectories for all φ, $0 \leq \varphi < 2T$, are separatrices of the singular cycle $x_0(t, \mu)$ tending to it as $t \to -\infty$, where the vector $u(t)$ is a solution of the system of Volterra integral equations of *the second kind*

$$u(t) = R(t + \varphi, \mu) \begin{pmatrix} e^{\beta_1(\mu)t} \\ 0 \end{pmatrix} +$$

$$R(t + \varphi, \mu) \int_{-\infty}^{t} e^{E(\mu)(t-\tau)} R^{-1}(\tau + \varphi, \mu) H(u(\tau), Q(\tau + \varphi, \mu), \mu) d\tau. \quad (4.63)$$

Proof. Since any solution of the autonomous system tends to a cycle with some asymptotic phase, it follows that desired separatrices of the cycle are the exact solutions of system (4.59) generated by solutions $u(t)$ of system (4.62) such that $u(t) \to 0$ for some φ as $t \to -\infty$. It follows from the representation (4.46) that the general solution of system (4.62) in the plane S can be written out in the form of

$$u(t) = R(t + \varphi, \mu) e^{E(\mu)t} \begin{pmatrix} C_1 \\ C_2 \end{pmatrix} +$$

$$R(t + \varphi, \mu) \int_{-\infty}^{t} e^{E(\mu)(t-\tau)} R^{-1}(\tau + \varphi, \mu) H(u(\tau), Q(\tau + \varphi, \mu), \mu) d\tau, \quad (4.64)$$

where C_1 and C_2 are arbitrary constants. Since $\beta_2(\mu) < 0$ for all $0 < \mu < \mu^*$, it necessarily follows from the condition $u(t) \to 0$ as $t \to -\infty$ that $C_2 = 0$. Further, without loss of generality, in view of the rotation of the vector $u(t)$ around the rotor O, one can assume that $C_1 > 0$, but in this case, the interval of the phase covering the entire domain of the separatrix surface of the cycle is $0 < \varphi < 2T$. Now we represent $C_1 e^{\beta_1 t}$ in the form $e^{\beta_1 s}$, where $s = t + (\ln C_1)/\beta_1 = t + a$, and rewrite system (4.64) in the form of

$$u(s - a) = R(s - a + \varphi) \begin{pmatrix} e^{\beta_1 s} \\ 0 \end{pmatrix} +$$

$$R(s - a + \varphi) \int_{-\infty}^{s} e^{E(s-\eta)} R^{-1}(\eta - a + \varphi) H(u(\eta - a), Q(\eta - a + \varphi)) d\eta,$$

by omitting the parameter μ. Since Q is a T-periodic function and R is a $2T$-periodic function, it follows from the last representation that, for $C_2 = 0$, the solution $u(t)$ of system (4.64) with a constant $C_1 > 0$ and a phase $0 \le \varphi < 2T$ at time t is also a solution of system (4.63) with a phase $(\varphi - a) \pmod{2T}$ at time $s = t + a$. Therefore, without loss of generality, in (4.64), one can set $C_1 = 1$ and $C_2 = 0$. Consequently, the separatrix surface of the singular cycle $x_0(t, \mu)$ of the set of systems (4.59) consists of solutions generated by solutions of the two-dimensional system of integral equations (4.63) for all $0 \le \varphi < 2T$. The proof of the lemma is complete$\square$

Corollary 4.4 *Since all solutions are continuously differentiable with respect to all arguments and hence continuously differentiable with respect to the phase φ, it follows that the separatrix surface constructed in Lemma 4.8 for the singular cycle is its continuously differentiable two-dimensional invariant unstable manifold W^u.*

Remark 4.2 *Obviously, the result of Lemma 4.8 and the assertion of Corollary 4.4 remain valid in the general case in which $z_1 \neq 0$ in the transformation $x(t + \varphi, \mu) = x_0(t + \varphi, \mu) + Q(t + \varphi, \mu)z(t)$. In this case, the two-dimensional separatrix surface is given by solutions of a system of three Volterra integral equations of the second kind.*

Theorem 4.3 *Any regular attractor (a stable limit cycle), any unstable cycle, and any singular attractor of the family of systems (4.59) in the three-dimensional phase space M for all $0 < \mu < \mu^*$ belong to the two-dimensional surface G that is the closure of the two-dimensional unstable invariant manifold W^u of the original singular cycle $x_0(t, \mu)$.*

Proof. First, let us consider an arbitrary periodic stable or unstable solution $x(t)$ of system (4.59) for some μ, $0 < \mu < \mu^*$. Let some point x_0 of its closed trajectory in the phase space M not belong to the surface G, i.e. $\mathrm{dist}(x_0, G) > 0$. We choose a sequence of times $t_n \to \infty$ as $n \to \infty$ such that $x(t_n) = x_0$ and hence $\mathrm{dist}(x(t_n), G) = \mathrm{const} > 0$. But, on the other hand, each solution of system (4.59), in particular, the cycle $x(t)$, is generated by one of solutions of the two-dimensional non-autonomous system (4.62) of the form (4.64). Since $\beta_2(\mu) < 0$, it follows that it tends to some solution of system (4.64) with constant $C_2 = 0$ and with some constant C_1 as $t \to \infty$. But, as was shown in Lemma 4.8, any solution of this kind of system (4.64) can be represented as a solution of system (4.63) with some different phase at some different time. Therefore, for any trajectory of the family of systems (4.59), there always exists a separatrix of the original singular cycle $x_0(t, \mu)$ that belongs to the surface G and is attracted to that trajectory. Consequently, $\mathrm{dist}(x(t_n), G) \to 0$ as $n \to \infty$. The resulting contradiction implies that any cycle belongs to the surface G. It also follows from the above-performed considerations that each singular attractor belongs to the surface G, since its arbitrary neighborhood contains an unstable limit cycle, which belongs to G. The proof of the theorem is complete. $\square$

The proved theorem plays an important role in understanding chaotic processes taking place in three-dimensional autonomous systems of nonlinear ordinary differential equations. It implies two important corollaries.

Corollary 4.5 *The fractal dimension of any singular attractor of system* (4.59) *cannot exceed* 2.

Corollary 4.6 *Any singular attractor of the family of autonomous three-dimensional systems* (4.59) *is the closure of a semistable nonperiodic trajectory and has one negative and two zero Lyapunov exponents.*

Proof. Any singular attractor is the closure of a nonperiodic trajectory lying on the two-dimensional surface G and such, that there exists a sequence of unstable limit cycles of the period doubling cascade tending to it. We show that this nonperiodic trajectory itself has one negative and two zero Lyapunov exponents, and, hence, it is a semistable nonperiodic trajectory. Indeed, its negative exponent is related to motion in a direction transversal to the surface G, and the one zero exponent is related to motion along the trajectory itself. The third exponent also vanishes: on one hand, it cannot be positive, since in this case the stable limit cycles of the period doubling cascade forming this nonperiodic trajectory and lying in its small neighborhood at lesser values of the bifurcation parameter must also have positive exponents, that is impossible for stable cycles. On the other hand, it cannot be negative, since in any neighbourhood of the trajectory, there exist trajectories not attracted to it (unstable limit cycles). $\square$

Let us now consider the mechanism of generation of singular attractors of autonomous three-dimensional systems of differential equations as a result of successive bifurcations of the originally stable singular cycle. As it was shown above, first a stable cycle of double period is generated from the original singular cycle $x_0(t, \mu)$ of the three-dimensional autonomous system for $\mu > 0$, which corresponds to the generation of a simple stable cycle without self-intersections in the two-dimensional plane S around the unstable rotor O of the non-autonomous two-dimensional system (4.62). The last cycle has a double period with respect to the original cycle. If the value of parameter $\mu > 0$ grows, then the trajectories of a two-dimensional system (4.62) corresponding to separatrices of the original cycle of the three-dimensional autonomous system untwist from the rotor and begin to intersect themselves and twist around the simple stable cycle of the system.

This corresponds to the beginning of passage of separatrices of the original cycle $x_0(t, \mu)$ of the three-dimensional autonomous system from the Mobius band containing it to the Mobius band containing the cycle of double period. Surface G wraps itself up and becomes two-sheeted. In terms of one-dimensional mapping $f(u_1)$, this corresponds to the appearance of a point of maximum on its graph in the domain $u_1 < 0$, which leads to the appearance of a double-valued inverse mapping $f^{-1}(u_1)$. In this case, there is a phase shift between separatrices of the original cycle, whose closures form different sheets of the surface G. Therefore, the loss of single-valuedness of the inverse one-dimensional mapping constructed in the plane S does not result in the loss of single-valuedness of solutions of the autonomous system in the three-dimensional space. The cycle of period 4 is generated even on the two-dimensional two-sheeted surface G of the phase three-dimensional space.

The subsequent mechanism of generation of increasingly complicated regular and singular attractors in the family of two-dimensional non-autonomous systems (4.62) and hence the family of three-dimensional autonomous systems (4.59) as a result of a subharmonic cascade of bifurcations is described in the following assertion.

Theorem 4.4 *Suppose that the period doubling bifurcation of a stable (for $\mu < 0$) singular cycle takes place in the family of autonomous three-dimensional systems of ordinary differential equations (4.59) for $\mu = 0$. Then the first stages of scenario of transition to chaos in the family of systems (4.59) for the growth of positive values of bifurcation parameter μ coincide with the stages of transition to chaos for iterations of a continuous self-mapping of the unit segment which has a double-valued inverse mapping. The Feigenbaum cascade of period doubling bifurcations of the generated stable limit cycle is realized first, and then a subharmonic cascade of bifurcations of stable cycles of arbitrary period takes place in accordance with the Sharkovskii order (4.47).*

Proof. Validity of the theorem follows from Theorem 4.2, which defines the order of transition to chaos in two-dimensional non-autonomous systems of differential equations with rotor type singular points and from Lemma 4.8, which provides a correspondence between trajectories of a three-dimensional autonomous system in a neighborhood of a singular cycle and trajectories of a two-dimensional non-autonomous system for which such a cycle is a rotor. In this case, a cycle of period k of a two-dimensional non-autonomous system corresponds to a cycle of period $2k$ of a three-dimensional autonomous system. □

Corollary 4.7 *There exists an infinite set of intervals of values of parameter μ for which the family of systems (4.59) has regular attractors (asymptotically orbitally stable periodic trajectories, even of a very large period). The family of systems (4.59) has singular attractors (closure of nonperiodic semistable trajectories) at infinitely many accumulation points of various infinite subcascades of period doubling bifurcations for various cycles. Obviously, the simplest is Feigenbaum attractor, the first attractor existing in the family of systems (4.59) for $\mu = \mu_\infty$, where the value μ_∞ corresponds to the limit of the sequence of values of parameter μ, for which period doubling bifurcations take place for the original singular cycle.*

As numerous examples show (see Chapter 3 and Sec. 4.5.3), that a subharmonic cascade of Sharkovskii bifurcations does not exhaust the entire complexity of transition to chaos in three-dimensional autonomous systems of ordinary differential equations. It can be continued at least by the Magnitskii homoclinic cascade of bifurcations. The transition to the two-dimensional plane S permits one to clearly detect the termination of a subharmonic cascade in plane S after the generation of a stable cycle of period 3 in this plane (a cycle of period 6 in the three-dimensional phase space) and a cascade of its period doubling bifurcations. For further growth of values of parameter μ, a homoclinic cascade occurs in plane S and turns to singular attractors and homoclinic cycles of a two-dimensional non-autonomous system approaching the rotor O. This, in terms of mapping $f(u_1)$, implies that $f(c)$ tends to zero when μ tends to μ_c. That corresponds to existence of a homoclinic separatrix loop of a rotor for a two-dimensional non-autonomous system and to an existence of a homoclinic separatrix loop of the original singular saddle cycle for the three-dimensional autonomous system at $\mu = \mu_c$ (see Fig. 2.20 in Chapter 2). But $f(u_1)$ cannot intersect the line $u_2 = 0$ at any point other than $u_1 = 0$, since this would imply the existence of a trajectory different from the original cycle but passing into it in a half-turn around the point O with some phase shift. The latter is impossible for autonomous systems of ordinary differential equations. Therefore, there appears a third branch in the mapping $f^{-1}(u_1)$ for $\mu > \mu_c$. This implies the appearance of a third sheet of the two-dimensional surface G; moreover, trajectories on the third sheet have a phase shift of more than a period as compared with trajectories lying on the first sheet. For further growth of values of parameter μ, generation of a stable cycle of the period 3 occurs in three-dimensional phase space of the autonomous system, that terminates the subharmonic cascade of bifurcations in this space. After that the homoclinic cascade of bifurcations occurs in three-dimensional phase space. Cycles of the homoclinic cascade converge, as a rule but not necessarily, to homoclinic saddle–focus separatrix loop. Further bifurcations cannot be described by the Sharkovskii and Magnitskii theory. Their theoretical analysis requires development of the theory of one-dimensional multi-valued mappings or mappings with multi-valued inverse mappings.

All known classical attractors of three-dimensional systems of ordinary differential equations, including Lorenz, Rossler, Chua, Magnitskii systems and other are complete or incomplete subharmonic or homoclinic singular attractors (see Chapter 3). Therefore, all of them lie on two-dimensional two- or three-sheeted surfaces in the phase space. However, attractors that are more complicated also exist in three-dimensional dissipative systems of ordinary differential equations. Such a system should have a singular cycle but cannot have a singular point of saddle–focus or saddle–node type. The complication of the structure of a two-dimensional unstable manifold of the singular cycle in such a system can be accompanied by the appearance of an arbitrary (either finite or infinite) number of sheets. An example of such many-sheeted system was constructed for the first time by N. Magnitskii. In this example and in other examples considered in the following section as two-sheeted, tree-sheeted and many-sheeted surfaces of subharmonic, homoclinic and more complex attractors are well observed both in plane S, and in original three-dimensional phase space M.

As follows from results proved above, complexity of singular attractors increases with growth of values of the bifurcation parameter. In the family of systems (4.59) there is at least one attractor, being presumably a fractal and having a dimension which is presumably equal to one. It is the elementary Feigenbaum attractor, not containing unstable limit cycles and coincident with the semi-stable nonperiodic trajectory which generated this attractor. At the same time, as follows from the theory of one-dimensional mappings, there are also attractors complex enough in three-dimensional systems which dimension presumably is equal to two. Such attractor is generated by a semi-stable nonperiodic trajectory, closure of which contains not only this trajectory but presumably some family of unstable limit cycles. Undoubtedly, the problem of classification of singular attractors on powers of sets of trajectories composing them, on their dimension or their measure is of great interest and of great importance, but is not considered in the present book.

4.5.3 *Some examples of three-dimensional autonomous systems with singular attractors*

Since the equations of cycles of particular nonlinear systems of ordinary differential equations can be known only in exceptional cases, it follows that the analysis of such systems has the main difficulty that is the construction of transformation $x(t+\varphi,\mu) = x_0(t+\varphi,\mu)+Q(t+\varphi,\mu)z(t)$ and its use for transition to a coordinate system related to the cycle. The first of examples considered below admits the construction of such a transformation in closed form with the vector $z(t) = (0, u_1(t), u_2(t))^{\mathrm{T}}$. This example is a complete, clear illustration of the entire above-represented theory of generation of singular attractors of nonlinear dissipative systems of ordinary differential equations. The second and third examples illustrate separate characteristic features of the theory.

Example 4.1   Consider the autonomous system of differential equations

$$\begin{aligned}
\dot{x}_1 &= -\omega x_2 + x_1 h(x_1, x_2, \mu, \omega), \\
\dot{x}_2 &= \omega x_1 + x_2 h(x_1, x_2, \mu, \omega), \\
\dot{x}_3 &= (x_2 + \omega/4)((x_1^2 + x_2^2) - 1) + 2(\mu - 1 - x_1)x_3,
\end{aligned} \tag{4.65}$$

where

$$h(x_1, x_2, \mu, \omega) = ((\mu - 1)\sqrt{x_1^2 + x_2^2} + x_1)(x_1^2 + x_2^2 - 1) + (2x_2 - \omega/2)x_3.$$

If $\mu < 1$ then system (4.65) has the singular point $(0, 0, \omega/8(\mu - 1))$ and a limit cycle $x_0(t, \mu) = (\cos \omega t, \sin \omega t, 0)^{\mathrm{T}}$, which lies in the plane of variables (x_1, x_2) and has the period $T = 2\pi/\omega$. We perform the change of variables

$$x(t + \varphi, \mu) = x_0(t + \varphi, \mu) + Q(t + \varphi, \mu)(0, u_1(t), u_2(t))^{\mathrm{T}}$$

with $2\pi/\omega$-periodic matrix $Q(t + \varphi, \mu)$:

$$Q(t) = \left(\dot{x}_0(t), x_0(t), (0,\ 0,\ 1)^{\mathrm{T}}\right) = \begin{pmatrix} -\omega \sin \omega t & \cos \omega t & 0 \\ \omega \cos \omega t & \sin \omega t & 0 \\ 0 & 0 & 1 \end{pmatrix}.$$

To avoid cumbersome manipulations, we omit the phase shift φ in forthcoming considerations. This change of variables reduces system (4.65) to coordinates attached to the cycle, i.e. directly to the two-dimensional non-autonomous system (4.41) with $2\pi/\omega$-periodic coefficients

$$\begin{aligned}
\dot{u}_1 &= 2(\mu - 1 + \cos \omega t)u_1 + (2 \sin \omega t - \omega/2)u_2 + h_1(u, t, \omega, \mu), \\
\dot{u}_2 &= (2 \sin \omega t + \omega/2)u_1 + 2(\mu - 1 - \cos \omega t)u_2 + h_2(u, t, \omega, \mu),
\end{aligned} \tag{4.66}$$

where $u = (u_1,\ u_2)^{\mathrm{T}}$,

$$h_1 = (\mu - 1 + \cos\omega t)((2u_1 + u_1^2)^2 + u_1^2) + ((2u_1 + 4)\sin\omega t - \omega/2)u_1 u_2,$$
$$h_2 = (2\sin\omega t + (u_1 + 1)\sin\omega t + \omega/4)u_1^2 - 2u_1 u_2 \cos\omega t.$$

The leading linear part of system (4.66) coincides with the leading linear part of system (4.48) and has the form (4.46) with $\beta_1(\mu) = 2\mu$, $\beta_2 = 2\mu - 4$. Consequently, for $\mu < 0$, the zero solution of system (4.66) and the cycle $x_0(t, \mu)$ of system (4.65) are stable. For $\mu > 0$, in system (4.66), there appears a stable solution with frequency $\omega/2$ (with the double period $4\pi/\omega$).

Complex numbers $\alpha_1 = 2\mu + i\omega/2$ and $\alpha_2 = 2\mu - 4 + i\omega/2$ are the Floquet exponents of the cycle $x_0(t, \mu)$ of the original system (4.65). For $\mu = 0$, the first of them passes through the imaginary axis from left to right, and the other remains in the left half-plane. The multipliers corresponding to the Floquet exponents are equal to

$$\lambda_1 = \exp((i\omega/2 + 2\mu)2\pi/\omega) = \exp(i\pi + 4\pi\mu/\omega),$$
$$\lambda_2 = \exp((i\omega/2 + 2\mu - 4)2\pi/\omega) = \exp(i\pi + 4\pi(\mu - 2)/\omega).$$

If $\mu = 0$, then, obviously, the first multiplier crosses the boundary of the unit circle at the point -1. In this case, the second multiplier is equal to $-\exp(-8\pi/\omega)$ and remains on the real axis inside the unit disk. Therefore, as bifurcation parameter μ passes through the value $\mu_1 = 0$, the singular stable cycle $x_0(t)$ of system (4.65) becomes a singular saddle cycle, and a stable cycle of double period is generated around it. The singular point $O(0,0)$ of the two-dimensional non-autonomous system (4.66) is a rotor.

All further bifurcations of the cycle can be observed in the three-dimensional phase space of the variables (x_1, x_2, x_3) as well as in the plane S of the variables (u_1, u_2), which rotates together with the cycle trajectory. One can also readily obtain the graph of a one-dimensional mapping $f(u_1)$ of the segment of line $u_2 = 0$ into itself for a half-turn of the trajectory of system (4.66) around the rotor O. In the plane S and in any Poincare section of the phase space by the plane $\varphi = \mathrm{const}$, one can clearly observe various sheets of the two-dimensional surface G on which all attractors of system (4.65) lie and to which all of its trajectories are attracted.

By integrating numerically the system (4.65), one can readily show that, for the growth of positive values of parameter μ and for a given value of parameter ω, the cascade of Feigenbaum period doubling bifurcations for the original stable limit cycle is first realized in the system (4.65), and a sub-harmonic cascade of bifurcations of stable cycles with an arbitrary period takes place in accordance with the Sharkovskii order. Thus, for example, for $\omega = 4$ in system (4.65), there appears a stable cycle of quadruple period for $\mu_2 \approx 0.079$, of period 8 for $\mu_3 \approx 0.0917$, of period 16 for $\mu_4 \approx 0.095$, of period 32 for $\mu_5 \approx 0.0951$, and of period 64 for $\mu_6 \approx 0.09517$. Obviously, the value of μ_∞ corresponding to the limit of the sequence of values of the parameter μ for which period doubling bifurcations of the original cycle take place, equals approximately to 0.0952. Here a Feigenbaum attractor is generated in the system (4.65). System (4.65) has a stable cycle of period 80, which corresponds to the cycle $5 \cdot 2^4$ in the Sharkovskii ordering for $\mu \approx 0.09523$ and a stable cycle of period $48 = 3 \cdot 2^4$ for $\mu \approx 0.09526$. The cycle of period $20 = 5 \cdot 2^2$ is generated for $\mu \approx 0.0962$, the cycle of period $12 = 3 \cdot 2^2$ is generated for $\mu \approx 0.0969$, the cycle of period $10 = 5 \cdot 2$ is generated for $\mu \approx 0.09986$, and the cycle of period 6 is generated for $\mu \approx 0.10295$. Generation of the last cycle implies the generation of a stable cycle of period 3 in the Sharkovskii order in the plane S of the variables (u_1, u_2). The subharmonic cascade of bifurcations terminates for $\mu \approx 0.1132$, when the eyes of singular attractors of system (4.66) in the plane S tend to zero. All cycles and singular attractors of the system (4.65) generated by that time lie on a two-dimensional two-sheeted surface (Fig. 4.21b), and the inverse of mapping $f(u_1)$ is double-valued (Fig. 4.21a).

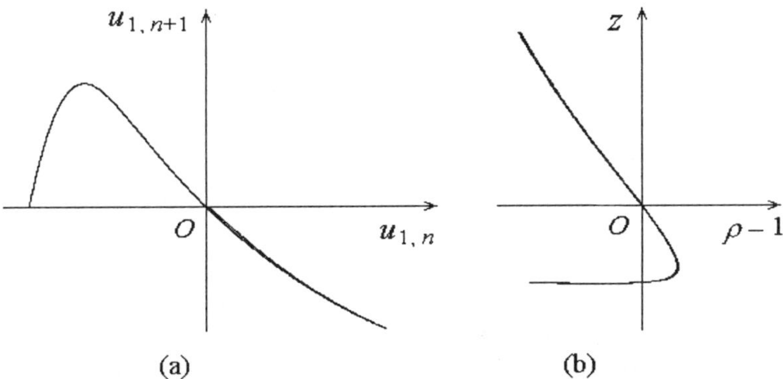

Fig. 4.21 Mapping $f(u_1)$ (a) and the Poincare mapping in the plane $\varphi = 1.3\pi$ (b) for the system (4.65) at $\omega = 4$ and $\mu = 0.1132$; $\rho = \sqrt{x_1^2 + x_2^2}$.

Further bifurcations of stable cycles and singular attractors of system (4.65) for $\omega = 4$ and for growing value of parameter μ, take place on a two-dimensional three-sheeted surface, and the inverse of $f(u_1)$ becomes a three-valued mapping (Fig. 4.22). An incomplete homoclinic cascade of bifurcations of stable cycles C_n tending to the homoclinic contour that is a separatrix loop of a singular point is realized in the system (4.65). But at a sufficiently large distance from the singular point for $\mu = \mu^* \approx 0.1274$ system (4.65) loses the dissipative property, and its attractors are destroyed.

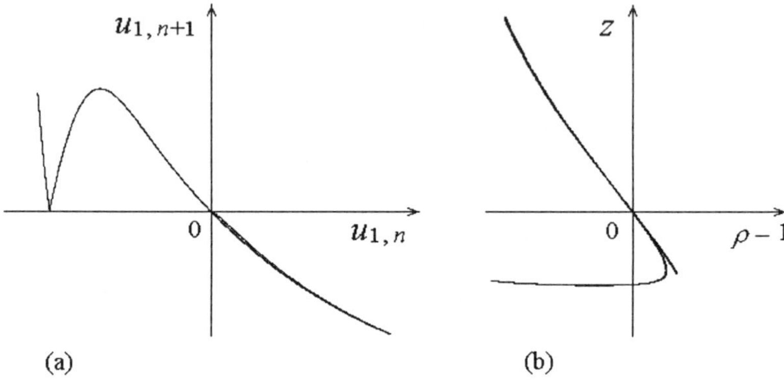

Fig. 4.22   Mapping $f(u_1)$ (a) and the Poincare mapping in the plane $\varphi = 1.3\pi$ (b) for the system (4.65) at $\omega = 4$ and $\mu = 0.127$.

One can readily see that, in the case under consideration, the singular point itself is a saddle–focus with a one-dimensional stable manifold and a two-dimensional unstable manifold, since, at this point, the linearization matrix of the system (4.65) has the eigenvalues

$$\lambda_1 = 2(\mu - 1) < 0, \quad \lambda_{2,3} = \frac{-\omega^2}{16(\mu - 1)} \pm i\omega.$$

For $\omega = 2$, in system (4.65) for a growing value of the parameter μ, even a complete homoclinic cascade of bifurcations of stable cycles converging to a homoclinic contour (which exists in the system for $\mu = \mu^* \approx 0.0625$ in the case under consideration) is realized. For example, there appears a stable cycle of the quadruple period for $\mu_2 \approx 0.0248$, of period 8 for $\mu_3 \approx 0.0302$, of period 16 for $\mu_4 \approx 0.0315$ and of period 32 for $\mu_5 \approx 0.0318$. The value μ_∞ corresponding to the limit of the sequence of values of the parameter μ for which period doubling bifurcations of the original cycle take place approximately equals 0.032. Here a Feigenbaum attractor is generated in the system. In system (4.65), there appears a stable cycle of period $10 = 5 \cdot 2$ for $\mu \approx 0.034$ and period 6 for $\mu \approx 0.0355$ (see Fig. 4.23). The subharmonic cascade of bifurcations is terminated for $\mu \approx 0.0415$. The stable cycle C_3 of the homoclinic cascade exists in the system for $\mu \approx 0.054155$ (see Fig. 4.24).

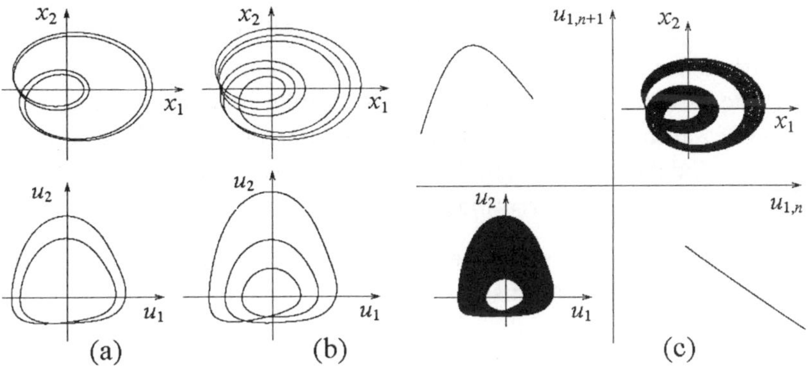

Fig. 4.23   The projection of the cycle with period 4 of system (4.65) and the cycle with period 2 of system (4.66) for $\mu = 0.0248$ (a); the projection of the cycle with period 6 of system (4.65) and the cycle with period 3 of system (4.66) for $\mu = 0.0355$ (b); the one-dimensional mapping $f(u_1)$, the Sharkovskii attractor of system (4.66) (at the bottom from the left) and the projection of the corresponding attractor of system (4.65) (at the top from the right) for $\mu = 0.0365$ (c).

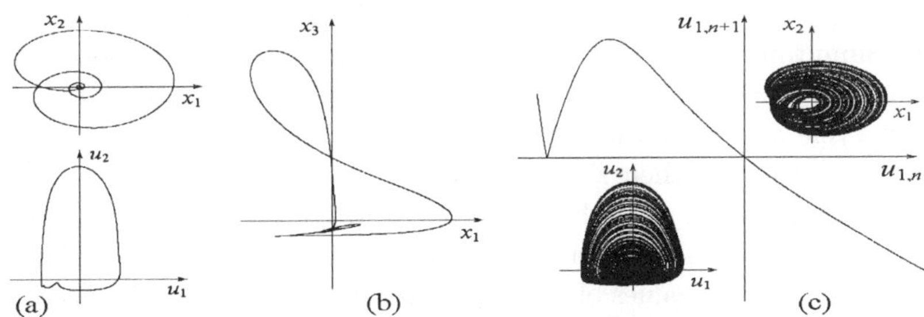

Fig. 4.24   The projection of the cycle with period 3 of system (4.65) and the cycle of system (4.66) for $\mu = 0.054155$ (a); the projection of the cycle with period 3 of system (4.65) for $\mu = 0.054155$ (b); the one-dimensional mapping $f(u_1)$, the singular attractor of system (4.66) (at the bottom from the left) and the projection of the corresponding attractor of system (4.65) (at the top from the right) for $\mu = 0.0454$ (c).

Numerical experiments show that, as μ tends to μ^*, the mapping $f^{-1}(u_1)$ becomes many-valued (for example, four-valued for $\mu \approx 0.0519$), but, in this case, the surface G remains three-sheeted, i.e. quite simple. This can be explained by the possible tangency of the two-dimensional invariant manifold W^u of the original singular cycle and the unstable two-dimensional invariant manifold of the saddle–focus.

We note that one can add a nonlinear member $-x_3^2$ in the third equation of the system (4.65) and easily obtain the system without any singular point for $0 < \mu < 2$ ($\omega = 4$) and with the same first stages of the scenario of transition to chaos as in the system (4.65) without, of course, a complete homoclinic cascade of bifurcations. This example shows that exactly singular cycle of autonomous system but not singular point generates a chaotic dynamics. Existence of a singular point can lead only to appearance of homoclinic cascade of bifurcations.

Example 4.2   As the second example we shall consider the system of ordinary differential equations

$$
\begin{aligned}
\dot{x}_1 &= \mu x_1 - \omega x_2 - x_1^2 x_3^2, \\
\dot{x}_2 &= \omega x_1 + \mu x_2 - x_1 x_2 x_3^2, \\
\dot{x}_3 &= x_1^2 + x_2^2 + x_3^2 - \sigma x_3,
\end{aligned}
\tag{4.67}
$$

dependent on three system parameters μ, $\omega > 0$ and $\sigma > 0$. The system (4.67) by its characteristics is closer to real autonomous systems of ordinary differential equations, than the modelling system (4.65). It has the zero singular point being a stable focus for $\mu < 0$, and a saddle–focus with one-dimensional stable and two-dimensional unstable manifolds for $\mu > 0$. Simultaneously at $\mu > 0$ a stable singular limit cycle is born in the system (4.67) as a result of Andronov–Hopf bifurcation. Characteristics of the last cycle are unknown to us. Such systems possess the most simple separatrix surfaces of their singular cycles.

Integrating system (4.67) by the Runge–Kutta method of the 4-th order, one can easily establish, that at fixed values of the parameters ω and σ and at growth of positive values of the parameter μ, the Feigenbaum cascade of period doubling bifurcations of the original singular limit cycle, and then the Sharkovskii subharmonic cascade of bifurcations and then the Magnitskii homoclinic cascade of bifurcations are realized in system (4.67), as well as in system (4.65).

So, for example, for $\omega = 1$, $\sigma = 5$ in system (4.67), the stable cycle of the double period appears for $\mu_1 \approx 0.02445$, of quadruple period for $\mu_2 \approx 0.02618$, of period 8 for $\mu_3 \approx 0.0265$, of period 16 for $\mu_4 \approx 0.02657$, of period 32 for $\mu_5 \approx 0.02658$, and of period 64 for $\mu_6 \approx 0.0265805$. The value μ_∞ corresponding to the limit of the sequence of values of the parameter μ for which the period doubling bifurcations of the original cycle take place, approximately equals 0.026581. Here the Feigenbaum attractor is generated in system (4.67). For $\mu \approx 0.0265829$ in system (4.67), there is already a stable cycle of the period 8 that corresponds to the cycle $3 \cdot 2^4$ in the Sharkovskii order. The cycle of period $12 = 3 \cdot 2^2$ is generated at $\mu \approx 0.026623$, the cycle of period $10 = 5 \cdot 2$ is generated at $\mu \approx 0.0267$, the cycle of period 6 is generated at $\mu \approx 0.02678$. The subharmonic cascade of bifurcations corresponding to the Sharkovskii order, comes to the end approximately at $\mu \approx 0.0273$.

At the further increasing of values of the parameter μ a complete homoclinic cascade of bifurcations of generation of stable homoclinic cycles C_n, converging to a homoclinic contour, is realized in the system (4.67) (Fig. 4.25). Cycle C_3 of the period three of homoclinic cascade is generated for $\mu \approx 0.0276$, cycle C_4 — for $\mu \approx 0.02814$, cycle C_5 — for $\mu \approx 0.02841$, and cycle C_6 — for $\mu \approx 0.028571$, *etc.* Homoclinic contour exists in the system (4.67) at $\mu = \mu^* \approx 0.0289659$ ($\omega = 1$, $\sigma = 5$).

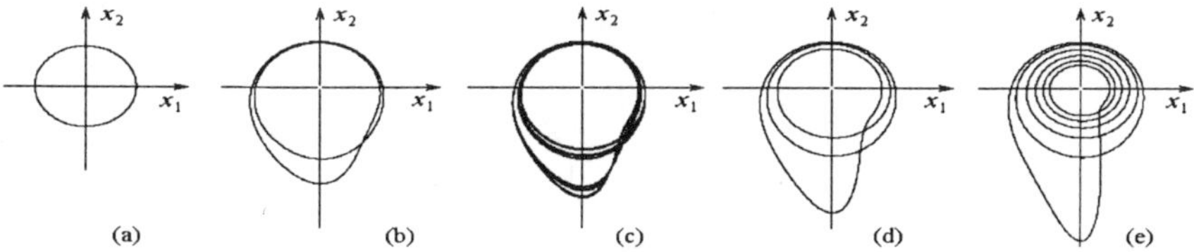

Fig. 4.25   Original singular cycle (a), a cycle of period 2(b) for $\mu = 0.025$, Feigenbaum attractor for $\mu = 0.026581$ (c), a cycle of the period 3 of the subharmonic cascade (d) for $\mu = 0.0276$ and a cycle C_6 of homoclinic cascade for $\mu = 0.028571$ (e) in autonomous three-dimensional system (4.67).

Not knowing the equations of a cycle, we have no opportunity to pass to a rotating plane S, transversal to a cycle. Nevertheless, the system (4.67) can be reduced to two-dimensional system

$$\dot{r} = \mu r - z^2 r^2 \cos(\omega t + \varphi), \qquad \dot{z} = r^2 - \sigma z + z^2$$

by change of variables

$$x_1 = r(t)\cos(\omega t + \varphi), \quad x_2 = r(t)\sin(\omega t + \varphi), \quad x_3 = z(t),$$

and then all its singular attractors can be observed clearly in Poincare's sections passing through axis z transversally to the original cycle, and also in a two-dimensional plane $\widetilde{S}$, given by variables (r, z) (Fig. 4.26a). Numerical experiments show, that all singular attractors of systems (4.67) lie on a two-sheeted surface. The original stable cycle lies on the bottom sheet.

After the period doubling bifurcation one turn of a cycle remains lying on the bottom sheet, and other turn passes to the top sheet. The cycle of the period 4 has in twos turns, lying on bottom and top sheets, *etc.* All turns of homoclinic cycles lie on the bottom sheet of surface G untwisting from the neighbourhood of a saddle–focus, and only the last turn, rising on a surface, passes on its top sheet and falls almost vertically in a neighbourhood of a stable one-dimensional manifold of the saddle–focus O, and then passes again to the bottom sheet. Thus, the system in which a singular cycle is born from a stable focus as a result of Andronov–Hopf bifurcation has singular attractors with the most simple structure. They lie on two-dimensional two-sheeted surfaces. The cause of such simplicity consists in coincidence of an unstable two-dimensional invariant manifold W^u of a singular cycle and unstable two-dimensional invariant manifold of a saddle–focus. For comparison the Poincare map of more complex three-sheeted classical Lorenz attractor is shown in Fig. 4.26b in coordinates $\rho = \sqrt{x^2 + y^2}$ and z.

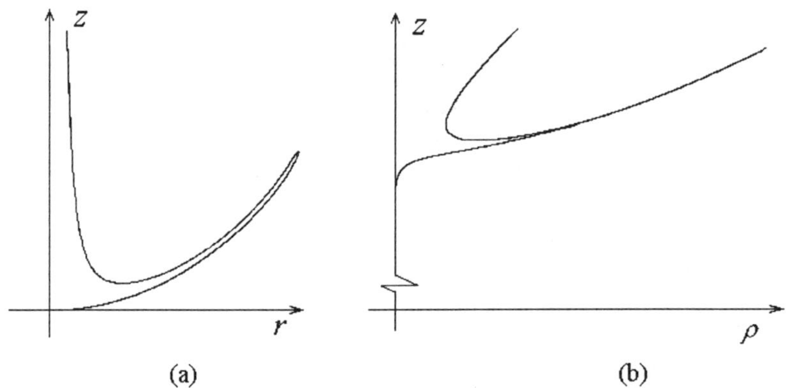

(a)   (b)

Fig. 4.26   Poincare mappings for the system (4.67) in the plane $\varphi = 0$ for $\mu = \mu^*$ (a) and for Lorenz system (3.1) in the plane $\varphi = 0.3\pi$ for $b = 8/3$, $\sigma = 10$, $r = 28$ (b).

Example 4.3   We shall show now, that there can be three-dimensional autonomous systems of ordinary differential equations having very complex structure of separatrix surfaces of their singular cycles. We shall consider an autonomous system of three differential equations

$$\dot{x}_1 = -\omega x_2 - \omega x_1 x_3/2 - ((\mu - 1)x_1 + 1 - 8x_3^2)(1 - x_1^2 - x_2^2),$$
$$\dot{x}_2 = \omega x_1 + 2x_3(1 - \omega x_2/4) - (\mu - 1)(1 - x_1^2 - x_2^2)x_2, \qquad (4.68)$$
$$\dot{x}_3 = 2(\mu - 1 - x_1)x_3 - (x_2 + \omega/4)(1 - x_1^2 - x_2^2).$$

The system (4.68), as well as system (4.65) of the Example 4.1, has the limit cycle $x_0(t) = (\cos\omega t, \sin\omega t, 0)^{\mathrm{T}}$, lying in the plane of variables (x_1, x_2). This cycle is stable for $\mu < 0$, and its period doubling bifurcation occurs at $\mu = \mu_1 = 0$. The system linearized on the cycle looks like:

$$\dot{y}_1 = 2\cos\omega t((\mu - 1)\cos\omega t + 1)y_1$$
$$+ (2((\mu - 1)\cos\omega t + 1)\sin\omega t - \omega)y_2 - \frac{\omega}{2}(\cos\omega t)y_3 + f,$$
$$\dot{y}_2 = (\omega + 2(\mu - 1)\cos\omega t \sin\omega t)y_1$$
$$+ 2(\mu - 1)(\sin^2\omega t)y_2 + 2(1 - \frac{\omega}{4}\sin\omega t)y_3 + g,$$
$$\dot{y}_3 = 2\cos\omega t\left(\frac{\omega}{4} + \sin\omega t\right)y_1$$
$$+ 2\sin\omega t\left(\frac{\omega}{4} + \sin\omega t\right)y_2 + 2(\mu - 1 - \cos\omega t)y_3 + h,$$

where expansions of functions $f(y_1, y_2, y_3)$, $g(y_1, y_2, y_3)$ and $h(y_1, y_2, y_3)$ in series at the point $(0, 0, 0)$ begin with members of the second order. We shall reduce the last system to coordinates connected with the cycle by change of variables $y(t) = Q(t)z(t)$ with the $2\pi/\omega$-periodic matrix $Q(t)$ described in the Example 4.1

$$\dot{z}_1 = -\frac{2}{\omega}(\sin\omega t)z_2 + \frac{2}{\omega}(\cos\omega t)z_3 + \tilde{f}(z_1, z_2, z_3),$$
$$\dot{z}_2 = 2(\mu - 1 + \cos\omega t)z_2 - \left(\frac{\omega}{2} - 2\sin\omega t\right)z_3 + \tilde{g}(z_1, z_2, z_3),$$
$$\dot{z}_3 = \left(\frac{\omega}{2} + 2\sin\omega t\right)z_2 + 2(\mu - 1 - \cos\omega t)z_3 + \tilde{h}(z_1, z_2, z_3).$$

Apparently from the last system, limit cycle $x_0(t)$ is a singular limit cycle, and analysis of all its bifurcations is reduced to the analysis of bifurcations of solutions of non-autonomous two-dimensional system with the same linear part, as in the system (4.66) from the first example. The difference is only that the system (4.68) has no saddle–focus type singular point and corresponding homoclinic contour. The last condition leads to extreme complication of structure of unstable two-dimensional invariant manifold W^u of the original singular cycle. It becomes many-sheeted. At different values of parameters ω and $\mu > 0$, the system (4.68) gives a great number of various many-sheeted attractors and many-valued mappings $f^{-1}(u_1)$, including rather exotic ones. All of them can be observed clearly in Poincare sections transversal to the original cycle, and also in a two-dimensional plane, asymptotically close to the plane S and given by variables $(\rho - 1, z)$.

At the fixed value of parameter ω and at growth of positive values of parameter μ in system (4.68), as well as in systems (4.65) and (4.67), the Feigenbaum cascade of period doubling bifurcations of the original singular limit cycle, and then the Sharkovskii subharmonic cascade of bifurcations of generation of stable cycles of arbitrary period are realized again. So, for example, for $\omega = 4$ in system (4.68), the stable cycle of quadruple period appears for $\mu_2 \approx 0.1135$, of period 8 for $\mu_3 \approx 0.1305$, of period 16 for $\mu_4 \approx 0.134$, of period 32 for $\mu_5 \approx 0.1346$. The value μ_∞ corresponding to the limit of sequence of values of parameter μ for which period doubling bifurcations of the original cycle take place approximately equals 0.135. Here the Feigenbaum attractor is generated in system (4.68). The system (4.68) has already a stable cycle of the period $10 = 5 \cdot 2$ in the Sharkovskii order for $\mu \approx 0.1406$, and a stable cycle of period 6 for $\mu \approx 0.1445$.

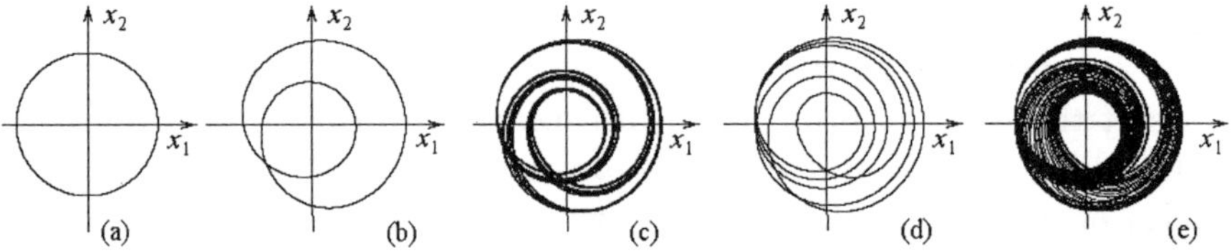

Fig. 4.27   Original singular cycle (a), a cycle of period 2(b) for $\mu = 0.1$, Feigenbaum attractor for $\mu = 0.135$ (c), a cycle of the period 6 of the subharmonic cascade (d) for $\mu = 0.1445$ and a singular subharmonic attractor for $\mu = 0.15$ (e) in autonomous three-dimensional system (4.68).

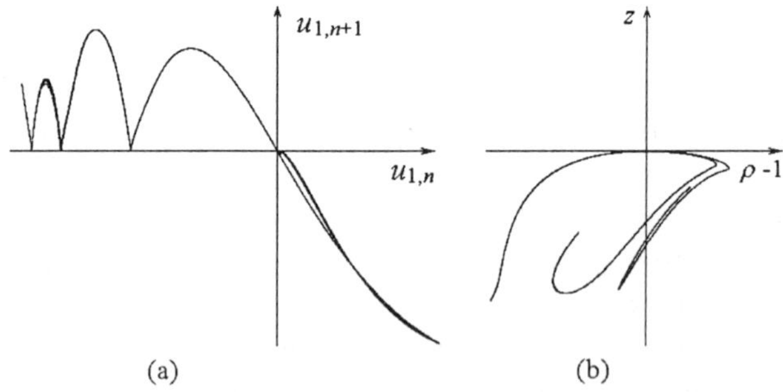

Fig. 4.28   The mapping $f(u_1)$ (a) and the Poincare mapping in the plane $\varphi = 0$ (b) for the system (4.68) at values $\omega = 4$ and $\mu = 0.26$.

The subharmonic cascade of bifurcations corresponding to the Sharkovskii order comes to the end approximately at $\mu \approx 0.155$ (Fig. 4.27).

At further increasing of values of the parameter μ, more complex than homoclinic cascade of bifurcations of stable cycles is realized in system (4.68) (Fig. 4.28). The mapping $f^{-1}(u_1)$ becomes many-valued (for example, it is five-valued for $\mu \approx 0.22$ and seven-valued for $\mu \approx 0.26$). The surface G thus becomes many-sheeted (it is four-sheeted for $\mu \approx 0.22$ and five-sheeted for $\mu \approx 0.26$). Hence, an existence is possible for much more structurally complex attractors in three-dimensional autonomous systems of ordinary differential equations than attractors of well-known classical chaotic systems including Lorenz, Rossler and Chua systems. Attractors of systems (4.65), (4.67) and (4.68) are shown in Fig. 4.29 a, b and c accordingly.

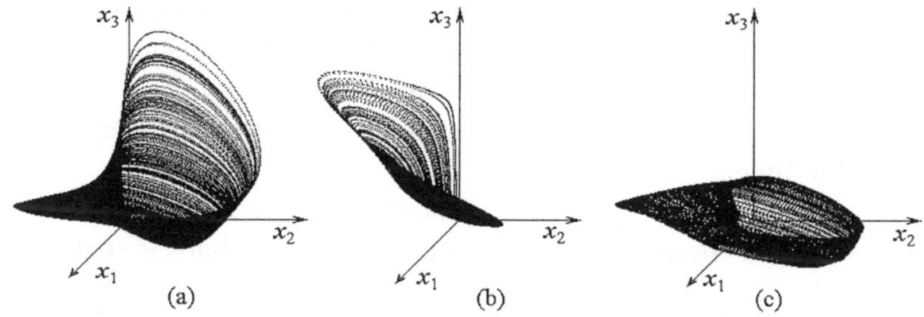

Fig. 4.29   Singular attractors of systems (4.65) (a), (4.67) (b) and (4.68) (c).

4.6   Final Remarks and Conclusions

In the present chapter we have proved that chaotic dynamics appears even in two-dimensional non-autonomous nonlinear systems of ordinary differential equations with periodic coefficients and rotor type singular points. We have proved that one universal Feigenbaum–Sharkovskii–Magnitskii scenario of transition to chaos is realized in all such systems. We have shown also that three-dimensional autonomous systems of ordinary differential equations with singular saddle limit cycles have the same universal scenario of transition to chaos and that the chaotic dynamics in such systems is induced by the dynamics of the corresponding two-dimensional non-autonomous systems written out in coordinates that are transverse to singular cycles and in which the cycles themselves are rotor type singular points. We have shown that transition to chaos in a wide class of three-dimensional autonomous nonlinear dissipative systems of ordinary differential equations with singular saddle limit cycles is performed in accordance with the following basic principles:

- any attractor (either periodic or singular) of the system lies on the two-dimensional (in general, many-sheeted) surface G that is the closure of a two-dimensional invariant unstable manifold (separatrix surface) of its singular saddle cycle;
- chaotic dynamics in the system appears owing to the phase shift between trajectories that form a separatrix surface of the singular cycle, and this leads to the possibility of appearance of continuous one-dimensional mapping with many-valued inverse mapping in the two-dimensional rotating surface S transverse to the cycle;
- any singular attractor of the system existing at accumulation points of values of the bifurcation parameter is the closure of some semistable nonperiodic trajectory, which belongs to G;
- any singular attractor has no positive Lyapunov exponent, and its fractal dimension does not exceed 2;
- in all systems of the class, at first the same universal Feigenbaum–Sharkovskii–Magnitskii scenario of transition to chaos is realized.

Therefore, in all systems of the above-considered class of two-dimensional non-autonomous systems and three-dimensional autonomous systems of ordinary differential equations, complete or incomplete subharmonic or homoclinic singular attractors are generated first during transition to chaos. As it has been shown in Chapter 3 and in the present chapter, the same attractors are also generated in all known classical two-dimensional non-autonomous and three-dimensional autonomous dissipative systems of nonlinear ordinary differential equations (including Duffing–Holmes and Matie equations, Lorenz, Rossler and Chua systems *etc.*). The same universal scenario of transition to chaos through a subharmonic and homoclinic cascades of bifurcations of stable cycles or stable two-dimensional tori is realized as well as in many-dimensional nonlinear dissipative systems of differential equations (including the complex Lorenz system and the Rikitaki system) and in infinite-dimensional nonlinear dissipative systems of ordinary differential equations with delay argument and partial differential equations (including the Mackey–Glass equation, Brusselator and Ginzburg–Landau equation considered in Chapter 5). Therefore, believable is the conjecture that the presented in this chapter theory of appearance of chaotic dynamics in nonlinear dissipative systems of differential equations is universal.

Chapter 5

Dynamical Chaos in Infinite-Dimensional Systems of Differential Equations

Chaotic systems of ordinary differential equations considered in previous chapters are in many cases finite-dimensional approximations of infinite-dimensional systems of partial differential equations. The classical system of Lorenz equations can be considered as a confirmation of this statement, because it is received in the form of finite-dimensional approximations of Navie–Stocks, continuity and heat conduction partial differential equations by *the Galerkin method* [Lorenz (1963); Schuster (1984)]. Therefore it is natural to assume, that complex irregular regimes of behaviour are inherent not only to finite-dimensional, but also to infinite-dimensional nonlinear systems of differential equations. At present the theory of dynamical chaos in infinite-dimensional nonlinear systems of differential equations is not practically developed. Among a few results received in this direction, it is necessary to distinguish the results of A. Samarskii, S. Kurdumov and their pupils about appearance of *nonstationary, spatially inhomogeneous and nonperiodic solutions (diffusion chaos)* of the Kuramoto–Tsuzuki equation describing behaviour of solutions of the system of reaction–diffusion equations in a neighbourhood of its stationary homogeneous state [Akhromeeva *et al.* (1992)]. A number of results concerning scenarios of occurrence of dynamical chaos in the system of equations, describing market economy, in systems of reaction–diffusion equations and in systems of differential equations with delay argument was received recently by authors in works [Magnitskii and Sidorov (1999); Magnitskii and Sidorov (2000); Magnitskii and Sidorov (2005b); Magnitskii and Sidorov (2005c)]. We show in the present chapter that in all considered cases of transition to dynamical chaos in three-dimensional systems of ordinary differential equations which are few-mode approximations for infinite-dimensional systems, the mechanism of formation of chaotic dynamics in such systems is described by the

261

universal Feigenbaum–Sharkovskii–Magnitskii theory stated in Chapter 4. Under the same *FSM scenario*, a transition to chaos occurs in nonlinear systems of ordinary differential equations with delay argument.

The scenarios for transition to diffusion chaos in systems of partial differential equations become less clear since not only cycles, but also two-dimensional tori begin to take part in such scenarios. We consider three systems of partial differential equations in the present chapter: the system of Brusselator equations as an example of a reaction–diffusion system, the Kuramoto–Tsuzuki equation and the Magnitskii system of market economy equations. We show that at least one universal scenario is realized in all examples of nonlinear partial differential equations considered by us. It is a subharmonic cascade of bifurcations of two-dimensional tori along one or both frequencies. This cascade can also be described by the universal Feigenbaum–Sharkovskii–Magnitskii theory. Considering the constructed model of a market economy, we discovered some more complex chaotic regimes, than in the Brusselator system or in the Kuramoto–Tsuzuki equation. We show in the present chapter that already homogeneous spatial solutions of the system of partial differential equations, describing variations of macroeconomic indices, possess a chaotic dynamics, the transition to which also occurs according to the theory presented in Chapter 4. Then the further complication of chaotic regimes in the market economy model of partial differential equations is realized through a subharmonic cascade of bifurcations of two-dimensional tori. Besides, we show that complexity of chaotic dynamics in partial differential equations depends essentially on the size of a spatial area.

5.1 Regular Dynamics and Diffusion Chaos in Reaction–Diffusion Systems

The wide class of physical, chemical and biological mediums strongly studied by nonlinear and chaotic dynamics, is described by *the reaction–diffusion system* of partial differential equations

$$
\begin{aligned}
u_t &= D_1 u_{xx} + f(u, v, \mu), \\
v_t &= D_2 v_{xx} + g(u, v, \mu), \\
&\quad 0 \leq x \leq l,
\end{aligned}
\tag{5.1}
$$

dependent on a scalar parameter μ. As a rule, there is a positive feedback on one of variables in the systems of a kind (5.1). Such variable is called *an activator*. The second variable which slows down the increase (development) of the activator, is called *an inhibitor*.

At studying systems of reaction–diffusion equations, one has the greatest interest in analysis of such boundary value problems, for which the system (5.1) has a stable stationary and homogeneous solution (U, V) for all values of scalar system parameter $\mu \leq \mu_0$. Such solution is named *a thermodynamic branch*. At $\mu > \mu_0$ the thermodynamic branch loses stability, and behaviour of solutions is defined by a spectrum of a boundary value problem linearized on a thermodynamic branch in a neighbourhood of bifurcation point μ_0. If one simple eigenvalue of the operator of linearization of a linearized boundary value problem on the solution (U, V) passes through zero at the value $\mu = \mu_0$ and other spectrum of this operator remains lying in the left half-plane, then stationary and spatial-inhomogeneous solution (*stationary dissipative structure*) appears in the reaction–diffusion system. For the first time such bifurcation was found by *A. Turing* at research of mathematical model of morphogenesis and got his name [Turing (1952)]. If two complex conjugate eigenvalues of linearization operator of a linearized boundary value problem on the thermodynamic branch pass from left to right through an imaginary axis at $\mu = \mu_0$ and other spectrum remain lying in the left half-plane, then the *Andronov–Hopf bifurcation* of birth of a cycle occurs [Hassard *et al.* (1981)]. In this case the thermodynamic branch loses stability at $\mu > \mu_0$, and points of segment $[0, l]$ begin to make periodic oscillations. The Turing and Andronov–Hopf bifurcations have obvious analogues among bifurcations of singular points of nonlinear ordinary differential equations considered in Chapter 2. They are the pitchfork bifurcation and the birth of a stable cycle bifurcation.

5.1.1 *Turing and Andronov–Hopf bifurcations in the Brusselator model*

Let us consider the Turing and Andronov–Hopf bifurcations in great detail by example of the first boundary value problem for the classical system of equations of a kind (5.1), offered for the first time by the Brussels school of I. Prigogine as a model of some self-catalyzed chemical reaction with diffusion and named as *Brusselator* [Lefever and Prigogine (1968); Hassard *et al.* (1981)]. One can find proofs for the theorems of common kind by reducing the infinite-dimensional system of the Eqs. (5.1) on two-dimensional or one-dimensional central manifold in the works [Marsden and McCracken (1976); Hassard *et al.* (1981)].

The system of the Brusselator equations considered on a segment $[0, l]$ looks like

$$u_t = D_1 u_{xx} + A - (\mu + 1)u + u^2 v,$$
$$v_t = D_2 v_{xx} + \mu u - u^2 v, \tag{5.2}$$
$$0 \le x \le l.$$

It is easily to see, that stationary spatially homogeneous solution (a thermodynamic branch) of the system (5.2) is the solution $u = A$, $v = \mu/A$. Therefore the first boundary value problem for Brusselator should satisfy the boundary conditions

$$u(0, t) = u(l, t) = A, \ v(0, t) = v(l, t) = \mu/A. \tag{5.3}$$

Let us linearize the problem (5.2)–(5.3) on the thermodynamic branch denoting $p = u - A$, $q = v - \mu/A$. We shall obtain

$$p_t = D_1 p_{xx} + (\mu - 1)p + A^2 q + h(p, q),$$
$$q_t = D_2 q_{xx} - \mu p - A^2 q - h(p, q) \tag{5.4}$$

with boundary conditions $u(0, t) = u(l, t) = v(0, t) = v(l, t) = 0$, where $h(p, q) = (\mu/A)p^2 + 2Apq + p^2 q$.

Operator of the linear part of the system (5.4) operating in *the Sobolev space* $H^2[0, l]$ with zero boundary values, can be represented in the form of $L = K + D\Delta$, where

$$K = \begin{pmatrix} \mu - 1 & A^2 \\ -\mu & -A^2 \end{pmatrix}, \quad D = \begin{pmatrix} D_1 & 0 \\ 0 & D_2 \end{pmatrix}, \quad \Delta = \frac{\partial^2}{\partial x^2}.$$

The operator Δ has eigenvalues $-\pi^2 n^2/l^2$ ($n = 1, 2, \dots$) responding to eigenfunctions $\sin(\pi n x/l)$. We shall expand components of eigenvector Ψ of the operator L, responding to its eigenvalue λ, on these functions $\sin(\pi n x/l)$.

Then $\qquad L\Psi = (K + D\Delta)\Psi = \lambda\Psi \qquad$ or

$$\sum_{n=1}^{\infty} \left[K \begin{pmatrix} a_n \\ b_n \end{pmatrix} + D \begin{pmatrix} a_n \\ b_n \end{pmatrix} \left(\frac{-\pi^2 n^2}{l^2} \right) \right] \sin \frac{\pi n x}{l} = \lambda \sum_{n=1}^{\infty} \begin{pmatrix} a_n \\ b_n \end{pmatrix} \sin \frac{\pi n x}{l}.$$

Hence,

$$\left(K - \frac{\pi^2 n^2}{l^2} D \right) \begin{pmatrix} a_n \\ b_n \end{pmatrix} = \lambda \begin{pmatrix} a_n \\ b_n \end{pmatrix}, \quad n = 1, 2, \dots$$

Therefore eigenvalues of the operator L are eigenvalues of matrices $G_n = K - \dfrac{\pi^2 n^2}{l^2} D$ and should satisfy the equations $\lambda^2 - \operatorname{tr} G_n + \det G_n = 0$ where

$$\operatorname{tr} G_n = \mu - 1 - A^2 - \frac{\pi^2 n^2}{l^2}(D_1 + D_2),$$

$$\det G_n = A^2 + \frac{\pi^2 n^2}{l^2}\left(A^2 D_1 + D_2 + \frac{\pi^2 n^2}{l^2} D_1 D_2 - \mu D_2\right).$$

Existence of the only pair of imaginary eigenvalues $\lambda = i\omega$, $\omega > 0$ is necessary for generation of a cycle birth bifurcation at some value of the parameter μ. That is a trace of one of matrices G_n should be equal to zero, and the determinant of this matrix should remain positive. Thus all other eigenvalues of the operator L should have negative real parts. In other words, the equation

$$\mu = 1 + A^2 + \frac{\pi^2 m^2}{l^2}(D_1 + D_2)$$

should to be necessary carried out at some $n = m$.

If $m > 1$ then there exists $n < m$ such, that

$$\operatorname{tr} G_n = (m^2 - n^2)\frac{\pi^2}{l^2}(D_1 + D_2) > 0.$$

Then at least one eigenvalue of the operator L, corresponding to this number n, will have a positive real part. Hence, it is necessary that $m = 1$,

$$\mu = \mu_0 = 1 + A^2 + \frac{\pi^2}{l^2}(D_1 + D_2),$$

and

$$\det G_n = A^2 + \frac{\pi^2 n^2}{l^2}\left(A^2(D_1 - D_2) + \frac{\pi^2}{l^2} D_2(n^2 D_1 - D_1 - D_2)\right).$$

It is easy to conclude from the last expression, that if coefficients of diffusion D_1 and D_2 of the system (5.2) satisfy the condition $D_1 \geq D_2 + \dfrac{\pi^2}{l^2}(D_2/A)^2$, then $\det G_n > \det G_1 = \omega_0^2 \geq A^2$ for all $n > 1$. Thus, conditions of the Andronov–Hopf theorem on a spectrum of the operator L are executed for $\mu = \mu_0$. A more detailed analysis shows [Hassard *et al.* (1981)], that stable periodic spatially inhomogeneous solutions of the system (5.2) have the following asymptotic representations for small $\varepsilon = (\mu - \mu_0)^{1/2}$:

$$u(x,t) = A + \varepsilon \cos \omega(\varepsilon)t \cdot \sin \frac{\pi x}{l} + O(\varepsilon^2),$$

$$v(x,t) = \frac{\mu}{A} + \varepsilon\gamma \cos \omega(\varepsilon)t \cdot \sin \frac{\pi x}{l} + \varepsilon\delta \sin \omega(\varepsilon)t \cdot \sin \frac{\pi x}{l} + O(\varepsilon^2),$$

where $\omega(\varepsilon) = \omega_0(1 + O(\varepsilon^2))$, $\quad \gamma$ and δ are some constants, and a kind of spatial harmonics is defined by boundary conditions of a problem, i.e. by eigenfunctions of the operator Δ for $n = 1$ in this case. Points of a segment make fluctuations with identical frequency and a constant gradient of a phase. The effect of "wave" running on a segment is created. We shall notice, that at small coefficients of diffusion the condition of birth of periodic solutions in the system (5.2) can be written down roughly in the form of $D_1 > D_2$. It can be interpreted as an interaction of a long-range activator and a short-range inhibitor in the Brusselator. A change of character of boundary conditions changes essentially the kind of periodic solutions of the system (5.2) born as a result of Andronov–Hopf bifurcation. So, for example, already functions $\cos(\pi n x/l)$, $n = 0, 1, 2, \ldots$, will be eigenfunctions of the operator Δ for the second boundary value problem on a segment with boundary conditions

$$u_x(0,t) = u_x(l,t) = v_x(0,t) = v_x(l,t) = 0. \tag{5.5}$$

And zero will be the least eigenvalue. Therefore the value $\mu_0 = 1 + A^2$ will be the bifurcation value for the parameter μ, and periodic solutions born at $\mu > \mu_0$ in the case of $D_1 > D_2$ will be spatially homogeneous.

Let us consider again the first boundary value problem for the system (5.2). For occurrence of Turing bifurcation in the system (5.2) at some value of parameter μ, it is necessary that a determinant of one of matrices G_n should be equal to zero, and the trace of this matrix should remain negative. And all other eigenvalues of the operator L should have negative real parts. Therefore traces of all matrices G_n should be negative, whence follows, that $\mu < \mu^* = 1 + A^2 + (\pi^2/l^2)(D_1 + D_2)$.

Let $\mu = \mu^* - \theta$, $\theta > 0$. Then

$$\det G_n = A^2 + \frac{\pi^2 n^2}{l^2}\left(A^2(D_1 - D_2) + \frac{\pi^2}{l^2}D_2(n^2 D_1 - D_1 - D_2) + \theta D_2\right).$$

Let us designate

$$y_n = \frac{\pi^2 n^2}{l^2}, \quad \alpha = A^2(D_1 - D_2) - \frac{\pi^2}{l^2}D_2(D_1 + D_2) + \theta D_2.$$

Then $\det G_n = D_1 D_2 y_n^2 + \alpha y_n + A^2$. Therefore one can choose a value of $\alpha < 0$ so that it will be executed $\det G_m = 0$ and $\det G_n > 0$ for all $n \neq m$ and for some values $\theta > 0$ and $m \geq 1$. Thus, conditions of the Turing theorem on a spectrum of the operator L will be executed at $\mu = \mu_0 = \mu^* - \theta$. A more detailed analysis shows, that born thus stable spatially inhomogeneous dissipative structures of the system (5.2) have the following asymptotic representations for small $\varepsilon = (\mu - \mu_0)^{1/2}$:

$$u(x, t) = A + \varepsilon \sin \frac{\pi m x}{l} + O(\varepsilon^2),$$

$$v(x, t) = \frac{\mu}{A} + O(\varepsilon) \sin \frac{\pi m x}{l} + O(\varepsilon^2),$$

where a kind of spatial structure is also defined by boundary conditions of the problem. We shall notice, that for small coefficients of diffusion, the condition of birth of stationary inhomogeneous spatial structure in the system (5.2) can be written down roughly in the form of $D_1 < D_2$. It can be interpreted as an interaction of a short-range activator and a long-range inhibitor in the Brusselator.

5.1.2 *Diffusion chaos for the Brusselator in a ring*

The above-presented simple linear analysis of the system (5.2) already evidently shows the complexity of systems of reaction–diffusion type. Behaviour of their solutions depends essentially on values of coefficients of diffusion and on relation between them, on the form of a spatial area and on its size, and also on the kind of boundary conditions. At bifurcation of a thermodynamic branch in such system, one can observe the appearance of periodic spatially homogeneous and inhomogeneous solutions and stationary dissipative structures. Numerical experiments show, that at further increasing of values of bifurcation parameter in the reaction–diffusion systems, one can observe spatially inhomogeneous nonperiodic solutions as well, i.e. diffusion chaos. However, the theory of transition to chaos in such systems is practically not developed now.

In the present section we consider the problem of a birth of chaotic regimes in the system (5.2) at growth of values of the parameter l which is the length of a spatial area. The fact, that complexity of solutions of the system (5.2) should increase at growth of the length of a spatial area, already follows from the above-presented linear analysis. At increase in l, the frequency of born periodic solutions decreases and, hence, the period and complexity of born periodic solutions grows. The value m grows and, hence, the complexity of born stationary dissipative structures grows too. Besides, a greater number of harmonics is required for expansion of a solution in the Fourier series with the given accuracy. Hence, the dimension of any finite approximation of the infinite-dimensional system (5.2) increases too. As a result of numerical experiments spent in the work [Dernov (2001a)] with the second boundary value problem for the system (5.2) in a ring, it was possible to find out some types of attractors at various values of length of the ring l. First spatially homogeneous oscillations are observed after the loss of stability of a thermodynamic branch. At increase in l oscillations become spatially inhomogeneous. At further increase in the length of a ring, a stable two-dimensional torus is generated, filled by a trajectory everywhere dense (Fig. 5.1a). The projection of its Poincare section to a two-dimensional plane looks like a closed curve. At greater values of l, different attractors are generated having more complex structures. This confirms the presence of diffusion chaos in the second boundary value problem for the Brusselator system (5.2) (Fig. 5.1b).

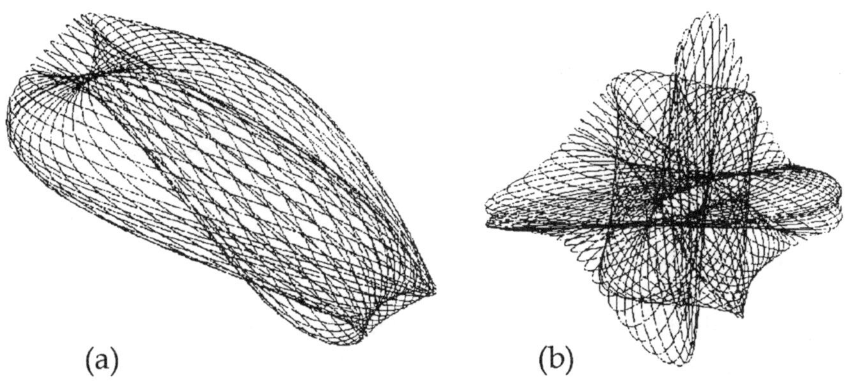

(a)   (b)

Fig. 5.1　Stable two-dimensional torus (a) and a more complex solution (b) in the Brusselator model.

Results stated in the present section give the basis to confirm that there exist complex nonstationary nonperiodic solutions (diffusion chaos) in systems of the reaction–diffusion equations of a kind (5.1). Moreover, there are also strong reasons to assume, that two-dimensional tori and, quite possible, all subharmonic cascade of two-dimensional tori participate in scenarios for transition to chaos just as it takes place in the five-dimensional system of the complex Lorenz equations (see Sec. 3.2).

5.1.3 *Diffusion chaos in the Brusselator on a segment*

In the present section, we consider scenarios for transition to diffusion chaos in the system of Brusselator equations on a segment. We show that this transition also occurs in accordance with the Feigenbaum–Sharkovskii–Magnitskii theory through a subharmonic cascade of bifurcations of stable cycles or stable two-dimensional tori.

In the beginning we shall consider the first boundary value problem (5.2)–(5.3). This problem was solved on the interval $l = \pi$ at the parameter value $A = 2$ and at various coefficients of diffusion. In the case of $D_1 = 0.04 > D_2 = 0.01$ and for the value of bifurcation parameter $\mu_0 = 5.05$, a stable spatially inhomogeneous periodic solution is generated for the problem (5.2)–(5.3) as a result of Andronov–Hopf bifurcation. This periodic solution is presented in the form of a closed curve (limit cycle) in the phase space. This cycle saves stability up to the value of $\mu \approx 5.43$ at which it generates a two-dimensional stable invariant torus as a result of the second Andronov–Hopf bifurcation. The torus exists for $\mu \approx 6$ and at further increase of the bifurcation parameter μ the torus disappears and stable stationary dissipative structure arises.

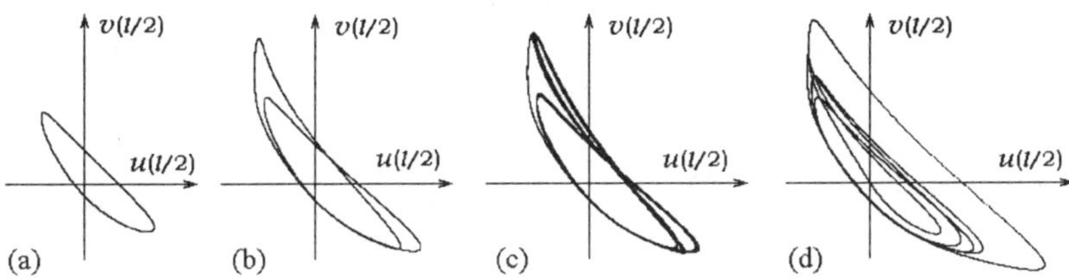

Fig. 5.2   Singular cycle (a) for $\mu = 6$, a cycle of the double period (b) for $\mu = 6.4$, the Feigenbaum attractor (c) for $\mu = 6.5905$ and a cycle of period 5 of the Sharkovskii subharmonic cascade (d) for $\mu = 6.825$.

For other coefficients of diffusion $D_1 = 0.15 < D_2 = 0.3$, another scenario of transition to chaos takes place in the same first boundary value problem, that is the period doubling cascade of bifurcations of stable limit cycles. A stable periodic spatially inhomogeneous solution, a stable limit cycle is born in the problem (5.2)–(5.3) as a result of Andronov–Hopf bifurcation for the value $\mu = 5.45$ (Fig. 5.2a). It remains stable up to the value of $\mu \approx 6.11$. A stable cycle of the double period (Fig. 5.2b) is observed in the region of values $\mu \in [6.12, 6.57]$, a cycle of period 4 is observed for $\mu = 6.58$, a cycle of period 8 is observed for $\mu = 6.587$. The period doubling cascade of bifurcations comes to the end with formation of the Feigenbaum singular attractor for $\mu \approx 6.5905$ (Fig. 5.2c).

For $\mu > 6.591$ stable limit cycles having the periods according to the Sharkovskii order are observed in the problem (5.2)–(5.3). For example, there is a stable cycle of the period $6 = 3 \cdot 2$ for $\mu = 6.64$, a stable cycle of the period 7 for $\mu = 6.765$. The subharmonic cascade of bifurcations comes to the end here with a cycle of the period 5 for $\mu = 6.825$ (Fig. 5.2d). The appearance of a stationary structure at the value of $\mu \approx 6.90$ terminates the subharmonic cascade for the given set of other fixed parameters.

Now we shall consider the second boundary value problem for the system (5.2) with boundary conditions (5.5) at the interval $l = \pi$. We did not find chaotic solutions in the region of parameter values $0 < A \leq 3$ and for various coefficients of diffusion. However, for $A = 3$ and for $D_1 = 0.04$, $D_2 = 0.01$, a bifurcation of birth of a two-dimensional invariant torus is found for the parameter value $\mu \approx 10.06$. And two more consecutive period doubling bifurcations of this torus on internal (basic) frequency were found for the values $\mu \approx 10.1224$ and $\mu \approx 10.14687$. However, at the value $A = 3$, the cascade of period doubling bifurcations of two-dimensional tori is incomplete. A two-dimensional invariant torus of the quadruple period on internal frequency passes into a two-dimensional invariant torus of period 2 on internal frequency for the value $\mu \approx 10.156$ and again turns into the usual two-dimensional invariant torus for $\mu \approx 10.16$.

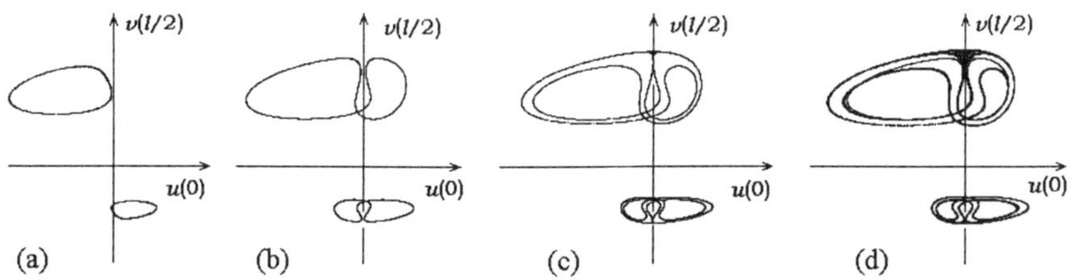

Fig. 5.3 Projections of the Poincare mapping in the section $u(l/2) = 0$ on a plane $(u(0), v(l/2))$ of two-dimensional torus (a), two-dimensional torus of double period on internal frequency (b), two-dimensional torus of period 4 (c) and the Feigenbaum singular attractor (d) in the problem (5.2), (5.5).

Singular attractors were found out in the second boundary value problem for Brusselator for the parameter value $A = 4$ and for coefficients of diffusion $D_1 = 0.1$, $D_2 = 0.02$. At these fixed parameters and for length of an interval $l = \pi$, there exists a stable limit cycle in the problem (5.2), (5.5) in the region of parameter values $\mu \in (17.00, \ 17.06)$. For

$\mu \approx 17.06$ a two-dimensional invariant torus is born in the phase space of the system (5.2) (Fig. 5.3a). This torus saves its stability up to the value of $\mu \approx 17.07392$ at which its period doubling bifurcation on internal frequency occurs (Fig. 5.3b). The existence of stable two-dimensional invariant tori with periods of 4 and 8 on internal basic frequency is established further for parameter values $\mu = 17.076$ and $\mu \approx 17.0798$ accordingly. For the value of $\mu \approx 17.0801$, a Feigenbaum singular attractor exists in the second boundary value problem (5.2), (5.5) formed as a result of the period doubling bifurcation cascade of two-dimensional invariant tori on internal frequency (Fig. 5.3d).

5.2   Transition to Spatio-Temporal Chaos in the Kuramoto–Tsuzuki Equation

It was shown in the paper [Kuramoto and Tsuzuki (1975)] that any solution of the system of reaction–diffusion equations (5.1), arising in the neighbourhood of a thermodynamic branch as a result of its bifurcation at $\mu > \mu_0$, can be expressed through some complex-valued function $W(r, \tau)$ satisfying the equation

$$W_\tau = W + (1 + ic_1)W_{rr} - (1 + ic_2)W|W|^2, \qquad (5.6)$$

where $r = \varepsilon x$, $\tau = \varepsilon^2 t$, c_1 and c_2 are some real constants with values defined by coefficients D_1, D_2, functions $f(u, v, \mu)$, $g(u, v, \mu)$ and their derivatives calculated at a thermodynamic branch, and $\varepsilon = (\mu - \mu_0)^{1/2}$ is a small parameter. The Eq. (5.6), named by *Kuramoto–Tsuzuki* or Time Dependent *Ginzburg–Landau equation*, plays the important role in studying and understanding of processes occurring in nonlinear dissipative diffusion type mediums. One can be easily to be convinced by direct substitution that function $W(\tau) = \exp(-i(c_2\tau + \varphi))$ is the homogeneous solution of Eq. (5.6) for any phase φ. Hence, each element of the medium (5.6) makes harmonic oscillations with frequency c_2 in the established regime, and this regime is stable in some large area with change of parameters c_1 and c_2. Such mediums are accepted to be named as *self-oscillating mediums*.

For finding spatially inhomogeneous solutions, the second boundary value problem of the Eq. (5.6) is usually considered. It has been shown in a cycle of works by A. Samarskii, S. Kurdumov and their pupils [Akhromeeva *et al.* (1992)] that an automodel solution of a kind $W(r, \tau) = F(r)\exp(i(\omega\tau + a(r)))$ is stable in some area of change of parameters c_1

and c_2. If $a(r) = kr$, then oscillations of the neighboring elements of the medium occur with a constant phase shift that corresponds to a movement of a phase wave. In a two-dimensional case the equation (5.6) has such solutions as spiral waves and leading centers, i.e. sequences of running up concentric phase waves. Areas of change of parameters c_1 and c_2 were also found, in which the second boundary value problem for the Eq. (5.6) on a segment $[0, R]$ has nonstationary nonperiodic and inhomogeneous solutions, that is a diffusion or *spatio-temporal chaos*.

The author's of work [Akhromeeva *et al.* (1992)] idea for explanation of nature of such solutions consisted in use of Galerkin few-mode approximations for reduction of Eq. (5.6) to a more simple finite-dimensional (three-dimensional) system of ordinary differential equations. They have shown, that when the length of a segment is insignificant ($R \sim \pi$) and Fourier coefficients of solutions quickly decrease with growth of their numbers, then not only qualitative, but in some cases quantitative conformity exists between solutions of the original Eq. (5.6) and two-mode (tree-dimensional) system of ordinary differential equations. And it concerns not only to elementary regular (periodic), but also to irregular attractors and to scenarios of their occurrence at change of parameters c_1 and c_2. It allowed them to find from the beginning of an area of dynamical chaos in the simplified three-dimensional system of ordinary differential equations, and then simply enough to find out a possible area of diffusion chaos in the Kuramoto–Tsuzuki Eq. (5.6). However, only areas of existence of stable singular points, simple stable cycles and cycles of double period of the simplified three-dimensional system of ordinary differential equations were found in the space of parameters (c_1, c_2). All more complex regular (periodic) and irregular attractors of the simplified three-dimensional system were simply referred to one class to which the remained areas of the space of parameters correspond. Therefore the approach offered in [Akhromeeva *et al.* (1992)] did not give an opportunity to explain mechanisms and to define natural scenarios for generation of chaotic dynamics as in a simplified three-dimensional system of ordinary differential equations and especially in the original Kuramoto–Tsuzuki Eq. (5.6). The main problem of relation between spatio-temporal chaos in diffusion type systems and dynamical chaos in dissipative systems of ordinary differential equations was not solved. As it was shown in Chapters 3 and 4, three-dimensional nonlinear dissipative systems of ordinary differential equations have a common universal scenario of transition to chaos through the Feigenbaum cascade of period doubling bifurcations of stable cycles and then through the Sharkovskii subharmonic cascade of bifurcations of stable cycles of all periods up to a cycle of period 3 and then through the Magnitskii homoclinic cascade of bifurcations of stable cycles tending to homoclinic contours. But in systems of ordinary differential equations of higher dimension, the scenario of transition to chaos contains two-dimensional tori and possibly the entire subharmonic cascade of bifurcations of two-dimensional tori with respect to one or to both frequencies, just as it is in the five-dimensional system of complex Lorenz equations (see Section 3.2). So, there are solid arguments to doubt in correctness of generalization of results valid for a few-mode (three-dimensional) systems to infinite-dimensional systems of partial differential equations.

In the present section, on the basis of numerical solution it is shown that the transition to chaos in the space of few-mode approximations for the Kuramoto–Tsuzuki equation occurs in accordance with *the Feigenbaum–Sharkovskii–Magnitskii (FSM) scenario*, but transition to spatio-temporal chaos in the phase space of solutions of the Kuramoto–Tsuzuki equation occurs through the cascades of Feigenbaum–Sharkovskii–Magnitskii bifurcations of two-dimensional invariant tori with respect to internal as well as external frequencies, and hence this scenario also can be described by *the FSM theory* presented in Chapter 4.

5.2.1 *Scenario of transition to chaos in system of few-mode approximations*

Following to the work [Akhromeeva *et al.* (1992)], let us consider the second boundary value problem for the Kuramoto–Tsuzuki equation

$$W_t = W + (1 + ic_1)W_{xx} - (1 + ic_2)W|W|^2,$$
$$W_x(0, t) = W_x(l, t) = 0, \quad W(x, 0) = W_0(x), \ 0 \leq x \leq l, \ 0 \leq t < \infty, \tag{5.7}$$

where $W = W(x, t) = u(x, t) + iv(x, t)$ is a complex-valued function. The use of Galerkin few-mode approximations

$$W(x, t) \approx \xi^{1/2}(t) \exp(i\theta_1(t)) + \eta^{1/2}(t) \exp(i\theta_2(t)) \cos kx, \ k = \pi/l,$$

permits one to reduce the infinite-dimensional problem (5.7) to the simpler three-dimensional system of nonlinear ordinary differential equations

$$\dot{\xi} = 2\xi - 2\xi(\xi + \eta) - \xi\eta(\cos\theta + c_2\sin\theta),$$
$$\dot{\eta} = 2\eta - 2\eta(2\xi + 3\eta/4) - 2\xi\eta(\cos\theta - c_2\sin\theta) - 2k^2\eta, \tag{5.8}$$
$$\dot{\theta} = c_2(2\xi - \eta/2) + (2\xi + \eta)\sin\theta + c_2(2\xi - \eta)\cos\theta + 2c_1k^2$$

for variables ξ, η and $\theta = \theta_2 - \theta_1$ which possess chaotic dynamics.

We show in this section that all irregular attractors of the three-dimensional system (5.8) are singular attractors, and that transition to chaos in the system (5.8) of few-mode approximations also occurs in accordance with *Feigenbaum–Sharkovskii–Magnitskii (FSM) theory* through the Feigenbaum cascade of period doubling bifurcations of stable cycles, then through the Sharkovskii subharmonic cascade of bifurcations of stable cycles of all periods up to a cycle of period 3 and then through the Magnitskii cascade of bifurcations of stable homoclinic cycles.

Indeed, set, say, $k = 1$ and $c_1 = 1.3$ and consider a scenario of transition to chaos in the system (5.8) as parameter c_2 varies. Attractors of the system (5.8) will be observed in a three-dimensional phase space with coordinates $x = \xi \cos \theta$, $y = \xi \sin \theta$, $z = \eta$. By integrating system (5.8) by a fourth-order Runge–Kutta method, one can readily show that if parameter c_2 is negative and decreases, then in system (5.8), just as in all other chaotic systems first cascade of Feigenbaum period doubling bifurcations is realized for the singular stable limit cycle existing in the system for $c_2 = -5$ and then a subharmonic cascade of bifurcations of stable cycles with arbitrary period in accordance with the Sharkovskii order is realized (Fig. 5.4). For example, a stable cycle of double period is generated in the system (5.8) for $c_2 \approx -6.782$, of quadruple period for $c_2 \approx -7.92$, of period 8 for $c_2 \approx -8.15$, of period 16 for $c_2 \approx -8.2$, of period 32 for $c_2 \approx -8.21$ and of period 64 for $c_2 \approx -8.211$.

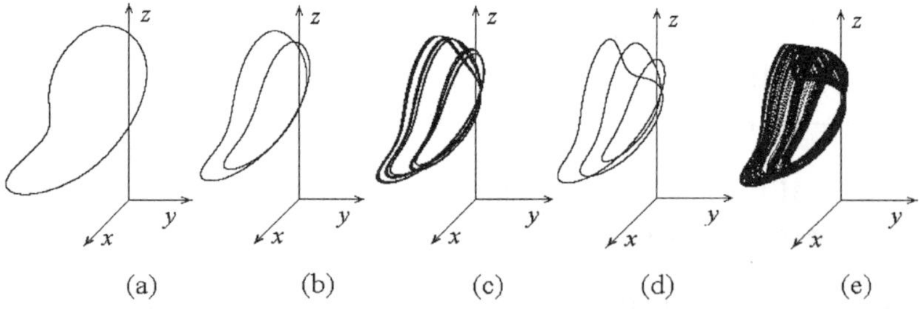

Fig. 5.4   Singular original cycle (a), period two cycle (b), Feigenbaum attractor (c), period three cycle (d) and some more complex subharmonic singular attractor (e) in the simplified system (5.8) for $c_1 = 1.3$ and when c_2 varies.

The Feigenbaum attractor appears in the system (5.8) for $c_2 \approx -8.2111$. If $c_2 \approx -8.2155$, then in the system (5.8) there is a stable cycle of period 40, which corresponds to the cycle of period $5 \cdot 2^3$ in the Sharkovskii order.

A cycle of period $20 = 5 \cdot 2^2$ is generated for $c_2 \approx -8.2509$, a cycle of period $14 = 7 \cdot 2$ is generated for $c_2 \approx -8.2754$; a cycle of period $10 = 5 \cdot 2$, for $c_2 \approx -8.2949$; a cycle of period $6 = 3 \cdot 2$, for $c_2 \approx -8.348$. Generation of the last cycle implies generation of a stable cycle of period 3 in some two-dimensional plane transversal to the original cycle. For $c_2 \approx -8.564$, in system (5.8), there appears a stable cycle of period 7; for $c_2 \approx -8.668$, a stable cycle of period 5; and for $c_2 \approx -9.0$, a stable cycle of period 3. All generated cycles undergo their own cascades of Feigenbaum period doubling bifurcations.

It is important to note that, in the three-dimensional phase space (x, y, z) of system (5.8), for some parameter values, there can simultaneously exist several distinct stable cycles with their attraction domains. Each cycle of this kind can generate its own cascade of bifurcations and its own set of complete or incomplete singular subharmonic attractors. For example, we fix the parameter value $c_2 = 9.0$ and vary the parameter c_1. One can readily show that the stable cycle of period 3 existing in the system (5.8) for $c_1 = 1.3$ results from a subharmonic cascade of bifurcations of the originally stable singular cycle as the parameter c_1 decreases from the value $c_1 = 1.43$. The projection of this original cycle onto the plane (y, z) makes four rotations around some conventional center. A stable cycle of period 8 is generated in the system for $c_1 \approx 1.425$; a stable cycle of period 6, for $c_1 \approx 1.366$; and a stable cycle of period 5, for $c_1 \approx 1.33$. If $c_1 \approx 1.282$, then in system (5.8), there appears another stable cycle of period 6, which is a double-period cycle for a cycle of period 3. For further decrease of the parameter c_1, a cascade of its period doubling bifurcations occurs.

In addition, as parameter c_1 increases from the value $c_1 \approx 1.21$, in the system (5.8), there is another subharmonic cascade of bifurcations, which starts from a simple stable singular cycle. The attraction domain of this cycle contains, for example, the initial point $\xi = 0.1$, $\eta = 0.01$, $\theta = -0.2$. The cascade continues with a stable cycle of double period for $c_1 \approx 1.215$, of quadruple period for $c_1 \approx 1.232$, of period 8 for $c_1 \approx 1.237$, of period 16 for $c_1 \approx 1.2375$, and so on. A Feigenbaum attractor is generated here for $c_1 \approx 1.238$. Further, a stable cycle of period 6 is generated in the system for $c_1 \approx 1.2415$; a cycle of period 5, for $c_1 \approx 1.252$; a new stable cycle of period 3, for $c_1 \approx 1.255$; and then a cascade of its period doubling bifurcations occurs as c_1 increases. In the domain $1.26 \leq c_1 \leq 1.28$, there exist numerous singular attractors that are not described by the Sharkovskii theory but are limiting attractors for cascades of bifurcations of stable cycles. Some of these cycles of periods 22, 14, and 12 have been detected for the parameter values $c_1 = 1.268$, $c_1 = 1.273$ and $c_1 = 1.279$, respectively. The singular subharmonic attractor of the system (5.8) for $c_1 = 1.27$ is represented in Fig. 5.5a. Note that if $c_1 = 1.21$, then in the neighborhood of the original stable cycle of the cascade, there is another stable cycle of double period whose attraction domain contains, for example, the initial point $\xi = 0.1$, $\eta = 0.01$, $\theta = 0$. This cycle does not generate singular attractors.

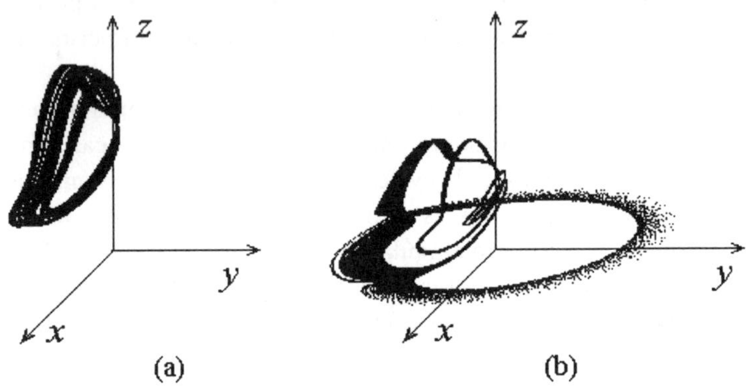

(a)   (b)

Fig. 5.5   Singular attractors of the system (5.8) in coordinates (x, y, z) for $c_2 = -9.0$ and $c_1 = 1.27$ (a), $c_1 = 1.57$ (b).

The third subharmonic cascade of bifurcations is generated by a stable cycle for $c_1 = 1.43$ if parameter c_1 increases. Here a stable cycle of double period is generated for $c_1 \approx 1.52$, of quadruple period for $c_1 \approx 1.5581$, and so on. The set of singular subharmonic attractors generated by this cycle is observed approximately in the domain $1.56 \le c_1 \le 1.59$ (see Fig. 5.5b). The fourth and fifth subharmonic cascades of bifurcations are generated by two stable singular cycles for $c_1 = 1.81$ as parameter c_1 decreases and increases. Attraction domain of the first cycle contains, for example, the initial point $\xi = 0.1$, $\eta = 0.01$, $\theta = 20$. Here a stable cycle of double period is generated for $c_1 \approx 1.802$, of quadruple period for $c_1 \approx 1.7875$, and so on. The set of singular subharmonic attractors generated by this cycle is observed approximately in the domain $1.76 \le c_1 \le 1.77$. Attraction domain of the second cycle contains, for example, the point $\xi = 0.1$, $\eta = 0.01$, $\theta = -0.2$. Here a stable cycle of double period is generated for $c_1 \approx 10.35$, of quadruple period for $c_1 \approx 11.67$ and so on. The set of singular subharmonic attractors generated by this cycle is observed approximately in the domain $12 \le c_1 \le 15$. Singular attractors of the system (5.8) for $c_2 = -9.0$, $c_1 = 1.765$ and $c_1 = 13$ are shown in Fig. 5.6. Singular attractors of the system (5.8) for some other values of parameters c_1 and c_2 are shown in Fig. 5.7.

The above-represented results imply that scenarios of transition to chaos in a three-dimensional system of few-mode Galerkin approximations for the Kuramoto–Tsuzuki equation do not differ from the scenario considered and theoretically justified in Chapter 4. However, this does not permit one to make a firm conclusion that chaotic dynamics of the infinite-dimensional system (5.7) is identical to chaotic dynamics of its three-dimensional few-mode approximation (5.8) considered in this section. This problem requires additional investigation which will be represented in the forthcoming sections of the present chapter.

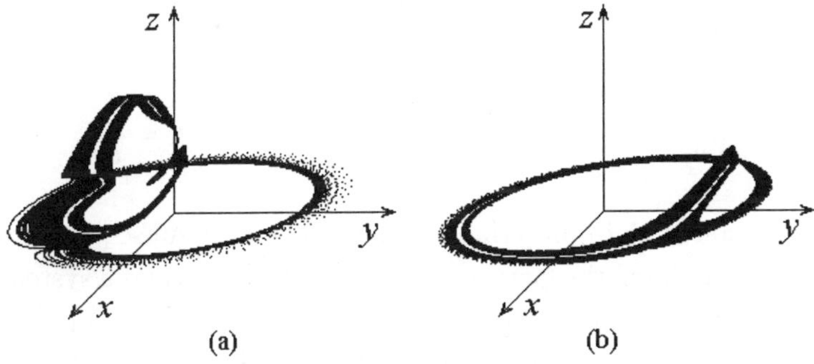

Fig. 5.6   Singular attractors of system (5.8) for the parameter values: $c_2 = -9.0$; $c_1 = 1.765$ (a); $c_1 = 13$ (b).

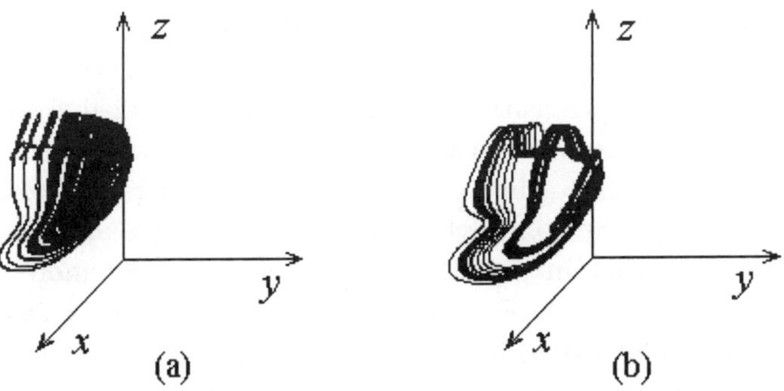

Fig. 5.7   Singular attractors of system (5.8) for the parameter values: $c_1 = 1.4125$, $c_2 = -11.5$ (a); $c_1 = 1.512$, $c_2 = -10$ (b).

5.2.2 *Transition to chaos in the space of Fourier coefficients*

The above-considered few-mode approximation of the solution $W(x,t) = u(x,t) + iv(x,t)$ of the Kuramoto–Tsuzuki equation is based on the Fourier expansion. For the second boundary value problem (5.7), this expansion has the form of

$$u(x,t) = \sum_{m=0}^{\infty} a_m(t) \cos \frac{\pi m x}{l}, \quad v(x,t) = \sum_{m=0}^{\infty} b_m(t) \cos \frac{\pi m x}{l}. \qquad (5.9)$$

Since the Fourier coefficients are rapidly decaying as their index increases, it follows that only the first harmonics are included in the few-mode approximation, i.e.

$$\begin{pmatrix} u(x,t) \\ v(x,t) \end{pmatrix} \approx \begin{pmatrix} a_0(t) \\ b_0(t) \end{pmatrix} + \begin{pmatrix} a_1(t) \\ b_1(t) \end{pmatrix} \cos kx, \quad k = \frac{\pi}{l}. \qquad (5.10)$$

Substitution of (5.10) into the system (5.7) and the subsequent omission of all terms containing the factors $\cos \dfrac{\pi m x}{l}$, $m > 1$, provides a system of ordinary differential equations for variables $a_0(t)$, $a_1(t)$, $b_0(t)$, and $b_1(t)$. System (5.8) is obtained with the use of the substitution $\rho_i^2(t) = a_i^2(t) + b_i^2(t)$, $i = 0, 1$, and then $\xi = \rho_0^2$, $\eta = \rho_1^2$, $\theta = \varphi_0 - \varphi_1$ with regard to the relations $a_i = \rho_i \cos \varphi_i$, $b_i = \rho_i \sin \varphi_i$. Therefore, the simplified system (5.8) essentially reflects the solutions not of problem (5.7) but of some other finite-dimensional system of equations in *the space of Fourier coefficients*. From this viewpoint, the system (5.8) is similar to the well-known Lorenz system of three nonlinear ordinary differential equations. In this connection, we note that transition to chaos in the extended five-dimensional Lorenz system occurs via cascade of bifurcations of two-dimensional invariant tori with respect to the external frequency rather than to limit cycles. The scenario of transition to chaotic modes in the five-dimensional system is simpler than that in the three-dimensional system; more precisely, it contains only cascades of Feigenbaum bifurcations, that is, cascades of period doubling bifurcations for invariant tori, and subharmonic bifurcations of stable invariant two-dimensional tori of arbitrary period in accordance with the Sharkovskii order. Thus it seems to be necessary to consider the scenario of transition to chaos in the space of Fourier coefficients without reducing the problem to a simplified few-mode system of the form (5.8).

To this end the Fourier coefficients for $m = 0, 1, \ldots$

$$\begin{pmatrix} a_0(t) \\ b_0(t) \end{pmatrix} = \frac{1}{l} \int_0^l \begin{pmatrix} u(x,t) \\ v(x,t) \end{pmatrix} dx, \qquad \begin{pmatrix} a_m(t) \\ b_m(t) \end{pmatrix} = \frac{2}{l} \int_0^l \begin{pmatrix} u(x,t) \\ v(x,t) \end{pmatrix} \cos \frac{\pi m x}{l} dx,$$

were computed with the use of functions $u(x,t)$ and $v(x,t)$ found numerically from the second boundary value problem (5.7) for the Kuramoto–Tsuzuki equation on the interval $[0, l]$. The problem was solved with the use of a purely implicit finite-difference scheme by the Thomas matrix method. The scheme had second-order approximation with respect to the time and space variables. Setting of boundary conditions also provided second-order approximation with respect to the space variable. The time integration step was chosen to be equal to $\tau = 0.005$, and the increment with respect to the space variable was determined by the number $n = 30$ of grid points in the interval $[0, l]$. To analyze the scenario of transition to chaos in the space of Fourier coefficients, we have chosen the variables $\rho_i = \sqrt{a_i^2 + b_i^2}$ and considered solutions in projection (ρ_0, ρ_1), just as was done in Sec. 5.2.1.

Let us compare the solutions found in the previous Section 5.2.1 for a simplified few-mode system (5.8) with solution of the original second boundary value problem (5.7) represented in the space of Fourier coefficients. To this end, we shall take the same parameter value $c_1 = 1.3$ and consider the scenario of transition to chaos as parameter c_2 varies in the domain of negative values. For parameter values $c_2 \in (-1.8, 0)$, the system has homogeneous periodic solutions. This state of problem (5.7) corresponds to the fixed point $(\rho_0, 0, 0, \ldots, 0)$ in the space of Fourier coefficients. In the phase space of variables (u, v), for the same range of parameter c_2, the trajectory is a circle, which implies that oscillations of variables $u(t)$ and $v(t)$ have the same amplitude.

For the parameter value $c_2 \approx -1.81$, in problem (5.7), there is a bifurcation after which the homogeneous periodic solution loses stability and another stable solution, spatially inhomogeneous and periodic in time, is generated. In the space of Fourier coefficients, this bifurcation corresponds to the loss of stability of the fixed point $(\rho_0, 0, 0, \ldots, 0)$ and the appearance of another fixed point with nonzero values of variables ρ_i. That fixed point in the space of Fourier coefficients is related to a circular orbit in the projection $(u(x_0, t), v(x_0, t))$ of the phase space, where $x_0 \in [0, l]$. Therefore, the loss of stability of the homogeneous periodic solution in the problem (5.7) is followed by generation of another inhomogeneous stable solution periodic in time and space.

This spatially inhomogeneous solution remains stable as parameter c_2 decreases to value of $c_2 = -2.66$ for which generation of a limit cycle is observed in the projection (ρ_0, ρ_1). For the value of $c_2 = -4.617$, there is a period doubling bifurcation for this cycle, and a cascade of Feigenbaum bifurcations is started. For example, a cycle of quadruple period is observed for $c_2 = -4.8$, of period 8 for $c_2 = -4.815$, of period 16 for $c_2 = -4.820$, and so on. The cascade of period doubling bifurcations generated by this cycle comes to the end with a Feigenbaum attractor for the parameter value of $c_2 = -4.8225$ (Fig. 5.8).

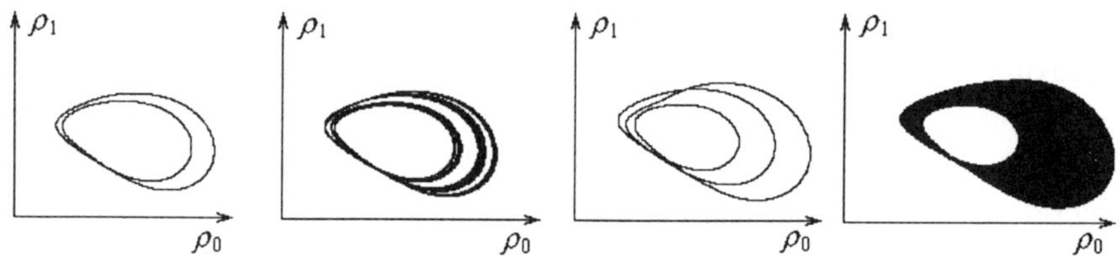

Fig. 5.8 Projections of period two cycle, Feigenbaum attractor, period three cycle and some more complex subharmonic singular attractor in many-dimensional system of Fourier coefficients for the Kuramoto–Tsuzuki equation (5.7) for $c_1 = 1.3$ and when c_2 varies.

For smaller parameter values, solutions have a complicated chaotic character in the space of Fourier coefficients. However, the existence of a solution with period 5 for $c_2 = -4.894$ and a solution with period 3 for $c_2 = -4.955$ implies that, in problem (5.7), there exists a subharmonic cascade of Sharkovskii bifurcations. In addition, note that the same cycles with periods of 3 and 5 are again registered for parameter values $c_2 = -5.20$ and $c_2 = -5.26$, respectively. This implies that in the space of Fourier coefficients of the system (5.7), just as in three-dimensional dissipative systems of ordinary differential equations, there exists an accumulation point, that is, a value of bifurcation parameter for which the singular attractor has most complicated structure. As was shown in [Kaloshin *et al.* (2003)], the scenario of generation of a chaotic attractor is largely the same on both sides of the accumulation point. If parameter c_2 decreases further, then solutions of system (5.7) are simplified in inverse order to the Feigenbaum–Sharkovskii cascade until the appearance of a stable fixed point for $c_2 = -6.42$.

Therefore, results obtained in the spaces of the same variables permit one to state the following important conclusions about the simplified system (5.8) and the original problem (5.7) for the Kuramoto–Tsuzuki equation.

First, transition to chaotic behavior in the Kuramoto–Tsuzuki equation for solutions in the space of Fourier coefficients follows the same scenarios as in nonlinear dissipative systems described by ordinary differential equations, more precisely, through cascades of Feigenbaum period doubling bifurcations of stable cycles and further through subharmonic cascades of bifurcations of stable cycles in accordance with the Sharkovskii order. Second, there are substantial quantitative differences between the values of bifurcation parameters in cascades of bifurcations of stable cycles in the system of few-mode approximation (5.8) and in solution of the second boundary value problem (5.7) in the space of Fourier coefficients.

5.2.3 *Scenario of transition to chaos in the phase space of the Kuramoto–Tsuzuki equation*

Let us now consider a scenario of transition to chaos in the second boundary value problem for the Kuramoto–Tsuzuki equation (5.7) in the phase space of variables (u, v). To this end, we use the cross-section of this space by plane $u(l/2) = 0$ and consider the Poincare mapping in projection to the coordinates $(u(0), v(l/2))$. For the fixed variable, we again take $c_1 = 1.3$ and vary the variable c_2 in the same domain as in Sec. 5.2.2. The initial conditions for solution of the second boundary value problem (5.7) are given to be homogeneous.

We note again that, in the range of $c_2 \in [-1.8, 0]$, the second boundary value problem (5.7) has a homogeneous periodic solution with equal amplitudes of oscillations of variables $u(x, t)$ and $v(x, t)$. For $c_2 \approx -1.81$, this homogeneous solution loses stability, and there appears another stable periodic but inhomogeneous solution, which also has equal amplitudes of oscillations with respect to the variables $u(x, t)$ and $v(x, t)$. For $c_2 \approx -2.66$, the periodic inhomogeneous solution also becomes unstable, and there appears a stable two-dimensional invariant torus, which is justified by the Poincare mapping. For $c_2 \approx -3.549$, there is a period doubling bifurcation of a two-dimensional invariant torus with respect to the basic (internal) frequency.

Note that for the value of $c_2 = -4.8$ the Poincare mapping is represented by a two-dimensional invariant torus with period 2 with respect to both internal and external frequencies (see Fig. 5.9a). Further, for $c_2 = -4.815$, one can observe a two-dimensional invariant torus of period 2 with respect to the internal frequency and of period 4 with respect to the external frequency; for $c_2 = -4.820$, one has a two-dimensional invariant torus with period 2 with respect to the internal frequency and period 8 with respect to the external frequency. Therefore, for problem (5.7), we have a cascade of period doubling bifurcations with respect to the external frequency for two-dimensional invariant tori of period 2 with respect to the internal frequency. This cascade is finished by generation of the Feigenbaum attractor for $c_2 \approx -4.8225$. The form of Feigenbaum attractor (which is induced by a cascade of period doubling bifurcations with respect to the external frequency for two-dimensional invariant tori of period 2 with respect to the internal frequency) in the cross-section $u(l/2) = 0$ for the Kuramoto–Tsuzuki equation at the point $(c_1, c_2) = (1.3, -4.8225)$ is shown in Fig. 5.9b.

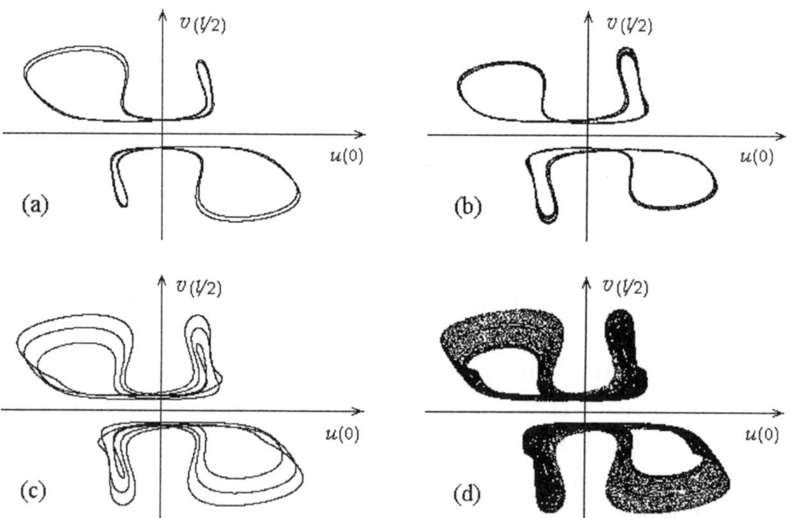

Fig. 5.9   Projections of the Poincare mapping in the cross-section $u(l/2)$ onto the coordinate plane $(u(0), v(l/2))$ for the parameter value $c_1 = 1.3$: a two-dimensional torus of period 2 with respect to the internal frequency and period 2 with respect to the external frequency (a) for $c_2 = -4.8$, a Feigenbaum attractor on a two-dimensional torus of period 2 with respect to the internal frequency (b) for $c_2 = -4.8225$, a two-dimensional torus of period 2 with respect to the internal frequency and period 3 with respect to the external frequency (c) for $c_2 = -4.955$, a singular attractor on a two-dimensional torus of period 2 with respect to the internal frequency (d) for $c_2 = -5.05$.

For values $c_2 = -4.894$ and $c_2 = -4.955$ there are two-dimensional invariant tori of periods 5 and 3 with respect to the external frequency and period 2 with respect to the basic internal frequency (see Fig. 5.9c). Existence of stable two-dimensional invariant tori of periods 5 and 3 implies that there is a subharmonic cascade of bifurcations of two-dimensional invariant tori in the scenario of transition to chaos in the Kuramoto–Tsuzuki equation (5.7). The form of *a singular subharmonic toroidal attractor* with period 2 with respect to the internal frequency, that is, the singular attractor resulting from the subharmonic cascade of bifurcations of two-dimensional invariant tori of period 2 with respect to the internal frequency in accordance with the Sharkovskii order, is shown in Fig. 5.9d for $c_2 = -5.05$.

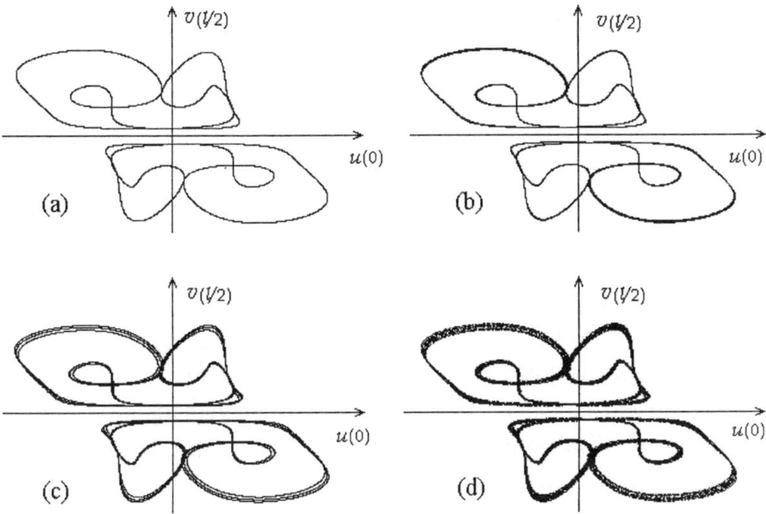

Fig. 5.10   Projections of the Poincare mappings in the cross-section $u(l/2)$ onto the coordinate plane $(u(0), v(l/2))$ for the parameter value $c_1 = 1.3$: a two-dimensional torus of period 4 with respect to the internal frequency (a) for $c_2 = -5.79$, a Feigenbaum attractor with respect to the external frequency on a two-dimensional torus of period 4 with respect to the internal frequency (b) for $c_2 = -5.843$, a two-dimensional torus of period 4 with respect to the internal frequency and period 3 with respect to the external frequency (c) for $c_2 = -5.85888$, a singular attractor with respect to the external frequency on a two-dimensional torus of period 4 with respect to the internal frequency (d) for $c_2 = -5.8605$.

Analysis of solutions of the second boundary value problem (5.7) for smaller negative values of the parameter c_2 shows that there also exists a cascade of period doubling bifurcations with respect to the basic internal frequency for two-dimensional tori. For $c_2 = -5.79$, the two-dimensional torus has period 4 with respect to the basic (internal) frequency (Fig. 5.10a);

and if the parameter c_2 is decreasing further, then the cascade of period doubling bifurcations with respect to the external frequency starts for this two-dimensional torus. This torus has period 2 with respect to the external frequency for $c_2 = -5.838$, period 4 for $c_2 = -5.840$, and so on (Fig. 5.10b). A Feigenbaum attractor with respect to the external frequency on a two-dimensional torus of period 4 with respect to the internal frequency exists for $c_2 = -5.843$. For the parameter value $c_2 = -5.8589$, this torus in the Poincare mapping has period 3 with respect to the external frequency (Fig. 5.10c), which implies again that there exists a subharmonic cascade of bifurcations of two-dimensional tori with respect to the external frequency in the scenario of transition to a chaotic regime which appears in the system for the parameter value $c_2 \approx -5.8605$.

The above-represented results imply that, in the Kuramoto–Tsuzuki equation (5.7), there may exist a subharmonic cascade of bifurcations of two-dimensional tori with respect to external as well as basic internal frequency. Such a cascade has been detected for fixed parameter value $c_1 = 2.5$ as the parameter c_2 decreases in the domain of negative values.

In the range of $c_2 \in [-1.85, 0]$, the second boundary value problem (5.7) has homogeneous periodic solutions with equal amplitudes of oscillations of variables $u(x, t)$ and $v(x, t)$. For the parameter value $c_2 \approx -1.851$, this homogeneous solution loses stability, but there appears another stable solution periodic with respect to time and inhomogeneous with respect to space, which has equal amplitudes of oscillations with respect to variables $u(x_0, t)$ and $v(x_0, t)$, where $x_0 \in [0, l]$. For $c_2 \approx -2.803$, periodic inhomogeneous solution also becomes unstable, and in problem (5.7) there appears a stable two-dimensional invariant torus, which is justified by the Poincare mapping. The period doubling bifurcation for a two-dimensional torus with respect to the basic (internal) frequency occurs for $c_2 \approx -3.134$.

This bifurcation starts a cascade of period doubling bifurcations for two-dimensional invariant tori with respect to the internal frequency. The solution of problem (5.7) is given by a two-dimensional invariant torus of quadruple period with respect to the internal frequency for $c_2 \in [-3.6186, -3.537]$, a torus of period 8 with respect to the internal frequency for $c_2 \in [-3.6409, -3.6187]$, a torus of period 16 with respect to the internal frequency for $c_2 \in [-3.64623, -3.6410]$, and so on. The cascade of period doubling bifurcations for two-dimensional tori with respect to the internal frequency finishes by generation of the Feigenbaum attractor for $c_2 \approx -3.655$ (Fig. 5.11). For subsequent decrease of parameter c_2, we have a subharmonic cascade of bifurcations of two-dimensional torus with respect to the internal frequency in accordance with the Sharkovskii order with generation of various singular toroidal attractors, one of which is shown in Fig. 5.11d for $c_2 = -3.75$.

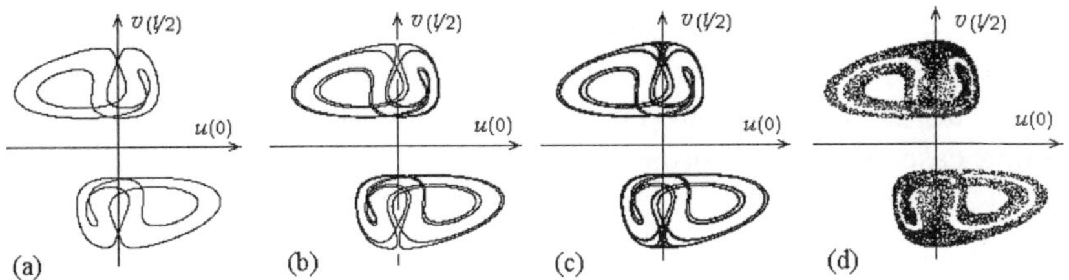

Fig. 5.11   Projections of the Poincare mappings in the cross-section $u(l/2)$ onto the coordinate plane $(u(0), v(l/2))$ for the parameter value $c_1 = 2.5$: a two-dimensional torus of period 4 with respect to the internal frequency (a) for $c_2 = -3.6$, the two-dimensional torus of period 8 with respect to the internal frequency (b) for $c_2 = -3.635$, the Feigenbaum attractor with respect to the internal frequency (c) for $c_2 = -3.655$, a singular attractor on the two-dimensional torus (d) for $c_2 = -3.75$.

The bifurcation diagram showing the existence of various subharmonic cascades of bifurcations of two-dimensional invariant tori of the second boundary value problem (5.7) for the Kuramoto–Tsuzuki equation in the space of parameters (c_1, c_2) is given in Fig. 5.12. This Fig. 5.12 represents a partition of the plane of parameters (c_1, c_2) into domains with various solutions of the problem (5.7). The curve L_0 separates the periodic homogeneous solution (the domain D_0) and the periodic spatially inhomogeneous solution (the domain D_1) of the second boundary value problem (5.7). On the set $(c_1, c_2) \in L_1$, there is a bifurcation of birth of simple two-dimensional invariant tori. The dashed lines in the domain D_2 represent the sets of values of parameters c_1 and c_2 corresponding to period doubling bifurcations for two-dimensional invariant tori with respect to the internal frequency. The set of parameter values lying to the right of the point A corresponds to termination of the complete cascade of period doubling bifurcations for the torus T_0 with respect to the internal frequency, while at points lying on the curve L_2 to the left of the point A, singular attractors are formed after incomplete cascades of period doubling bifurcations for tori with respect to the internal as well as external frequency.

In the domain D_4, there is a stable two-dimensional torus T_1 of another form, which generates a cascade of period doubling bifurcations with respect to the internal frequency. Therefore, singular attractors lying in the part of the domain D_3 between the domains D_2 and D_4 are generated by subharmonic cascades of bifurcations of stable two-dimensional invariant tori whose period multiplicity with respect to the internal frequency is determined by the Sharkovskii order. Singular attractors lying below the domain D_4 are generated by subharmonic cascades of bifurcations (with respect to the external frequency) of tori T_2, T_3, and T_4, which are stable in domains D_5, D_6 and D_7, respectively. Dotted curves in the domain D_7 show the set of values of parameters c_1 and c_2 corresponding to the period doubling cascade for the torus T_4 with respect to the external frequency.

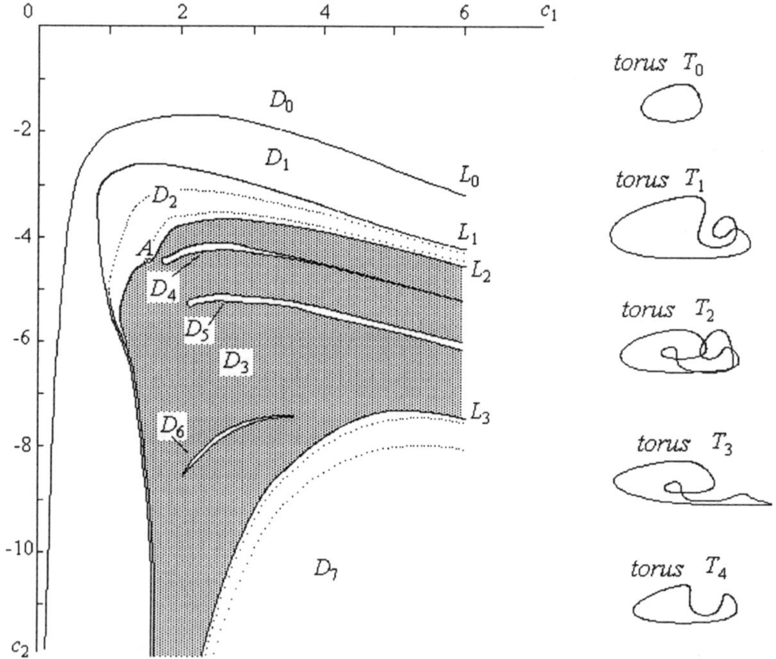

Fig. 5.12 The bifurcation diagram of solutions of the second boundary value problem for the Kuramoto–Tsuzuki equation on the interval $l = \pi$.

The results of numerical analysis of solutions of the Kuramoto–Tsuzuki equation in the three-dimensional space of few-mode approximations and in the space of Fourier coefficients as well as for the original system in the space of phase variables of this equation permit us to make the following conclusions.

(1) Transition to chaotic modes in the Kuramoto–Tsuzuki equation is based on the same mechanisms as in nonlinear dissipative systems of

ordinary differential equations (see Chapter 4); more precisely, for solutions in the space of few modes approximations these mechanisms include cascades of period doubling bifurcations (Feigenbaum cascades) and subharmonic cascades of cycle bifurcations in accordance with Sharkovskii order. For solutions in the infinite-dimensional phase space, these mechanisms include cascades of Feigenbaum bifurcations and subharmonic cascades of bifurcations of invariant tori; in addition, it was shown that cascades of period multiplication for invariant tori take place with respect to internal as well as external frequency.

(2) It is not adequate to use three-dimensional few-mode approximations to describe spatio-temporal or diffusion chaos in diffusion-type equations. In three-dimensional few-mode systems, singular attractors are generated only by bifurcations of limit cycles, while in the corresponding diffusion equations, generation of singular attractors is caused by cascades of bifurcations of at least two-dimensional (no other have been found yet) invariant tori. Moreover, the corresponding bifurcation diagrams of solutions have substantial differences.

(3) The following conjecture is likely to be true: appearance of spatio-temporal chaos in systems of partial differential equations is caused by cascades of bifurcations of two-dimensional invariant tori rather than the destruction of a three-dimensional torus with the generation of some hypothetical strange attractor, as was assumed in modern publications following [Ruelle and Takens (1971)].

5.3 Dynamical Chaos in Differential Equations with Delay Argument

It is be shown in the present section, that one of the basic scenarios of transition to chaos in infinite-dimensional nonlinear differential *equations with delay argument* is also the subharmonic cascade of bifurcations of stable cycles. We shall consider the equation

$$\dot{x} = f(x(t), x(t - \tau)), \ t \geq 0, \tag{5.11}$$

where $x(t)$, $f(\cdot)$ are scalar functions, $\tau > 0$ is a constant delay. Some continuous function $\varphi(\vartheta)$, given on an interval $-\tau \leq \vartheta \leq 0$, is an initial condition for the Eq. (5.11). Parameter τ in the Eq. (5.11) can be a bifurcation parameter, that is at its change there can be a complication of structure of attractors of Eq. (5.11) down to occurrence of chaotic dynamics in this equation. Simple regular attractors, such as stable stationary states and periodic solutions, can be observed directly at numerical integration of the Eq. (5.11), for example, by Runge–Kutta method of the fourth order. However, for the analysis of more complex regular and, especially, irregular attractors of the Eq. (5.11), one needs a transition to some finite-dimensional phase space. One of the possibilities of such transition is an approximation of the Eq.(5.11) by some finite-dimensional system of ordinary differential equations. For this purpose we shall divide an interval $[-\tau; 0]$ into m identical parts and designate

$$x_t(0) = y_0, \quad x_t(-\tau) = y_m, \quad x_t(-i\tau/m) = y_i, \quad i = 1, \ldots, m-1,$$

where $x_t(\vartheta) = x(t + \vartheta)$. Then, using, for example, a finite-difference approximation of the derivative, we obtain

$$\dot{y}_0 = f(y_0, y_m, \tau), \quad \dot{y}_i = (m/\tau)(y_{i-1} - y_i), \quad i = 1, \ldots, m. \tag{5.12}$$

Eq. (5.11)) is thereby reduced to the $(m+1)$-dimensional system of ordinary differential equations

$$\dot{y} = F(y, \tau), \tag{5.13}$$

where the vector $y = (y_0(t), y_1(t), \ldots, y_m(t))^{\mathrm{T}}$ determines a vector function

$$\varphi_t^{[m]}(\vartheta) = y_i + \frac{(y_{i-1} - y_i)m}{\tau}\vartheta, \quad \frac{-\tau i}{m} \leq \vartheta \leq \frac{-\tau(i-1)}{m}, \quad i = 1, \ldots, m.$$

Each coordinate of this vector-function linearly approximates the function $x_t(\vartheta)$ on an interval of the length $h = \vartheta_{i-1} - \vartheta_i$ on the basis of two values of the function at the nodes y_i and y_{i-1} with error, not exceeding $O(h^2)$ [Samarskii and Gulin (1989)]. Obviously, for a sufficiently high order m, the function $\varphi_t^{[m]}(\vartheta)$ is arbitrarily close to $x_t(\vartheta)$ on the interval $[-\tau; 0]$. Thus the solution $x(t)$ of the Eq. (5.11) corresponds to the values of the coordinate $y_0(t)$ of system (5.13), and the trajectory of the Eq. (5.11) in the expanded phase space $\mathbb{R} \times C[-\tau; 0]$ corresponds to the trajectory of the system (5.13) in the phase space $\mathbb{R}^{m+1}$.

Let us notice, that the error of approximation of the Eq. (5.11) by the system (5.13) is defined, mainly, by accuracy of calculation of derivatives $\dot{y}_i$, $i = 1, \ldots, m$ in nodes of the net function $y_i = x_t(-im/\tau)$, $i = 1, \ldots, m$. It is shown in work [Magnitskii and Sidorov (2000)], that for difference scheme of the first order (5.12) for calculation of derivative, the order of the system (5.13) should be equal to $m \approx 10^3$ for obtaining the errors of approximation, comparable with the accuracy of the Runge–Kutta numerical integration method of the fourth order at comprehensible steps of integration. Essential reduction of the order of system up to $m = 20 - 50$ can be realized by use of interpolation of function $x_t(\vartheta)$ on the interval $[-\tau, 0]$ by cubic splines. But in any case the Eq. (5.11) is not reduced to a few-mode system of ordinary differential equations. So, the problem of comparison of scenarios of transition to chaos in the original equation with delay argument (5.11) and in many-dimensional approximate system of ordinary differential Eqs. (5.13) is represented as very interesting.

Example 5.1   As an example we shall consider well-known *Mackey–Glass equation* describing the process of haemopoiesis [Mackey and Glass (1977)]:

$$\dot{x} = -ax(t) + \frac{\beta_0 \theta^n x(t - \tau)}{\theta^n + x^n(t - \tau)}, \tag{5.14}$$

where a, β_0, θ and n are positive constants such that $\beta_0 > a > 0$, $nB > 2$, $6aB > \beta_0$, $B = (\beta_0 - a)/\beta_0$. The Eq. (5.14) has a unique stationary state $x = \theta \sqrt[n]{\dfrac{\beta_0 - a}{a}}$, which loses stability for $\tau > \dfrac{\arccos(-a/b)}{\sqrt{b^2 - a^2}}$, $b = a(nB - 1)$, as a result of Andronov–Hopf bifurcation. We use the following values of parameters in the Eq. (5.14): $a = 1$, $\beta_0 = \theta = 2$, $n = 10$. For these parameter values and for $0 < \tau < 0.4708$, the Eq. (5.14) has the stationary stable solution $x = 2$, then a bifurcation of birth of a stable periodic solution (or a limit cycle in the expanded phase space) occurs for $\tau = 0.4708$, then there are period doubling bifurcations for further growth of τ. So a stable cycle of the double period is born for $\tau \approx 1.32$, a cycle of the quadruple period is born for $\tau \approx 1.57$, *etc.* Thus each bifurcation results in the loss of stability of the previous limit cycle. For $\tau > 1.608$, the Eq. (5.14) has chaotic oscillations. Then a stable cycle of period 6 is born for the parameter value $\tau \approx 1.677$, a stable cycle of period 5 is born for $\tau \approx 1.766$, and a stable cycle of period 3 is born for $\tau \approx 1.874$.

The Eq. (5.14) has been approximated by a 20-dimensional system of Eqs. (5.13), and projections of phase portraits of different attractors of the last system on a plane (y_0, y_m) were considered. It has been established, that at growth of parameter values τ, the subharmonic cascade of bifurcations of stable cycles is realized in the system (5.13), and the first bifurcation values for the system (5.13) and the Eq. (5.14) practically coincide (Fig. 5.13). For example, a stable cycle is born in system (5.13) as a result of Andronov–Hopf bifurcation for $\tau \approx 0.47$, a cycle of double period is born for $\tau \approx 1.32$, and a cycle of quadruple period is born for $\tau \approx 1.57$. Domains of the parameter τ for which other stable periodic solutions of system (5.13) and Eq. (5.14) practically coincide too (Fig. 5.14).

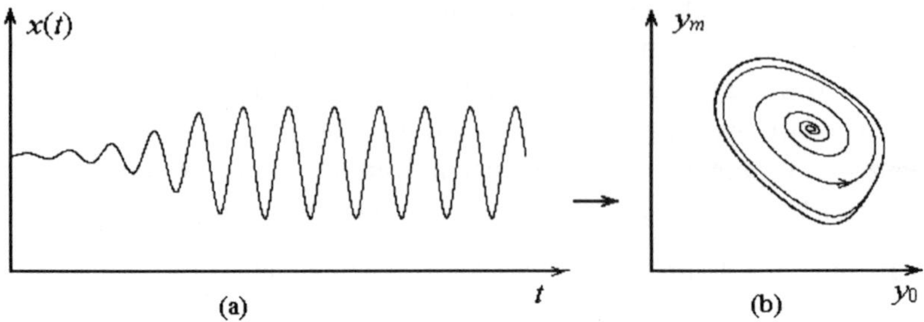

Fig. 5.13   The stable periodic solution (a) of the equation (5.14) and a projection of a stable limit cycle of system (5.13) for $\tau = 0.745$ (b).

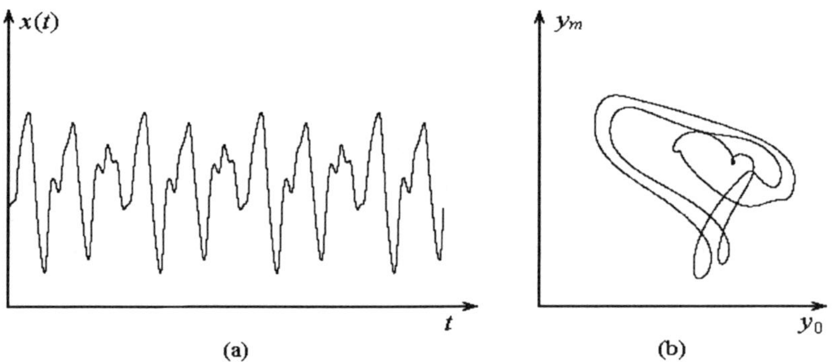

Fig. 5.14   The stable periodic solution of the equation (5.14) (a) and a projection of a stable cycle of period 3 of the system (5.13) (b) for $\tau = 1.874$.

Thus, it is possible to draw a conclusion that one of scenarios of transition to chaos in systems of differential equations with delay argument is the scenario of transition to chaos through the Feigenbaum cascade of period doubling bifurcations of stable cycles and then the Sharkovskii subharmonic cascade of bifurcations of stable cycles of arbitrary period. This scenario can be also successfully described by the Feigenbaum–Sharkovskii–Magnitskii theory considered in Chapter 4. The question on, whether transition to chaos is possible in systems of differential equations with delay argument through the subharmonic cascade of bifurcations of two-dimensional or many-dimensional tori, remains opened.

5.4   Cycles and Chaos in Distributed Economic Systems

Another, essentially different example of formation of spatio-temporal chaos in the nonlinear mediums is the distributed model of a market economy offered by one of the authors in [Magnitskii (1991)] and developed then in [Magnitskii and Sidorov (2005b)]. The model is a system of three nonlinear differential equations, two of which describe the change and intensity of motion (diffusion) of capital and consumer demand in a technology space under the influence of change of profit rate. The last is described by the third ordinary differential equation. The obtained system possesses many remarkable properties. One of them is the presence of a sequence of bifurcations of birth of stable periodic spatial homogeneous solutions of arbitrary period forming a spatial homogeneous, but chaotic in time singular attractor. The last is impossible both for the system of reaction–diffusion equations, and for the Kuramoto–Tsuzuki equation. And transition to chaos is carried out by the same subharmonic cascade of bifurcations of stable cycles considered in Chapters 2–4. Therefore the presence of diffusion processes in economic system should also lead to existence more complex regimes of diffusion chaos in it in comparison with chaotic regimes considered in the previous sections of the present chapter. The distributed model of self-developing market economy offered by the authors is at present the only economic-mathematical model possessing such properties. Therefore we shall derive its equations in detail. Then we shall carry out an analysis of the model and obtain some interesting results following from this analysis, important both from the point of view of mathematics, and from the point of view of their various economic applications.

5.4.1   *Description of the model of self-developing market economy*

In the present section we construct a mathematical model of a self-developing economy whose development is characterized by spontaneous growth of capital and its movement in the technology space in response to differences in profitability. The model is a system of partial differential equations that describe formation of social wealth, including production, distribution, exchange, and consumption. A distinctive feature of the model is that distribution of profitability (profit rates) provides a main model stimulating economic development. Profit rates determine the direction and the intensity of motion (diffusion) of capital and its spontaneous growth through generation of added value, the magnitude of which is also determined by profit rates. This approach has enabled us to dispense with the traditional use of production functions, which are empirically not very suitable for description of self-development processes. Three economic agents having their own interests take part in economic processes that are employers, workers and government. In the model, self-development of a market economy involves movement and spontaneous growth of capital of employers, which in turn is the result of creation of added value by workers under a government control.

The model assumes an unstructured closed economic system that develops in a finite-dimensional Euclidean space $\mathbb{R}^n$, called *the technology space*. Each point $c \in \mathbb{R}^n$ corresponds to a certain production technology of some commodity and its coordinate c_i $(i = 1, 2, \ldots, n)$ is the consumption of resource i per unit output. The following functions are used to describe the main characteristics of the economic system:

- $\widetilde{K}(t, c)$ — distribution density of capital at time t in the technology space, i.e. the value of capital (total value of productive capital, commodity capital, and financial capital) used by the firms at time t to produce some consumer products by technology (with cost) c and to produce new means of production for these products;
- $C_T(t, c)$ — distribution density of productive capital of firms (both fixed $K(t, c)$ and variable $H(t, c)$);
- $Y(t, c)$ — distribution density of commodity capital of firms, equal to the value at time t of the commodity stocks of consumer product and means of its production that was produced at cost c;
- $M(t, c)$ — distribution density of the financial capital of the firms (the demand of the firms for production assets and labor required for production by technology c);
- $D_1(t, c)$, $D_2(t, c)$ and $D_3(t, c)$ — demand of the firms, the workers, and the government, respectively, for consumer goods produced by technology c;
- $u(t, c)$ — distribution of profit rate at time t in the technology space;
- $\rho_{C_T}(t, c, \cdot)$ and $\rho_M(t, c, \cdot)$ — flow density vectors of productive and financial capitals, respectively, i.e. the value of capital passing in unit time through a unit surface of some volume element in the technology space $\mathbb{R}^n$;
- $\rho_{D_1}(t, c, \cdot)$, $\rho_{D_2}(t, c, \cdot)$ and $\rho_{D_3}(t, c, \cdot)$ — flow density vectors of the demand of firms, workers, and the government, respectively, for con-

sumer goods, defined as the value of cash resources passing in unit time through a unit surface of some volume element in the technology space $\mathbb{R}^n$;

- $R_1(t, c)$, $R_2(t, c)$ and $R_3(t, c)$ — current consumption by firms, workers, and the government, respectively, of consumer goods produced by technology c.

The profit rate in production of consumer goods uniquely determines the profit rate in production of means of production of these goods. It is justified to use only one function $u(t, c)$ for each vector in the technology space.

Karl Marx's theory of added value, being based on rigorous rules of economic development that remain valid up to this day, suggests that self-development of a market economy involves movement and spontaneous growth of capital, which is the result of creation of added value by workers in the circulation process of capital.

Let us consider in detail the circulation process of capital, measuring time in units of the turnover cycle. In the first stage of capital turnover cycle the firms spend their financial assets M to acquire means of production (fixed capital ΔK) and labor (variable capital ΔH). Combining fixed and variable assets, the firms embark on the second stage of capital turnover cycle, in production of value and added value. Having undergone a transformation from financial assets M to production assets $C_T = K + H$, capital continues its movement in the production sphere. Here, the value of production capital is decreased by the value of variable assets ωH paid to workers in the form of wages and by the value of fixed capital μK lost through working capital and depreciation and obsolescence of fixed capital. At the same time, productive capital flows $\rho_T(t, c\cdot)$ appear in technology space from points with a lower profit rate to points with a higher profit rate. Thus, equation describing the change of productive capital in an arbitrary volume v in the space $\mathbb{R}^n$ has the form

$$\frac{\partial}{\partial t} \int_v C_T(t, c) dv = \int_v (-\omega H - \mu K + \Delta K + \Delta H) dv - \int_S \rho_T(t, c, \cdot) dS.$$

Integrating by volume in the last term, we obtain in virtue of arbitrary v

$$\frac{\partial C_T(t, c)}{\partial t} = -\operatorname{div} \rho_T(t, c, \cdot) + \Delta K - \mu K + \Delta H - \omega H. \tag{5.15}$$

Production thus transforms capital into stocks of goods form, and the value of newly produced goods is the sum of depreciation νK plus the new value created by workers, which in turn is the sum of variable capital ωH

and added value uC_T. The goods produced by technology c are purchased by the firms R_1, the workers R_2, and the government R_3. The change of commodity capital is thus described by the equation

$$\frac{\partial Y(t,c)}{\partial t} = \nu K + \omega H + uC_T - (R_1 + R_2 + R_3). \qquad (5.16)$$

Selling the produced goods in the market, the firm converts their value into cash. The capital is transformed from commodity to financial assets. The source of funds for increase of the firm's financial assets is the retained portion I of sales revenue earned from consumers; the use of funds is the cost of added productive capital $\Delta K + \Delta H$. At the same time, financial capital flows $\rho_M(t, c, \cdot)$ move in the technology space to points with higher profit rate. This reflects the willingness of lenders to invest their funds in development of firms that ensure the highest profit and thus pay the highest interest rate on loans. Applying the same approach as in Eqs. (5.15) and (5.16), we obtain an equation for the motion of financial assets:

$$\frac{\partial M(t,c)}{\partial t} = -\operatorname{div} \rho_M(t, c, \cdot) + I - \Delta H - \Delta K. \qquad (5.17)$$

Adding up Eqs. (5.15)–(5.17), we obtain the equation for movement of capital:

$$\frac{\partial \widetilde{K}(t,c)}{\partial t} = -\operatorname{div} \rho(t, c, \cdot) + (\nu - \mu)K + uC_T + I - R, \qquad (5.18)$$

where $R = R_1 + R_2 + R_3$, $\rho = \rho_T + \rho_M$, $\widetilde{K} = C_T + Y + M$.

Eq. (5.18) describes spontaneous growth of capital. If production is profitable, then capital grows, production expands, both fixed and variable capitals increase. But this process cannot continue indefinitely. As capital grows, certain factors reduce the profit rate, which ultimately leads to reduced production, unemployment, and economic crisis. These factors include periodic appearance of excess supply of consumer goods and inability to sell the produced goods in the market due to low consumer demand. Moreover, rapid production growth periodically produces situations when demand for means of production exceeds the limited supply of financial assets and the demand for labor exceeds its limited supply [Popov (1989)]. All these factors can be formalized using the demand function.

As we have noted previously, the portion I of funds R generated by the firms from the sale of consumer goods is used for capital accumulation. Other portions C_K and G are sources of demand D_1 and D_3 of the firms and the government for consumer goods produced by technology c. The sink of demand D_1 and D_3 at the point c in $\mathbb{R}^n$ is the value of purchased goods R_1 and R_3. At the same time, demand flows $\rho_{D_1}(t, c, \cdot)$ and $\rho_{D_2}(t, c, \cdot)$ of the firms and the government, respectively, redistribute the demand over the technology space in accordance with consumer prices of goods produced by various technologies. The demand equations of the firms and the government thus have the form

$$\frac{\partial D_1(t, c)}{\partial t} = -\operatorname{div} \rho_{D_1}(t, c, \cdot) + C_K - R_1, \tag{5.19}$$

$$\frac{\partial D_3(t, c)}{\partial t} = -\operatorname{div} \rho_{D_3}(t, c, \cdot) + G - R_3. \tag{5.20}$$

The source of the worker's demand C_L at the point c is their wages $C_L = \omega H$, the sink is the value of consumer goods that they purchase, R_2. Thus,

$$\frac{\partial D_2(t, c)}{\partial t} = -\operatorname{div} \rho_{D_2}(t, c, \cdot) + C_L - R_2. \tag{5.21}$$

Adding up Eqs. (5.19)–(5.21), we obtain the demand equation in the form of

$$\frac{\partial D(t, c)}{\partial t} = -\operatorname{div} \rho_D(t, c, \cdot) + C_L + C_K + G - R, \tag{5.22}$$

where $\rho_D = \rho_{D_1} + \rho_{D_2} + \rho_{D_3}$.

Overproduction does not necessarily imply that the goods cannot be sold and consumed. They simply cannot be sold at prices that ensure a certain profit rate to the firms. Therefore, when supply of consumer goods and labor $Y + H$ exceeds the demand D and the financial capital M, prices may drop, interest rate on loans may rise, and as an end result the profit rate will decrease:

$$\frac{\partial u(t, x)}{\partial t} = \alpha((D + M) - (Y + H)). \tag{5.23}$$

The full system of equations of a self-developing market economy is thus representable in the form of

$$\frac{\partial \widetilde{K}(t, c)}{\partial t} = -\operatorname{div} \rho(t, c, \cdot) + (\nu - \mu)K + uC_T + I - R,$$

$$\frac{\partial D(t, c)}{\partial t} = -\operatorname{div} \rho_D(t, c, \cdot) + C_L + C_K + G - R, \tag{5.24}$$

$$\frac{\partial u(t, c)}{\partial t} = \alpha((D + M) - (Y + H)),$$

where $\widetilde{K} = C_T + Y + M, \quad C_T = K + H, \quad R = C_K + G + I.$

Integrating the equations of system (5.24) over the positive orthant in technology space and noting that

$$\int_{\mathbb{R}^n_+} \operatorname{div} \rho(t, c, \cdot)dc = \int_{\mathbb{R}^n_+} \operatorname{div} \rho_D(t, c, \cdot)dc = 0,$$

we obtain a system of equations that describes the variation of macro-variables of a market economy over time:

$$\begin{aligned}
\frac{d\widetilde{K}(t,c)}{dt} &= (\nu - \mu)K + uC_T + I - R, \\
\frac{dD(t,c)}{dt} &= C_L + C_K + G - R, \\
\frac{du(t,c)}{dt} &= \alpha(D + M - Y - H), \\
\widetilde{K} &= C_T + Y + M, \quad C_T = K + H, \quad R = C_K + G + I.
\end{aligned} \qquad (5.25)$$

Note that if we add up the first two equations in system (5.25), then obtain an equation for the variation of value of all elements contributing to economic development:

$$\frac{d(\widetilde{K}(t,c) + D(t,c))}{dt} = (\nu - \mu)K + C_L + uC_T - R,$$

where $\nu K + C_L + uC_T$ is the aggregate social product, and $C_L + uC_T$ is the national income.

Systems (5.24), (5.25) are undetermined, because they contain six equations (of which three are differential equations) and twelve variables plus capital and demand flows. Systems (5.24) or (5.25) can be augmented with various behavioral equations (relationships) that have a clear economic interpretation in terms of market economy. We propose the following approach to augmenting system (5.24) and thus (5.25) (it is obviously not the only possible approach).

Assumption 1. Capital flow density vector is proportional to the gradient of profit rate. Proportionality coefficient is not constant and it depends both on the point c in the technology space and on the profit rate itself:

$$\rho(t, c, \cdot) = \kappa_1(c, \widetilde{K}, u) \operatorname{grad} u(t, c). \qquad (5.26)$$

Capital diffusion is determined by the coefficient of diffusion $\kappa_1(c, \widetilde{K}, u)$ characterizing the properties of economic environment. Depending on the form of κ_1 (in general κ_1 is a tensor), environment may be homogeneous or inhomogeneous, isotropic or anisotropic. Different economic mechanisms of capital diffusion may lead to different, including nonlinear, dependencies of κ_1 on $\widetilde{K}$ and u.

Assumption 2. Profit rate decreases primarily due to the decrease of demand for goods. We may accordingly assume that the demand flow density is proportional to the gradient of profit rate, and the proportionality coefficient depends on the point c in the technology space, the level of consumer demand, and the profit rate itself:

$$\rho_D(t, c, \cdot) = \kappa_2(c, D, u)\, \operatorname{grad} u(t, c). \tag{5.27}$$

Coefficient $\kappa_2(c, D, u)$ determines the diffusion of consumer demand and characterizes the economic environment in terms of consumer properties of the goods produced by various technologies, such as their quality, image appeal, fashionable style, *etc.*

Assumption 3. The value of consumer goods purchased in the market is proportional to the value of commodity capital and to consumer demand:

$$R = \beta Y D. \tag{5.28}$$

Assumption 4. The funds G available to satisfy the government's demand are a portion of the total funds earned by the firms from the sale of consumer goods (taxes, excises, custom duties, *etc.*):

$$G = \delta R. \tag{5.29}$$

Since $C_L \leq (1 - \delta)R$, we have $0 < \delta \leq \delta_0 < 1$.

Assumption 5. The funds C_K available to satisfy the firm's demand are a portion of the added value generated in production and remaining after payment of taxes to the government:

$$C_K = \varepsilon(1 - \widetilde{\delta})uC_T, \quad \varepsilon < 1. \tag{5.30}$$

Taking in (5.30) $\sigma = \varepsilon(1 - \delta)$ and $\widetilde{\delta} = \delta$, we obtain $C_K = \sigma u C_T$, where $\sigma < 1 - \delta$.

Finally, Eqs. (5.24)–(5.30) can be augmented with variables characterizing three components of the structure of capital: organic capital $\gamma = K/H$, productive capital $\theta = C_T/M$, and commodity capital $\eta = Y/M$. Under these behavioral assumptions, system (5.24) takes the form

$$\frac{\partial \widetilde{K}(t,c)}{\partial t} = -\operatorname{div}(\kappa_1(c,\widetilde{K},u)\operatorname{grad} u) + \frac{\gamma\theta(\nu-\mu)\widetilde{K}}{(\gamma+1)(1+\theta+\eta)}$$

$$+ \frac{\theta(1-\sigma)\widetilde{K}u}{1+\theta+\eta} - \frac{\beta\delta\eta\widetilde{K}D}{1+\theta+\eta},$$

$$\frac{\partial D(t,c)}{\partial t} = -\operatorname{div}(\kappa_2(c,D,u)\operatorname{grad} u) + \frac{\omega\theta\widetilde{K}}{(1+\gamma)(1+\theta+\eta)} \qquad (5.31)$$

$$+ \frac{\sigma\theta\widetilde{K}u}{1+\theta+\eta} - \frac{\beta(1-\delta)\eta\widetilde{K}D}{1+\theta+\eta},$$

$$\frac{\partial u(t,c)}{\partial t} = \alpha\Big(D - \frac{\theta+(\eta-1)(\gamma+1)}{(1+\theta+\eta)(\gamma+1)}\widetilde{K}\Big).$$

Note that the system of Eqs. (5.31) is a particular case of systems with multicomponent diffusion, where the activator (the variable providing positive feedback) is the capital and the inhibitor (the variable suppressing capital growth) is the consumer demand. Activator and inhibitor interact under control of yet another variable — the profit rate, and the coefficients of diffusion in general are nonlinear functions of the properties of economic environment. Thus, we may assume the existence of complex nonlinear structures in system (5.31), such as limit cycles, tori, autowaves, and dissipative structures. The presence of three equations in the system suggests the possibility of chaotic behavior and diffusion chaos.

We define new variables

$$x = \frac{\beta\eta}{1+\theta+\eta}\widetilde{K}, \quad y = \frac{\beta\eta(1+\gamma)}{\omega\theta}D, \quad z = \frac{1+\gamma}{\omega}u$$

and assume that in general $\nu = \mu$. System (5.31) is thus reduced to the form of

$$\frac{\partial x(t,c)}{\partial t} = -\operatorname{div}(d_1(c,x,z)\operatorname{grad} z) + bx((1-\sigma)z - \delta y),$$

$$\frac{\partial y(t,c)}{\partial t} = -\operatorname{div}(d_2(c,y,z)\operatorname{grad} z) + x(1 - (1-\delta)y + \sigma z), \qquad (5.32)$$

$$\frac{\partial z(t,c)}{\partial t} = a(y - dx),$$

where

$$b = \frac{\omega\theta}{(1+\theta+\eta)(1+\gamma)}, \quad a = \frac{\alpha\theta}{\beta\eta}, \quad d = \frac{\theta+(\eta-1)(1+\gamma)}{\omega\theta}. \qquad (5.33)$$

System (5.25) describing the variation of macroeconomic variables can be similarly reduced under Assumptions 1–5 to the form of

$$\dot{x}(t) = bx((1-\sigma)z - \delta y),$$
$$\dot{y}(t) = x(1 - (1-\delta)y + \sigma z), \qquad (5.34)$$
$$\dot{z}(t) = a(y - dx).$$

5.4.2 *Behavior of macroeconomic variables*

In this section, we investigate only the system of ordinary differential equations (5.34). Solutions of this system display chaotic behavior, but also have a clear economic interpretation. Fix parameters γ, θ, η, that determine respectively the organic, productive, and commodity structure of capital; also fix parameters α, β and ω. We take, for instance $\gamma = 1$, $\theta = 12$, $\eta = 2$, $\beta = 6\alpha/7$, $\omega = 1$ and examine the effect of two parameters α and δ on the qualitative behavior of solutions of the system (5.34) keeping values of the other three parameters fixed: $a = 7$, $b = 0.4$, $d = 1.17$. Note that for all values of parameters σ and δ satisfying the relationship $\sigma < 1 - \delta$, system (5.34) has a stationary solution (a fixed point) with positive coordinates

$$O^*(x^*, y^*, z^*) = \left(\frac{1-\sigma}{d(1-\delta-\sigma)}, \ \frac{1-\sigma}{1-\delta-\sigma}, \ \frac{\delta}{1-\delta-\sigma} \right). \qquad (5.35)$$

Then eigenvalues of the Jacobi matrix of the right-hand side of the system (5.34)

$$J(x, y, z) = \begin{pmatrix} b((1-\sigma)z - \delta y) & -b\delta x & b(1-\sigma)x \\ 1 - (1-\delta)y + \sigma z & -(1-\delta)x & \sigma x \\ -ad & a & 0 \end{pmatrix},$$

evaluated at the point O^* satisfy the characteristic equation

$$\lambda^3 + (1-\delta)x^*\lambda^2 + ax^*(bd(1-\sigma) - \sigma)\lambda + ab(1-\sigma)x^* = 0. \qquad (5.36)$$

Applying the Routh–Hurwitz conditions

$$(1-\delta)x^* > 0, \ ax^{*2}(1-\delta)(bd(1-\sigma) - \sigma) > abx^*(1-\sigma) > 0,$$

that determine stability of the polynomial (5.36), we obtain the stability region G^s of the point O^*, where all three roots of the characteristic equation have negative real parts: $G^s = G_1^s \cup G_2^s$, where $G_1^s = \{\sigma | \sigma < 0\} \cap \{(b, d, \delta) | bd < (1-\delta)/\delta\}$, $G_2^s = \{\sigma | \sigma > 0\} \cap \{(b, d, \delta) | bd > (1-\delta)/\delta\}$. The remaining part of the parameter region $\{(\sigma, \delta) | \sigma < 1 - \delta, \ 0 < \delta \leq \delta_0 < 1\}$

is the region of instability G^u of the point O^*. It also consists of two subregions: $G^u = G_1^u \cup G_2^u$, where $G_1^u = \{\sigma|\sigma < 0\} \cap \{(b,d,\delta)|bd > (1-\delta)/\delta\}$, $G_2^u = \{\sigma|\sigma > 0\} \cap \{(b,d,\delta)|bd < (1-\delta)/\delta\}$. Regions G^s and G^u in the parameter plane (σ, δ) are shown in Fig. 5.15, where $(1-\delta^*)/\delta^* = bd = 0.468$ and thus $\delta^* \approx 0.681$, and $\delta_0 = 0.8$.

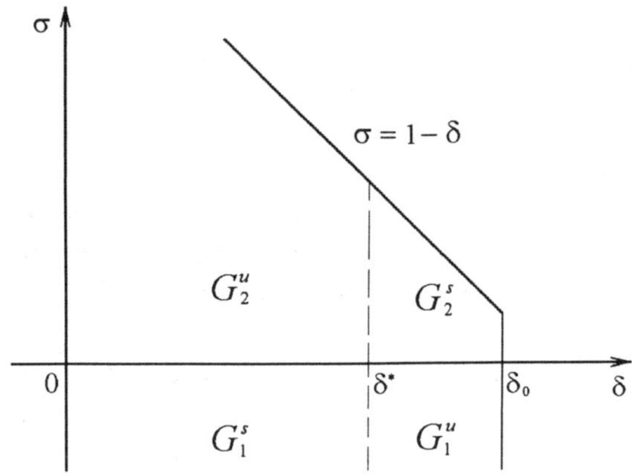

Fig. 5.15 Regions of stability (G_1^s and G_2^s) and instability (G_1^u and G_2^u) of the fixed point O^* of system (5.34).

Let us consider a number of natural and most relevant scenarios of economic development when parameters σ and δ change.

Scenario 1: $\sigma < 0$. This scenario corresponds to a government with an extremely strong administrative command system of economic controls, when the authorities force the firms to surrender their obtained profit to the government. If the government "pressure" on business is high ($\delta > \delta^*$, $(\delta, \sigma) \in G_1^u$), then this economy may exist for some time with low levels of capitalization and consumption, but ultimately it will be destroyed as the consumer demand drops to zero (Fig. 5.16a).

As δ decreases keeping $\sigma = -1$ constant, the system moves in the parameter space from G_1^u to G_1^s, and for $\delta < \delta^*$ the fixed point O^* of the system (5.34) becomes a stable fixed point (a focus) surrounded by an unstable saddle cycle. This means that as the government "pressure" on business is relaxed, the administrative command economy may exist in a virtually stationary state (in essence, stagnation) at low levels of capitalization and consumption, without breaking up for a long time (Fig. 5.16b).

Scenario 2: $\delta < \delta^*$. The previous case $\sigma < 0$ corresponds to a strong administrative command style of economic control. When parameter σ

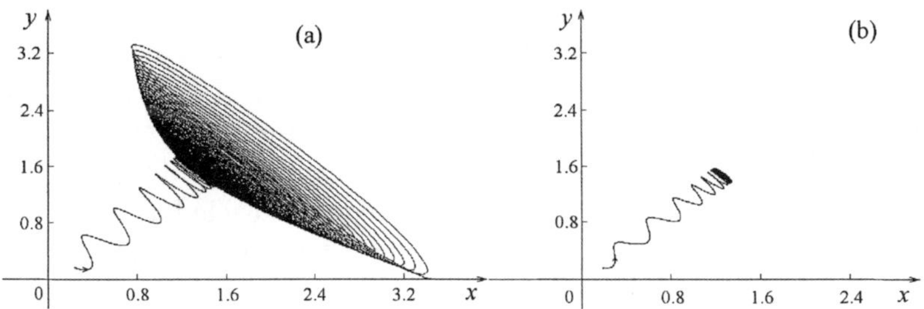

Fig. 5.16   Projection on the plane (x, y) (capital, consumer demand) of the solution of system (5.34) for $\delta = 0.7$ and $\sigma = -1$ (a) and $\delta = 0.65$ and $\sigma = -1$ (b).

crosses the boundary $\sigma = 0$ from region G_1^s to region G_2^u, the system experiences a bifurcation that reverses stability of the fixed point and its encircling cycle: the point O^* becomes unstable and the limit cycle becomes stable. Mathematically, this bifurcation occurs when two complex-conjugate roots of the characteristic equation (5.36) cross the imaginary axis from left to right. The third root has a negative real part. Indeed, for $\sigma = 0$ the characteristic equation (5.36) takes the form

$$(\lambda + (1 - \delta)x^*)(\lambda^2 + abdx^*) = 0$$

and thus has the roots $\lambda_{1,2} = \pm i\sqrt{abdx^*}$, $\lambda_3 = -(1 - \delta)x^* = -1/d < 0$. Differentiating the characteristic equation (5.36) with respect to parameter σ, we find that

$$\Re\left\{\frac{d\lambda(\sigma)}{d\sigma}\right\}\Bigg|_{\substack{\sigma=0 \\ \lambda=\lambda_1}} = \frac{a\delta}{2(1 - \delta)(abd^2 + 1 - \delta)}\left(\frac{1 - \delta}{\delta} - bd\right) > 0,$$

because $(1 - \delta)/\delta > bd$ for $\delta < \delta^*$.

Note that the conditions listed above are not sufficient for the Andronov–Hopf bifurcation, i.e. bifurcation that results in "soft" creation of a limit cycle from a stable fixed point. To ensure sufficiency we need yet another condition: for $\sigma = 0$ the fixed point O^* remains a stable focus. Yet in our case, for $\sigma = 0$ the point O^* is a center.

In economic terms, this bifurcation corresponds to a transition from an administrative command economy to a market economy. For $\sigma > 0$ a new class of employers emerge: these employers are relatively independent of the government and can spend their profits as they fit, including on private consumption. In our case $(\delta < \delta^*)$, such an economy may develop only cyclically. The complexity of periodic oscillations of capital and consumer demand increases with the increase of σ (keeping δ constant) until chaotic oscillations appear; if, however, we increase δ keeping σ constant, oscillations are much simpler. It is notable that for each $\delta < \delta^*$ there is always some $\sigma < 1 - \delta$ such that the economy is ultimately destroyed due to a global crisis, when capital and the consumer demand drop to zero. For instance, for $\delta = 0.4$ model (5.34) has a unique stable cycle for $0 < \sigma < 0.05$ (Fig. 5.17a), but this cycle is destroyed for $\sigma > 0.05$. For $\delta = 0.65$, increasing σ in model (5.34) produces a subharmonic cascade of bifurcations of transition to chaos.

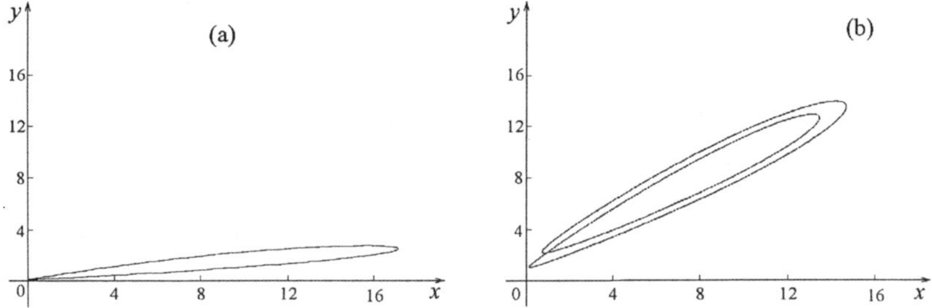

Fig. 5.17 Cyclic development of economy for $\delta = 0.4$ and $\sigma = 0.05$ (a) and for $\delta = 0.65$ and $\sigma = 0.266$(b).

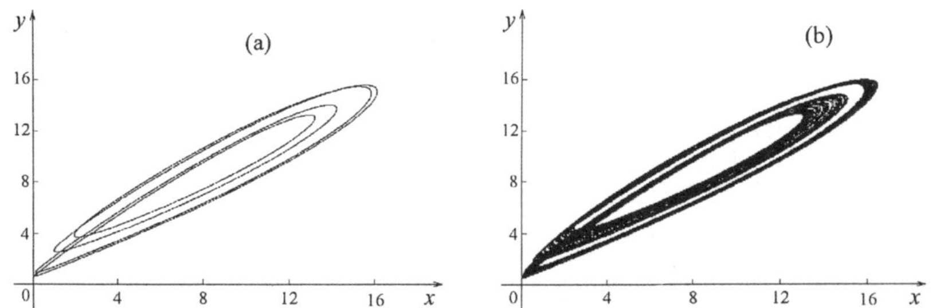

Fig. 5.18 Projection on the (x, y) plane of the quadruple-period cycle for $\delta = 0.65$, $\sigma = 0.276$ (a) and the Feigenbaum singular attractor for $\delta = 0.65$ and $\sigma = 0.278$ (b).

Thus, a double-period cycle appears for $\sigma = 0.266$ (Fig. 5.17b), a quadruple-period cycle appears for $\sigma = 0.275$ (Fig. 5.18a), and the Feigenbaum singular attractor terminating the infinite cascade of period doubling bifurcations appears for $\sigma = 0.278$ (Fig. 5.18b). A cycle of period 5 appears for $\sigma = 0.2802$ (Fig. 5.19a); it is followed by the next cascade of period doubling bifurcations, and a symbolic cycle of period 3 appears for $\sigma = 0.284$

(Fig. 5.19b). According to Sharkovskii theorem (see Chapter 4), this cycle signals the existence of cycles of arbitrary period in system (5.34). For $\sigma > 0.284$ the cyclic behavior disappears and the economic system breaks up. These results indicate that uncontrolled growth of personal consumption of the employers (corresponding to the increase of the parameter σ in a model (5.34)) inevitably leads to chaos and ultimate destruction of economy. A vivid illustration of our rigorous mathematical results is provided by the heavy consequences of the decade of "wild" capitalism in Russia following the improvised reforms of Yeltsin–Gaidar, in contrast to the economic success in China, where market relations were developing during the same period under government control and regulation.

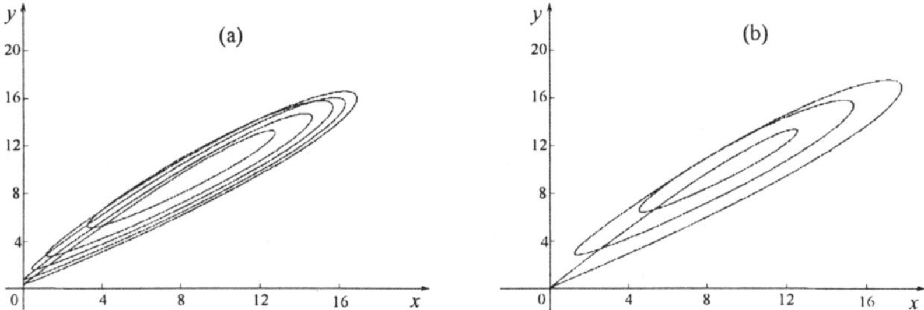

Fig. 5.19   Projection on the (x, y) plane of a cycle of period 5 for $\delta = 0.65$, $\sigma = 0.2802$ (a) and a cycle of period 3 for $\delta = 0.65$ and $\sigma = 0.284$ (b).

Scenario 3: $\sigma > 0$. As we have noted previously, complexity of periodic oscillations of capital and consumer demand decreases with the increase of δ and increases with the decrease of δ (keeping σ fixed). Therefore, economy is destroyed for small δ, just as for large σ. If we take, for instance, $\sigma = 0.284$, then for all $\delta < 0.65$ the economy is destroyed due to a global crisis. For $\delta = 0.65$, as noted previously, the system (5.34) has a cycle of period 3 (Fig. 5.19b), and for $\delta > 0.65$ we observe a bifurcation cascade which is the reverse of the cascade in Scenario 2. Specifically, at first we have chaos, then a cycle of period 4 appears for $\delta = 0.656$, followed by a cycle of period 2 for $\delta = 0.657$, and finally, for $\delta = 0.662$, the solution of system (5.34) is a simple stable limit cycle (Fig. 5.20.a). When parameter δ is increased to $\delta^* = 0.681$, this limit cycle contracts to a point and disappears through an inverse Andronov–Hopf bifurcation (Fig. 5.20b).

The last scenario leads to a second important conclusion, namely that a higher government demand for consumer goods ensures a more stable development of a market economy, which becomes less susceptible to various crisis phenomena. Conversely, a low government demand destroys the economic system.

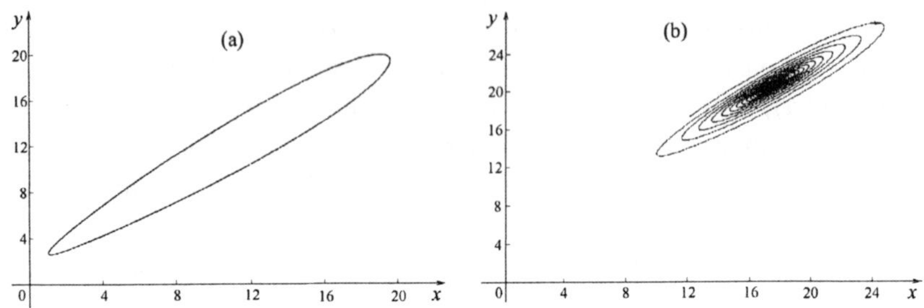

Fig. 5.20 Projection on the (x, y) plane of a stable cycle of system (5.34) for $\delta = 0.662$, $\sigma = 0.284$ (a) and fixed point for $\delta = \delta^* = 0.681$, $\sigma = 0.284$ (b).

An illustration of this rigorous mathematical result is provided on one hand by crisis and bankruptcy of many Russian enterprises which have not been provided by government orders, and on the other hand by decisions of President George Bush to withdraw from the ballistic missile control agreement and to begin the war in Iraq: these decisions were dictated by cleanly economic reasons that is by the need to inject government orders into the ailing U.S. economy.

We can see from the model (5.34) that the same conclusion relates also to the worker's demand, which is mainly determined by their wage earnings. The worker's demand is characterized by the coefficient γ reflecting the organic structure of capital. It is thus totally false to assert that the Russian people will be able to get a decent wage only when the economy begins to function normally. The opposite is true: the economy will begin to function normally only when everybody gets a decent wage and thus contributes to a stable demand for consumer goods produced in the economy.

5.4.3 *Behavior of economic variables in the presence of diffusion of capital and consumer demand*

Every stable solution of the system of ordinary differential Eqs. (5.34) obtained in the previous section can be continued homogeneously to the entire interval $[0, l]$. This continuation is obviously a stable solution of the second boundary value problem for the system (5.34) with homogeneous initial conditions. We should naturally explore the stability of these solutions under small spatial perturbations of the homogeneous initial conditions. So, in this section we consider the second boundary value problem on an interval for the system of Eqs. (5.32) with constant positive diffusion coefficients of capital and consumer demand for $\sigma > 0$:

$$\frac{\partial x(t,c)}{\partial t} = -d_1\,\frac{\partial^2 z}{\partial c^2} + bx((1-\sigma)z - \delta y),$$

$$\frac{\partial y(t,c)}{\partial t} = -d_2\frac{\partial^2 z}{\partial c^2} + x(1-(1-\delta)y + \sigma z),$$

$$\frac{\partial z(t,c)}{\partial t} = a(y - dx), \tag{5.37}$$

$$0 \le c \le l,\ 0 \le t < \infty,$$

$$x(0,c) = x_0(c),\ y(0,c) = y_0(c),\ z(0,c) = z_0(c),$$

$$\frac{\partial x(t,0)}{\partial c} = \frac{\partial x(t,l)}{\partial c} = \frac{\partial y(t,0)}{\partial c} = \frac{\partial y(t,l)}{\partial c} = \frac{\partial z(t,0)}{\partial c} = \frac{\partial z(t,l)}{\partial c} = 0.$$

We start with analysis of stability of the thermodynamic branch of system (5.37), which is the homogeneous time-independent solution $x(t,c) = x^*$, $y(t,c) = y^*$, $z(t,c) = z^*$ of this system. We linearize problem (5.37) in the neighborhood of the thermodynamic branch, setting

$$u(t,c) = (u_1, u_2, u_3)^{\mathrm{T}} = (x(t,c) - x^*, y(t,c) - y^*, z(t,c) - z^*)^{\mathrm{T}}.$$

Then $\dot{u} = Lu + O(\|u\|^2)$,   where the operator L has the form

$$L = J(x^*, y^*, z^*) + \begin{pmatrix} 0 & 0 & -d_1 \\ 0 & 0 & -d_2 \\ 0 & 0 & 0 \end{pmatrix} \frac{\partial^2}{\partial c^2}$$

$$= \begin{pmatrix} 0 & -b\delta x^* & b(1-\sigma)x^* \\ 0 & -(1-\delta)x^* & \sigma x^* \\ -ad & a & 0 \end{pmatrix} + \begin{pmatrix} 0 & 0 & -d_1 \\ 0 & 0 & -d_2 \\ 0 & 0 & 0 \end{pmatrix} \frac{\partial^2}{\partial c^2}.$$

The operator $\frac{\partial^2}{\partial c^2}$ with boundary conditions of the second kind on the interval $[0, l]$ has the eigenvalues $\nu_n = -(\pi n/l)^2$, $n = 0, 1, \ldots$. Thus, expanding the eigenfunctions of the operator L in the eigenfunctions $\cos(\pi n c/l)$ of the operator $\frac{\partial^2}{\partial c^2}$, we obtain that the eigenvalues λ_n of the operator L are the eigenvalues of the matrices

$$G_n = \begin{pmatrix} 0 & -b\delta x^* & b(1-\sigma)x^* + d_1\dfrac{\pi^2 n^2}{l^2} \\ 0 & -(1-\delta)x^* & \sigma x^* + d_2\dfrac{\pi^2 n^2}{l^2} \\ -ad & a & 0 \end{pmatrix}, \quad n = 0, 1, \ldots,$$

which satisfy the characteristic equations

$$\lambda^3 + (1 - \delta)x^* \lambda^2 + a\left[bdx^*(1 - \sigma) - \sigma x^* + (dd_1 - d_2)\frac{\pi^2 n^2}{l^2}\right]\lambda$$

$$+ ax^*\left[b(1 - \sigma) + d((1 - \delta)d_1 - b\delta d_2)\frac{\pi^2 n^2}{l^2}\right] = 0, \quad n = 0, 1, \ldots. \quad (5.38)$$

We write the Routh–Hurwitz stability conditions for Eqs. (5.38):

$$(1 - \delta)x^* > 0,$$

$$(1 - \delta)x^* a\left[bdx^*(1 - \sigma) - \sigma x^* + (dd_1 - d_2)\frac{\pi^2 n^2}{l^2}\right]$$

$$> ax^*\left[b(1 - \sigma) + d((1 - \delta)d_1 - b\delta d_2)\frac{\pi^2 n^2}{l^2}\right],$$

$$ax^*\left[b(1 - \sigma) + d((1 - \delta)d_1 - b\delta d_2)\frac{\pi^2 n^2}{l^2}\right] > 0, \quad n = 0, 1, \ldots.$$

The first condition is always true, because $x^* > 0$, $\delta < 1$. The second condition may be written as

$$ax^*\left[bdx^*(1 - \sigma)(1 - \delta) - \sigma x^*(1 - \delta) - b(1 - \sigma)\right.$$

$$\left. + d_2\delta\frac{\pi^2 n^2}{l^2}\left(bd - \frac{1 - \delta}{\delta}\right)\right] > 0, \quad n = 0, 1, \ldots,$$

or

$$ax^*\delta\left(bd - \frac{1 - \delta}{\delta}\right)\left(\frac{\sigma(1 - \sigma)}{d(1 - \delta - \sigma)} + d_2\frac{\pi^2 n^2}{l^2}\right) > 0, \quad n = 0, 1, \ldots.$$

Thus, the second condition holds for all $n \geq 0$ in the region $\delta > \delta^*$ and does not hold for any n in the region $\delta < \delta^*$. The third condition is true for all n only if $(1 - \delta)d_1 - b\delta d_2 \geq 0$. Otherwise, there is always some n such that the third condition does not hold. The last condition may be rewritten as

$$bd \leq k(1 - \delta)/\delta, \quad \text{where} \quad k = dd_1/d_2.$$

Therefore, the thermodynamic branch of problem (5.37) is always unstable in the region $\delta < \delta^*$, and in the region $\delta > \delta^*$ it is stable only when $k \geq 1$ or $d_1 \geq d_2/d$.

The thermodynamic branch is the only homogeneous solution of problem (5.37) in the region $\delta > \delta^*$. Therefore greater inertia or low mobility of capital in the technology space due to weak laws, criminalization of business, corruption, difficult access to credit, and other reasons leads to instability of the economic system and its ultimate destruction for $\delta > \delta^*$. For $\delta < \delta^*$, problem (5.37) has other homogeneous periodic or chaotic solutions alongside the unstable thermodynamic branch (see Figs. 5.17–5.19), but stability analysis of these solutions under small spatial perturbations is a difficult mathematical problem for which no adequate tools are available. Numerical calculations show, however, that at least some of these solutions (cycles for $\delta \lesssim \delta^*$), while unstable under small spatial perturbations for $d_1 < d_2/d$, remain stable under small perturbations for $d_1 > d_2/d$.

Our mathematical analysis suggests a third important conclusion that capital should be allowed to flow and respond sufficiently rapidly to all changes in demand for various consumer goods produced in the economic system.

Thus, we have shown that the system of three ordinary differential equations describing essentially the dynamics of macroeconomic variables may have stable time-independent solutions, stable cycles with an arbitrary period, and also chaotic solutions. We have also examined the solutions of the second boundary value problem for the system of partial differential equations with constant coefficients of diffusion of capital and consumer demand. Our analysis of several scenarios of economic development gave us possibility to draw a number of important conclusions:

- uncontrolled growth of personal consumption of the firms inevitably leads to chaos and ultimate destruction of economy;
- low demand from the side of the government (government orders, government support to business, *etc.*) and the workers (wages) also inevitably leads to chaos and destruction of economy;
- high inertia of capital, slowing down its response to changes in profit rates and consumer demand, also makes the economic system unstable and leads to its ultimate destruction.

We have received some more interesting results for the model (5.37) in a case of $d_2 = 0$ [Dernov and Magnitskii (2005)]. In this case operator L has one real negative eigenvalue and infinite number of pairs of complex conjugate eigenvalues simultaneously passing through an imaginary axis at reduction of parameter δ and its passing through the point $\delta = \delta^*$. However, in the given case a spatially homogeneous stable limit cycle is born as a result of infinite degenerate bifurcation of loss of stability of a thermodynamic branch in the problem (5.37). Numerical experiments have shown, that the size of a region of stability of the born cycle on parameter δ essentially depends on the value of diffusion coefficient d_1. If the diffusion coefficient is less, then the stability region is less too (Fig. 5.21a). At enough great values of diffusion coefficient when the system (5.37) on its properties is close enough to the concentrated system of ordinary differential equations (5.34), the subharmonic cascade of bifurcations of stable cycles is realized in it. However, at reduction of values of the diffusion coefficient d_1 in the system (5.37), another qualitative picture is observed that is illustrated in Fig. 5.21a. Spatially homogeneous cycle (area 1) loses stability and generates a spatially inhomogeneous cycle (area 2), which bifurcates with formation of stable two-dimensional torus (area 3). A projection of two-dimensional torus and a projection of its Poincare section are shown in Fig. 5.21b,c. Further two-dimensional torus bifurcates with formation of more complex attractors (area 4) which structure description demands carrying out of additional research.

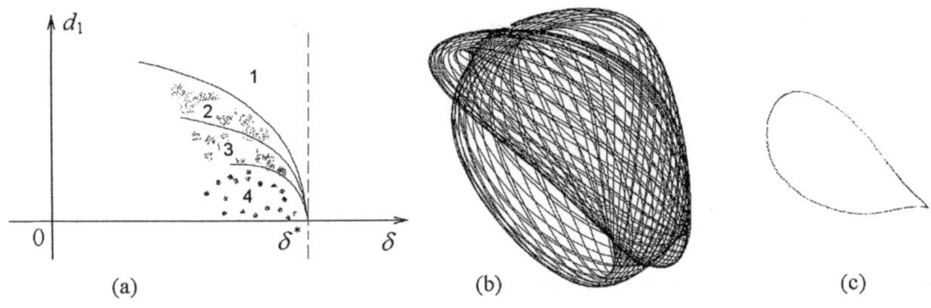

Fig. 5.21 Bifurcation diagram (a), projection of two-dimensional torus of the system (5.37) for $d_2 = 0$ (b) and projection of its Poincare section (c).

Thus, in spite of the fact that the system (5.37) possesses many essential differences from classical systems of reaction–diffusion type, transition to spatio-temporal chaos in it is carried out also according to the Feigenbaum–Sharkovskii–Magnitskii theory stated in Chapter 4.

In summary we shall note one more important direction in studying behaviour of solutions of system (5.37), research of the running waves connected with redistribution of capital and consumer demand in space of technologies. The first results of numerical experiments confirming existence of such waves in system (5.37), are resulted in [Dernov (2002a)]. Further study of the solutions of this system of equations in more general cases should allow for multidimensional spatial distribution of the variables and non-constant coefficients of diffusion of capital and consumer demand.

Chapter 6

Chaos Control in Systems of Differential Equations

Presence of chaos is an essential part of the majority of nonlinear dynamical systems describing complex enough physical, chemical, biological and social processes and phenomena. Chaotic systems are characterized by the increased sensitivity to small perturbations of system parameters and initial conditions. Owing to this fact the behaviour of such systems was considered unpredictable and uncontrollable for many years. There was an opinion that it is possible to obtain desirable behaviour of system only having suppressed a chaos in it let even big and expensive changes in the system, leading to a change in its dynamics as a whole. The task in view was reduced to a choice of control actions or in the opened form (program control), or in the form of a feedback on a state or an output with the purpose of reduction of the solution of system to the given periodic kind or with the purpose of synchronization of the solution of system with the solution of some other system possessing the necessary regular properties (see numerous links in [Chen and Dong (1998)] and in reviews [Andrievskii and Fradkov (2003); Loskutov (2001)]. In other words, the problem of stabilization of the given or desirable trajectory in a system with chaotic behaviour was solved.

However, understanding of a special role of chaos in self-organizing of various natural phenomena has come during last years. It has been realized, that the chaos not only does not prevent, but more likely it is an indispensable condition for efficiency of complex systems, such, for example, as a human brain [Prigogine and Stengers (1984); Sepulchre and Babloyantz (1994)]. Only owing to the presence of chaotic attractor, containing, as a rule, an infinite number of unstable periodic orbits (cycles), it is possible to achieve a qualitative change in dynamics of the system (passing from a neighbourhood of one cycle into a neighbourhood of another) by small perturbations of system parameters. In this connection in *a problem of chaos*

309

control, a problem has naturally appeared for stabilization of not *a priori* given or desirable trajectories of chaotic dynamical systems, but namely those unstable periodic trajectories, which infinite number is situated in a web of irregular attractor. And, any information about location of these trajectories in a phase space, about the periods and amplitudes of their oscillations is practically absent. Methods of solution of this last and the most interesting, from our point of view, problem are considered in the present chapter. This problem is reduced to *localization* (revealing) and *stabilization* of unstable periodic orbits (in particular, stationary states) of chaotic dynamical systems (including chaotic mappings).

The most known and widely quoted in the modern literature methods of solution of the problem are presented in Sec. 6.1. It is, first of all, the OGY-method based on linearization of the Poincare mapping, and Pyragas method based on construction of a feedback with delay, close to the period of the existing unstable periodic solution. The Magnitskii method is stated in Sec. 6.2, consisting in construction of feedback in an expanded phase space, i.e. in construction in space of greater dimension of some dynamical system for which the required unstable periodic orbit of original chaotic system is a projection of some its asymptotically (orbital asymptotically) stable periodic trajectory.

The scope of last method has appeared extremely wide: chaotic mappings, chaotic dynamical systems described by ordinary differential equations, distributed chaotic dynamical systems and dynamical systems with delay argument (see papers from [Magnitskii (1996)] to [Magnitskii (1997c)], from [Magnitski and Sidorov (1998)] to [Magnitskii and Sidorov (2001b)]). The problem of reconstruction of chaotic system in the trajectory of its irregular attractor is considered in Sec. 6.3. This problem is often connected with the problem of control of chaotic systems.

6.1 Ott–Grebogi–Yorke and Pyragas methods

Ott–Grebogi–Yorke method (OGY-method) is offered and developed, basically, in papers [Ott *et al.* (1990); Shinbort *et al.* (1993)]. It consists in stabilization of an unstable periodic solution of chaotic system of ordinary differential equations by application of discrete operating influences in the form of feedback in some Poincare section in a neighbourhood at a fixed point of the Poincare mapping, corresponding to a required cycle. The kind of feedback is defined by linearization of the Poincare mapping at a fixed point. The method offered by Pyragas in [Pyragas (1992)] uses a feedback with delay, and time of delay should be close by the period of the required unstable periodic solution.

6.1.1 *The OGY-method*

We shall consider a smooth family of nonlinear autonomous systems of ordinary differential equations

$$\dot{x} = F(x, \mu), \quad x \in M \subset \mathbb{R}^m, \ \mu \in L \subset \mathbb{R}^k, \ F \in C^\infty, \tag{6.1}$$

given in phase space M by smooth vector fields F, depending on coordinates of vectors of system parameters μ, lying in the region L of the space $\mathbb{R}^k$. Let the unstable limit cycle $x^*(t, \mu^*)$ be the required solution of family of systems (6.1) which has in addition a regular or singular attractor at the same parameter value $\mu = \mu^*$. Let us construct the Poincare section S, passing through the point $x_0 = x^*(0, \mu^*)$ of a cycle $x^*(t, \mu^*)$ transversally to it. We shall consider the Poincare control mapping $x \to P(x, \mu)$, in which $P(x, \mu)$ is a point of the first returning to surface S of the trajectory of the system (6.1) starting in point x at value of parameter vector μ, being in this case a vector of control parameters. (For complex cycle having several turns, it is necessary to consider a corresponding iteration of the mapping). Applying the sequence of such control mappings, we shall obtain a discrete dynamical system

$$x_{n+1} = P(x_n, \mu_n), \tag{6.2}$$

where $x_n = x(t_n)$, t_n is the moment of time of n-th crossing of the surface S, and μ_n is the value of vector of control parameters in an interval between t_n and t_{n+1}.

Let us replace now the mapping (6.2) by close to them mapping linearized at the point (x_0, μ^*)

$$y_{n+1} = Ay_n + Bu_n, \quad A = \frac{\partial P}{\partial x}(x_0, \mu^*), \quad B = \frac{\partial P}{\partial \mu}(x_0, \mu^*), \tag{6.3}$$

where $y_n = x_n - x_0$, $u_n = \mu_n - \mu^*$. For linear system (6.3), we shall choose stabilizing control u_n in the form of a linear feedback on a state: $u_n = -Cy_n$. Then we shall obtain from (6.3), that

$$y_{n+1} = (A - BC)y_n. \tag{6.4}$$

Thus, the fixed point x_0 of the Poincare map and, hence, a required unstable cycle $x^*(t, \mu^*)$ of the system (6.1) will be stabilized for matrix C such that modules of eigenvalues of the matrix $(A - BC)$ are less than unit.

The advantage of OGY-method is that stabilization of a fixed point of the Poincare mapping and a limit cycle of a system of differential equations can be achieved by small control influences during the discrete moments of time. However, a great shortcoming is that the fixed point of the Poincare mapping is unstable. Therefore for applicability of the method, it is necessary not only to know precisely the matrix A (that is possible only numerically for systems of differential equations), but also both its eigenvalues and eigenvectors corresponding to stable and unstable manifolds of the fixed point of the Poincare mapping. The trajectory thus should be corrected on each iteration aside stable manifold of the fixed point. A big problem is also the choice of an initial point. In an OGY-method, it is supposed implicitly, that the system (6.1) has a chaotic attractor in the sense that it is the closure of all periodic trajectories containing in it, whence follows, that any trajectory with any initial condition in due course will necessarily get in some small neighbourhood of a required cycle. But it is far not so. Many singular attractors, considered in the present book, do not possess this property. The elementary example is the Feigenbaum attractor, coexisting together with infinite number of unstable cycles and lying in finite distance from each of them. One can find numerous links to works, devoted to development of the OGY-method and its various modifications in reviews [Chen and Dong (1998); Andrievskii and Fradkov (2003)].

6.1.2   *The Pyragas method*

We shall consider smooth family of nonlinear control systems of ordinary differential equations

$$\dot{x} = F(x, \mu, u), \quad x \in M \subset \mathbb{R}^m, \ \mu \in L \subset \mathbb{R}^k, \ u \in U \subset \mathbb{R}^n, \ F \in C^\infty, \quad (6.5)$$

depending on a vector u of control parameters. Let it be required to stabilize an unstable limit cycle $x^*(t, \mu^*)$ of the period T, being the solution of the system of family (6.5) at $u = 0$ and $\mu = \mu^*$. Let at the same values of parameters $u = 0$ and $\mu = \mu^*$ the family (6.5) have a regular or singular attractor. Then the problem of stabilization of the cycle $x^*(t, \mu^*)$ can be solved in some cases by a choice of simple law of feedback with delay of a kind

$$u(t) = K(x(t) - x(t-T)),\tag{6.6}$$

where K is a matrix of transfer coefficients. If the initial condition $x(0)$ is chosen lying in enough small neighbourhood of an orbit of a cycle, then the solution $x(t)$ of the system

$$\dot{x}(t) = F(x(t), \mu^*, K(x(t) - x(t-T)))\tag{6.7}$$

with the feedback (6.6) at $\mu = \mu^*$ can converge to a required unstable cycle $x^*(t, \mu^*)$.

Analytical research of asymptotical properties of solutions of the closed system (6.7) is a problem enough difficult. Therefore until recently exclusively numerical and experimental results concerning properties and area of applicability of *the Pyragas method* (see the literature in the review [Andrievskii and Fradkov (2003)]) were known. The problem of finding of sufficient conditions guaranteeing applicability of the method, till now remains unresolved. Besides, greater lack of the law of control (6.6) is its sensitivity to a choice of time of delay. So, if the period T of a required cycle is unknown in advance, namely this situation is typical for chaotic systems of differential equations, then it is possible to obtain a required convergence only in unusual cases, successfully having estimated the period value by any heuristic methods.

6.2   The Magnitskii Method

Let us consider in detail *the Magnitskii method* of localization and stabilization of unstable singular points and periodic solutions of chaotic systems of nonlinear differential equations and discrete chaotic dynamical systems. The method was offered in papers from [Magnitskii (1996)] to [Magnitskii (1997c)] and developed then in papers from [Magnitski and Sidorov (1998)] to [Magnitskii and Sidorov (2001b)]. It is based on construction of a coordinate-parametrical feedback in the expanded space, that makes possible search of stable fixed points (unlike the OGY-method) or asymptotic orbitally stable periodic trajectories (unlike the Pyragas method). Except for that the method has no problems with choice of initial conditions and, unlike the OGY-method and the Pyragas method, it can be applied to chaotic systems of differential equations in case of absence of any information on the size of period and on position of a required unstable cycle in a phase space. The field of the method applicability includes chaotic mappings, chaotic systems of ordinary and partial differential equations, and also differential equations with delay argument. All algorithms of applicability of the method are proved analytically.

6.2.1   *Localization and stabilization of unstable fixed points and unstable cycles of chaotic mappings*

Let us consider a family of m-dimensional nonlinear smooth mappings

$$z_{n+1} = F(z_n, \mu), \quad z \in \mathbb{R}^m, \tag{6.8}$$

where μ is a scalar parameter and F is a smooth vector-function. Let points $z_1^*(\mu)$, $z_2^*(\mu)$, $\ldots$, $z_k^*(\mu)$ be k-periodic points of the mapping (6.8), for which equalities

$$z_2^*(\mu) = F(z_1^*(\mu), \mu), \ z_3^*(\mu) = F(z_2^*(\mu), \mu), \ldots, \ z_1^*(\mu) = F(z_k^*(\mu), \mu) \tag{6.9}$$

take place. If $k = 1$, then point $z_1^*(\mu)$ is the fixed point of the mapping (6.8). Usually there is a critical value μ_k^* of the system parameter such that the cycle (6.9) is a stable periodic trajectory of the mapping (6.8) in the domain $\mu \leq \mu_k^*$, whereas in the domain $\mu > \mu_k^*$ the cycle (6.9) is an unstable periodic trajectory of the mapping (6.8), which has in this case other regular or singular attractors. The problem is to localize (to define the position) and stabilize an unstable periodic trajectory (6.9) of the mapping (6.8) in the domain $\mu > \mu_k^*$ by small perturbations of the system parameter μ. Note that each point $z_i^*(\mu)$, $i = 1, \ldots, k$ of the cycle (6.9) is a stable fixed point of the mapping F^k in the domain $\mu \leq \mu_k^*$ and is an unstable fixed point for $\mu > \mu_k^*$.

Let us consider a $(m + 1)$-dimensional mapping

$$\begin{aligned} z_{n+1} &= F^k(z_n, \mu) + \varepsilon(q_n - \mu), \\ q_{n+1} &= Q(z_n, \mu) + \beta(q_n - \mu) + \mu, \end{aligned} \tag{6.10}$$

where $\varepsilon \in \mathbb{R}^m$, $\beta \in \mathbb{R}$, $Q(z_i^*(\mu), \mu) = 0$, $i = 1, \ldots, k$. Clearly, that if the cycle (6.9) is a periodic trajectory of the mapping (6.8), then each point $(z_i^*(\mu), \mu)$, $i = 1, \ldots, k$, is a fixed point of the mapping (6.10). Let us calculate the Jacobi matrix of the mapping (6.10) at the point $(z_i^*(\mu), \mu)$

$$J(\mu) = \begin{pmatrix} \partial F^k / \partial z & \varepsilon \\ \partial Q / \partial z & \beta \end{pmatrix}.$$

All eigenvalues of the matrix $J(\mu)$ at the point μ_k^* are equal to zero if and only if

$$\sum_{i=1}^{C_{m+1}^l} J_{li}(\mu_k^*) = 0, \quad l = 1, \ldots, m+1, \tag{6.11}$$

where $J_{li}(\mu)$ is the i-th principal minor of the order l of the matrix $J(\mu)$. Therefore, control parameters β and $\varepsilon = (\varepsilon_1, \ldots, \varepsilon_m)^{\mathrm{T}}$ in (6.10) should satisfy the system of $m+1$ linear Eqs. (6.11). In particular, it is obvious, that one of control parameters β of the mapping (6.10) can be calculated directly from the first equation $\operatorname{tr} J(\mu_k^*) = 0$ of the system (6.11)

$$\beta = -\operatorname{tr} J_0(\mu_k^*) = -\operatorname{tr}\left\{\frac{\partial F^k(z_i^*(\mu_k^*), \mu_k^*)}{\partial z}\right\}.$$

Therefore, in view of smoothness of the mapping F, we claim that if system (6.11) has a solution, then there exists a domain $\mu_k^* \leq \mu \leq \mu_{k1}^*$ such that for each $\mu \in [\mu_k^*, \mu_{k1}^*]$ the absolute values of all eigenvalues of the matrix $J(\mu)$ are less than 1. Hence we obtain the following statement.

Theorem 6.1 *If the determinant D of system of linear Eqs. (6.11) is not equal to zero, then there exists a value $\mu_{k1}^* > \mu_k^*$ of the system parameter such that for each $\mu \in [\mu_k^*, \mu_{k1}^*]$ the point $(z_i^*(\mu), \mu)$ is an asymptotically stable fixed point of the mapping (6.10) and can be localized and stabilized in the domain of $\mu \in [\mu_k^*, \mu_{k1}^*]$ by means of the iterative process (6.10) with initial conditions $q_0 = \mu$, $z_0 = z_i^*(\mu_k^*)$. Any other k-periodic point $z_j^*(\mu)$, $j \neq i$ of the mapping (6.8) also can be localized and stabilized in the domain of $\mu \in [\mu_k^*, \mu_{k1}^*]$ by means of the Eq. (6.9).*

Using the value μ_{k1}^* as a new critical value of the system parameter and calculating the new Jacobi matrix $J(\mu_{k1}^*)$, we can refine the values of control parameters ε_1, ..., ε_m, β in (6.10) and once more localize and stabilize the point $z_i^*(\mu)$ of the cycle (6.9) in the new domain of $\mu \in [\mu_{k1}^*, \mu_{k2}^*]$. This process can be continued for the entire interval on which the periodic trajectory (6.9) of the mapping (6.8) exists.

In particular, mapping $Q(z, \mu)$ in (6.10) can be chosen in the form of

$$Q(z, \mu) = a_1(F_1^k(z, \mu) - z_1) + \cdots + a_m(F_m^k(z, \mu) - z_m),$$

where parameters a_i, $i = 1, \ldots, m$, are set to either zero or unity so as to satisfy the condition $D \neq 0$ for the determinant D of the linear system

(6.11). The general approach stated above, can be concretized in case when the mapping F in (6.8) is one- or two-dimensional mapping.

6.2.1.1 *Case of one-dimensional mapping*

Theorem 6.2 *Let $z_i^*(\mu)$ be a k-periodic point of one-dimensional chaotic mapping (6.8) and μ_k^* be a critical value of the system parameter such that*

$$\nu = \partial F^k(z_i^*(\mu_k^*), \mu_k^*)/\partial z \neq 1.$$

Then there exists $\mu_{k1}^* > \mu_k^*$ such that for any $\mu \in [\mu_k^*, \mu_{k1}^*]$ the point $(z_i^*(\mu), \mu)$ is an asymptotically stable fixed point of a two-dimensional mapping

$$
\begin{aligned}
z_{n+1} &= F^k(z_n, \mu) + \varepsilon(q_n - \mu), \\
q_{n+1} &= F^k(z_n, \mu) - z_n + \beta(q_n - \mu) + \mu,
\end{aligned}
\tag{6.12}
$$

where $\beta = -\nu$, $\varepsilon = \nu^2/(1 - \nu)$.

Let us notice, that the value of

$$\nu = \frac{\partial F}{\partial z}(z_1^*(\mu_k^*), \mu_k^*) \times \ldots \times \frac{\partial F}{\partial z}(z_k^*(\mu_k^*), \mu_k^*)$$

is the same for all points $z_i^*(\mu)$ of the cycle (6.9) of the mapping (6.8).

Example 6.1 By way of example, we consider the logistic mapping

$$z_{n+1} = \mu z_n(1 - z_n), \quad 3 < \mu \leq 4. \tag{6.13}$$

The fixed point $z^*(\mu) = 1 - 1/\mu$ of the mapping (6.13) is an unstable stationary point in the domain $\mu > \mu^* = 3$. Moreover, if $\mu > 3.57$, then the mapping (6.13) exhibits a chaotic behavior. Numerical experiments showed, that the fixed point $z^*(\mu)$ of the mapping (6.13) can be localized and stabilized at the intervals $3 \leq \mu \leq 3.6$ with the help of iterative process (6.12) with $\beta = -F_z = 1$ and $\varepsilon = 0.5$, calculated at the point $\mu^* = 3$. Then this fixed point can be localized and stabilized at the interval $3.6 \leq \mu \leq 4.25$ with parameters $\beta = 1.6$ and $\varepsilon = 1$, calculated at the point $\mu_1^* = 3.6$.

At the latter interval the mapping (6.13) has a chaotic behavior. The bifurcation diagram of the Eq. (6.13) and the system (6.12) in case of $k = 1$ with the above-mentioned two sets of control parameters β and ε is represented in Fig. 6.1a.

The points

$$z_1^*(\mu) = \frac{(\mu+1) - \sqrt{(\mu+1)(\mu-3)}}{2\mu},$$
$$z_2^*(\mu) = \frac{(\mu+1) + \sqrt{(\mu+1)(\mu-3)}}{2\mu} \tag{6.14}$$

are points of an unstable cycle of the period 2 of the mapping (6.13) for $\mu > \mu^* = 1 + \sqrt{6}$.

Numerical experiments showed, that it is possible to localize and stabilize a double cycle (6.14) of the mapping (6.13) at the interval $\mu^* \leq \mu \leq 3.58$ by means of iterations (6.12) for $k = 2$ with $\beta = -\nu = 1$, $\varepsilon = 0.5$, calculated at the point μ^* (Fig. 6.1b). We shall remind, that the mapping (6.13) has a chaotic behaviour near to the right end of this interval (for $\mu > 3.57$).

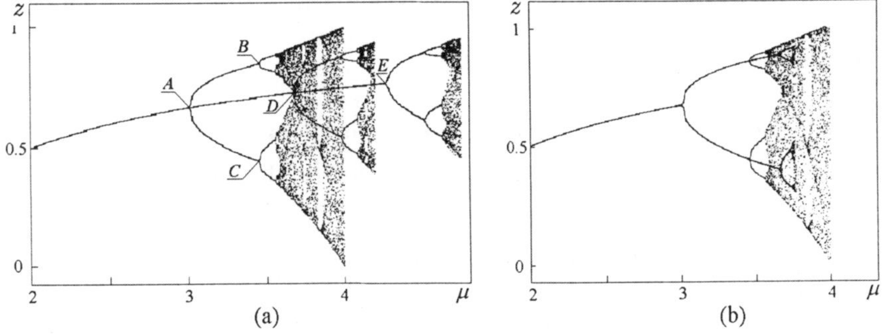

Fig. 6.1   Bifurcation diagram of logistic mapping (6.13) and the system (6.12) stabilizing it for $k = 1(a)$ and for $k = 2(b)$. Points A, B, C correspond to the period doubling bifurcations of the mapping (6.13), and points D, E correspond to the period doubling bifurcations of the stabilizing mapping (6.12). Parts of a trajectory $A - D$ and $D - E$ are obtained at the first and second iterations of the process of stabilization of the mapping (6.12) accordingly at intervals $3 \leq \mu \leq 3.6$ and $3.6 \leq \mu \leq 4.25$.

Stabilization by this method of unstable cycles of logistic mapping of the big period is spent in the paper [Dernov (2001b)]. In the paper [Kaloshin (2001)] the method is applied to search and stabilization of unstable cycles of the Lorenz system, lying in a neighbourhood of a saddle–node separatrix loop (homoclinic butterfly), and it is applied for search and stabilization of saddle cycles in the Lorenz system and in the Chua system in the paper [Dernov (2002b)].

6.2.1.2 *Case of two-dimensional mapping*

Theorem 6.3 *Let $(x_i^*(\mu), y_i^*(\mu))$ be a k-periodic point of two-dimensional chaotic mapping (6.8) with $F = \begin{pmatrix} f_1(x_n, y_n, \mu) \\ f_2(x_n, y_n, \mu) \end{pmatrix}$ and let μ_k^* be a critical value of the system parameter such that a determinant*

$$D \equiv [a_1^2 f_{1y}^k - a_1 a_2 (f_{1x}^k - f_{2y}^k) - a_2^2 f_{2x}^k](1 - \operatorname{tr} J_0 + \det J_0) \neq 0,$$

where $J_0 = D_z F^k$ is the Jacobi matrix of the mapping F^k, and f_{1x}^k, f_{1y}^k, f_{2x}^k, f_{2y}^k are partial derivatives of the mapping $F^k = (f_1^k, f_2^k)^T$ at the point $\left(x_i^*(\mu_k^*), y_i^*(\mu_k^*)\right)$. Then there exists $\mu_{k1}^* > \mu_k^*$ such that for any $\mu \in [\mu_k^*, \mu_{k1}^*]$ point $(x_i^*(\mu), y_i^*(\mu), \mu)$ is an asymptotically stable fixed point of a three-dimensional mapping

$$\begin{aligned}
x_{n+1} &= f_1^k + \varepsilon_1(q_n - \mu), \\
y_{n+1} &= f_2^k + \varepsilon_2(q_n - \mu), \\
q_{n+1} &= a_1[f_1^k - x_n] + a_2[f_2^k - y_n] + \beta(q_n - \mu) + \mu,
\end{aligned} \qquad (6.15)$$

where $\beta = -\operatorname{tr} J_0$,

$$\begin{aligned}
\varepsilon_1 = -\big\{ &[a_1 f_{1y}^k - a_2 f_{1x}^k + a_2(\operatorname{tr} J_0 - 1)]\operatorname{tr} J_0 \det J_0 \\
&+ (a_1 f_{1y}^k - a_2 f_{1x}^k + a_2 \det J_0)[\det J_0 - (\operatorname{tr} J_0)^2]\big\}/D,
\end{aligned}$$

$$\begin{aligned}
\varepsilon_2 = \big\{ &[a_2 f_{2x}^k - a_1 f_{2y}^k + a_1(\operatorname{tr} J_0 - 1)]\operatorname{tr} J_0 \det J_0 \\
&+ (a_2 f_{2x}^k - a_1 f_{2y}^k + a_1 \det J_0)[\det J_0 - (\operatorname{tr} J_0)^2]\big\}/D.
\end{aligned}$$

Equality $c \equiv 1 - \operatorname{tr} J_0 + \det J_0 = 0$ means, that one or two eigenvalues of the Jacobi matrix of the mapping F^k are equal to $+1$ at the point $\mu = \mu^*$. If $c \neq 0$ and $f_{1y}^k \neq 0$ or $f_{2x}^k \neq 0$, then it is possible to choose $a_1 = 1$, $a_2 = 0$ or $a_1 = 0$, $a_2 = 1$ in the mapping (6.15). If $c \neq 0$ and $f_{1y}^k = f_{2x}^k = 0$, and $f_{1x}^k \neq f_{2y}^k$, then it is possible to fix $a_1 = a_2 = 1$ in (6.15).

Example 6.2 As an example we shall consider the Henon mapping

$$\begin{aligned}
x_{n+1} &= \mu + 0.3y_n - x_n^2, \\
y_{n+1} &= x_n.
\end{aligned} \qquad (6.16)$$

The fixed point $x^*(\mu) = y^*(\mu) = -0.35 + \sqrt{0.1225 + \mu}$ of the mapping (6.16) is an unstable stationary point in the domain $\mu > \mu^* = 0.3675$. For $\mu = \mu^*$ any of eigenvalues of the Jacobian of the mapping (6.16) is not equal to +1. And as $k = 1$ and $f_{2x} \equiv 1$, then it is possible to fix $a_1 = 0$, $a_2 = 1$ in (6.15).

The mapping (6.16) results from the mapping (1.7) of Chapter 1 by replacement $ax_n \to x_n$; $ay_n/b \to y_n$; $a \to \mu$. Therefore, as it was noted in Chapter 1, it has a strange attractor for $\mu = 1.4$. Numerical experiment showed, that, using iterative process (6.15), it is possible to localize and stabilize the fixed point $\left(x^*(\mu),\, y^*(\mu)\right)$ of the mapping (6.16) at the interval $\mu^* \leq \mu \leq 0.75$ at $\beta = 0.7$, $\varepsilon_1 = -0.376$, $\varepsilon_2 = 0.414$, calculated at the point μ^*. Then by means of iterative process (6.15), it is possible to stabilize a fixed point of the mapping (6.16) at the interval $0.75 \leq \mu \leq 1.1$ with new control parameters $\beta = 1.168$, $\varepsilon_1 = -0.964$, $\varepsilon_2 = 0.7$, calculated at the point $\mu_1^* = 0.75$. At last, it is possible to localize and stabilize a fixed point $\left(x^*(\mu),\, y^*(\mu)\right)$ of the Henon mapping (6.16) at the interval $1.1 \leq \mu \leq 1.43$, using iterative process (6.15) with new control parameters $\beta = 1.512$, $\varepsilon_1 = -1.622$, $\varepsilon_2 = 0.96$, calculated at the point $\mu_2^* = 1.1$. The last interval contains the value $\mu = 1.4$ of the system parameter, for which the Henon mapping has a well-known strange attractor (see, for example, Chapter 1, and also [Henon (1976); Schuster (1984)]). The two-dimensional diagram of the Henon mapping (6.16) and its stabilizing system (6.15) for $k = 1$ is represented in Fig. 6.2.

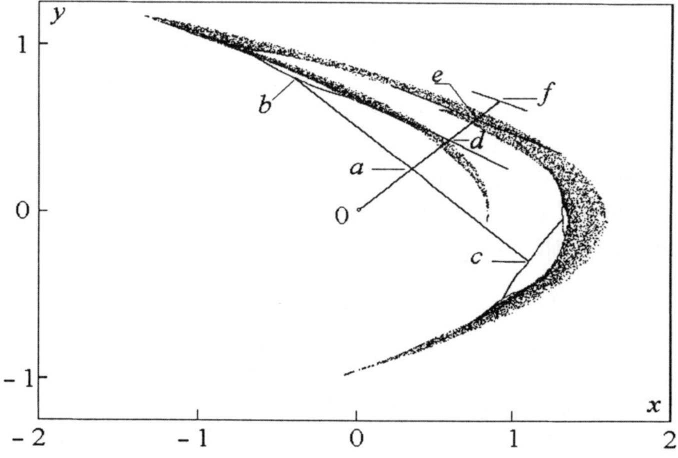

Fig. 6.2 Two-dimensional diagram of the Henon mapping (6.16) and its stabilizing system (6.15). The points a, b, c correspond to period doubling bifurcations of the mapping (6.16) and the points d, e, f correspond to period doubling bifurcations of the stabilizing mapping (6.15).

Points

$$x_1^* = 0.35 + \sqrt{\mu - 0.3675}, \ y_1^* = 0.35 - \sqrt{\mu - 0.3675},$$
$$x_2^* = 0.35 - \sqrt{\mu - 0.3675}, \ y_2^* = 0.35 + \sqrt{\mu - 0.3675} \tag{6.17}$$

form an unstable cycle of period 2 of the Henon mapping in the domain of $\mu > \mu^* = 0.9125$. If $\mu = \mu^*$, then each of eigenvalues of the Jacobian of the mapping F^2 is not equal to $+1$. As $f_{2x}^2 \neq 0$, then it is possible to fix values $a_1 = 0$, $a_2 = 1$ in (6.15) for $k = 2$.

Numerical experiments showed, that the cycle (6.17) of the mapping (6.16) can be localized and stabilized at the interval $\mu^* \leq \mu \leq 0.97$ by iterations (6.15) for values of control parameters β, ε_1, ε_2, calculated at the point μ^*. It is similarly possible to stabilize the cycle (6.17) of the mapping (6.16) at the interval $0.97 \leq \mu \leq 1.1$ by iterations (6.15) with new values of control parameters β, ε_1, ε_2, calculated at the point $\mu_1^* = 0.97$. The last interval contains values of parameter μ for which the Henon mapping already has a strange attractor.

Example 6.3   An interesting illustration of the presented method is the example of stabilization of a fixed point in the simplified two-dimensional *Lorenz gas model*. This model describes the trajectory of a material point at elastic collisions with fixed disks, and it is a special case of a scattering *Sinai billiard* [Sinai (1970); Machta (1983)]. In the Sinai billiard, trajectories close in the beginning eventually quickly diverge, and an angle between the neighbouring trajectories of originally parallel beam increases after each reflection from boundary. As the billiard area is finite, trajectories start to be crossed repeatedly in time that leads to chaos.

For a simplified plane Lorenz gas model in which the movement of material point is considered inside the area, limited by three circles of unit radius touching each other, equations describing the trajectory of a particle after n of its collisions with boundary would look like

$$\varphi_{n+1} = \varphi_n + \psi_n - \pi/6 + \arcsin(2\cos(\varphi_n + \psi_n) - \cos\psi_n),$$
$$\psi_{n+1} = \pi/2 + \arcsin(2\cos(\varphi_n + \psi_n) - \cos\psi_n), \tag{6.18}$$

where φ is the angular distance from the point of collision of a particle with a disk up to the point of contact of disks, ψ is an angle which is formed by the trajectory before collision with the trajectory after collision with a disk. It is obvious, that angles φ_n and ψ_n uniquely determine the position of a particle inside the area after n-th collision. The fixed point $z^* = (\varphi^*, \psi^*) = (\pi/6, \pi/3)$ is unstable. It is important to note, that the mapping (6.18) does not contain any parameters, and position of the point z^* is independent.

For solution of the problem of stabilization at such fixed point, we shall enter artificially in the system (6.18) some parameters μ and r and shall consider the mapping

$$\begin{aligned}
\varphi_{n+1} &= \varphi_n + \psi_n - \pi/6 \\
&\quad + \arcsin\big((1+\mu)\cos(\varphi_n + \psi_n) - \mu\cos\psi_n + r(\mu - 1)\big), \\
\psi_{n+1} &= \pi/2 \\
&\quad + \arcsin\big((1+\mu)\cos(\varphi_n + \psi_n) - \mu\cos\psi_n + r(\mu - 1)\big).
\end{aligned} \tag{6.19}$$

It is obvious, that for $\mu = 1$ the mapping (6.19) coincides completely with the original mapping (6.18). We shall note, that the fixed point $z^* = (\alpha(\mu_0),\ \pi/3)$ is a stable fixed point of the mapping (6.19) for $\mu_0 = 0$ and $r_0 = 0.5 + \sqrt{13}/4$. And there exists a value $0 < \mu_1^* < 1$ such that the given point is stable for $\mu \in [0, \mu_1^*]$ and it is unstable for $\mu > \mu_1^*$. We fix the value $r = r_0$. Then the problem consists in stabilizing the fixed point $z^* = (\alpha(\mu),\ \pi/3)$ of the mapping (6.19) at $\mu = 1$, where

$$\alpha(\mu) = \arccos\frac{(\mu - 1)(0.5 - r)}{\mu + 1} - \pi/3.$$

Using the iterative process (6.15), it is numerically shown, that for values of control parameters $\varepsilon_1 = 0.014$, $\varepsilon_2 = 0.67$, $\beta = 1.29$, calculated at $\mu_1^* = 0.2863$, it is possible to localize and stabilize this fixed point at an interval $[0.2863;\ 0.51]$. Then for $\varepsilon_1 = 0.236$, $\varepsilon_2 = 1.127$, $\beta = 1.82$, calculated for $\mu_2^* = 0.51$, the fixed point will be stabilized at an interval $[0.51;\ 0.77]$, and values $\varepsilon_1 = 0.493$, $\varepsilon_2 = 1.54$, $\beta = 2.29$, calculated at $\mu_3^* = 0.77$, provide stability of the fixed point z^* of the mapping (6.19) for $\mu \in [0.77;\ 1]$.

6.2.2 *Localization and stabilization of unstable fixed points of chaotic dynamical systems*

Let us consider a nonlinear dynamical system

$$\dot{x} = F(x, \mu), \quad x \in \mathbb{R}^m, \tag{6.20}$$

given by the family F of smooth mappings. Let $x^*(\mu)$ be a fixed point of system (6.20), where μ is a scalar parameter. Just as in Sec. 6.2.1, we assume that there exists a critical value μ^* of the system parameter such that point $x^*(\mu)$ is the stable fixed point of system (6.20) for $\mu \leq \mu^*$, and is an unstable fixed point for $\mu > \mu^*$. In the latter case, the system has other regular or chaotic attractors. The problem is to localize and stabilize the unstable fixed point $x^*(\mu)$ of system (6.20) for values $\mu > \mu^*$ by means of small perturbations of the parameter μ. Let us consider the $(m+1)$-dimensional dynamical system

$$
\begin{aligned}
\dot{x} &= F(x, \mu) + \varepsilon(q - \mu), \\
\dot{q} &= Q(x, \mu) + \beta(q - \mu),
\end{aligned}
\tag{6.21}
$$

where $Q(x^*(\mu), \mu) = 0$, and $\varepsilon = (\varepsilon_1, \ldots, \varepsilon_m)^{\mathrm{T}}$ and $\beta \in \mathbb{R}$ are control parameters of the system. Clearly, if point $x^*(\mu)$ is a fixed point of system (6.20), then point $(x^*(\mu), \mu)$ is the fixed point of system (6.21). Let us calculate the Jacobian of the right-hand side of the mapping (6.21) at the point $(x^*(\mu), \mu)$

$$
J(\mu) = \begin{pmatrix} \partial F/\partial x & \varepsilon \\ \partial Q/\partial x & \beta \end{pmatrix}_{x=x^*(\mu),\, q=\mu} .
$$

The characteristic polynomial of the matrix $J(\mu)$ has the form

$$
\begin{aligned}
P(\lambda, \mu) &= \det[J(\mu) - \lambda E] \\
&= (-\lambda)^{m+1} + b_1(-\lambda)^m + \cdots + b_m(-\lambda) + b_{m+1},
\end{aligned}
\tag{6.22}
$$

where $b_k = \sum_{i=1}^{C_{m+1}^k} J_{ki}(\mu)$, J_{ki} is the principal minor of the order k of the matrix $J(\mu)$. In particular, $b_1 = \operatorname{tr} J(\mu)$ is the trace of the matrix $J(\mu)$, $b_{m+1} = \det J(\mu)$.

We require that all roots λ of the polynomial (6.22) are equal to some negative number $d < 0$ at the point μ^*. In this case $P(\lambda, \mu^*) = (d - \lambda)^{m+1}$, and $b_k = C_{m+1}^k d^k$, $k = 1, \ldots, m+1$. Consequently, the control parameters $\varepsilon_1, \ldots, \varepsilon_m, \beta$ should satisfy the system of $m+1$ linear algebraic equations

$$
\sum_{i=1}^{C_{m+1}^k} J_{ki}(\mu^*) = C_{m+1}^k d^k, \quad k = 1, \ldots, m+1.
\tag{6.23}
$$

In particular, the control parameter β is defined directly from the equation $\operatorname{tr} J(\mu) = (m+1)d$:

$$
\beta = -\operatorname{tr}\left\{ \frac{\partial F(x^*(\mu^*), \mu^*)}{\partial x} \right\} + (m+1)d = -\operatorname{tr} J_0 + (m+1)d.
$$

Since the family F occurring in (6.20) is smooth, we have the following statement.

Theorem 6.4 *If the determinant D of the system of linear Eqs. (6.23) is nonzero, then there exists a domain $\mu^* \leq \mu \leq \mu_1^*$ such that for any $\mu \in [\mu^*, \mu_1^*]$ real parts of all eigenvalues of matrix $J(\mu)$ are negative, and the point $(x^*(\mu), \mu)$ is an asymptotically stable fixed point of system (6.21). Hence, the fixed point $x^*(\mu)$ of the dynamical system (6.20) can be localized and stabilized in the domain $\mu \in [\mu^*, \mu_1^*]$ with the help of system (6.21) with initial conditions $x_0 = x^*(\mu^*)$, $q_0 = \mu$.*

Using μ_1^* as a new critical value of the system parameter and calculating the new Jacobian $J(\mu_1^*)$, we can again refine the values of control parameters β, ε_1, ..., ε_m in (6.21) and again localize and stabilize the fixed point $x^*(\mu)$ of system (6.20) for new values $\mu \in [\mu_1^*, \mu_2^*]$. This procedure can be continued to the entire interval on which the fixed point of system (6.20) exists. In a special case, the mapping $Q(x, \mu)$ in (6.21) can have the form

$$Q(x, \mu) = a_1 F_1(x, \mu) + \cdots + a_m F_m(x, \mu),$$

where the parameters $a_i \in \{0, 1\}$, $i = 1, \ldots, m$, are taken to be either zero or unity so that the determinant D of the system of $m + 1$ linear Eqs. (6.23) is nonzero.

Remark 6.1 *The condition $\det J_0 \neq 0$ is a necessary condition of applicability of system (6.21) for localization and stabilization of fixed points of system (6.20). It follows from the last equation $\det J(\mu^*) = d^{m+1}$ of the system (6.23), which can be presented in the form of*

$$\det J_0 \left(\beta - \sum_{i=1}^{m} a_i \varepsilon_i \right) = d^{m+1}.$$

Indeed, if $\det J_0$ is equal to zero, then the determinant D of the linear system (6.23) also vanishes.

Example 6.4 To illustrate the method, we consider the Rossler chaotic dynamical system (see Chapters 1–3)

$$\begin{aligned}
\dot{x} &= -(y + z), \\
\dot{y} &= x + ay, \\
\dot{z} &= b + z(x - \mu).
\end{aligned} \tag{6.24}$$

If $a = 0.5$ and $b = 0.75$, then the fixed point $x^*(\mu)/a = -y^*(\mu) = z^*(\mu) = \mu/2 - \sqrt{\mu^2/4 - ab}$ of system (6.24) is unstable for values $\mu > \mu^* = 1.375$. If $\mu > 2.35$, then the system (6.24) has a singular attractor, which is well-known as the Rossler attractor. We localize and stabilize a fixed point of the system (6.24) with the help of the system

$$
\begin{aligned}
\dot{x} &= -(y + z) + \varepsilon_1(q - \mu), \\
\dot{y} &= x + ay + \varepsilon_2(q - \mu), \\
\dot{z} &= b + z(x - \mu) + \varepsilon_3(q - \mu), \\
\dot{q} &= -a_1(y + z) + a_2(x + ay) + a_3\big(b + z(x - \mu)\big) + \beta(q - \mu),
\end{aligned}
\tag{6.25}
$$

where $d = -1$, $a_1 = 1$, $a_2 = a_3 = 0$. Numerical experiment showed, that the system (6.25) enables to localize and stabilize a fixed point of the system (6.24) at the interval $\mu^* \le \mu \le 4$ for values of parameters $\beta = -5$, $\varepsilon_1 = -1$, $\varepsilon_2 = 7/3$, $\varepsilon_3 = 1/6$, calculated at the point μ^*. The interval of localization and stabilization covers all domain of chaotic behaviour of the Rossler system.

6.2.3 *Localization and stabilization of unstable cycles of chaotic dynamical systems*

This problem is one of the most important parts of the chaos control problem. Let us consider a nonlinear dynamical system

$$
\dot{x} = F(x, \mu), \quad x \in \mathbb{R}^m, \quad \mu \in \mathbb{R},
\tag{6.26}
$$

given by a family F of smooth mappings. Let $x^*(t, \mu)$ be a closed periodic trajectory (a limit cycle) of the system (6.26), depending on the system parameter μ. Without loss of generality, we assume that there exists a critical value of the system parameter μ^* such that the trajectory $x^*(t, \mu)$ is an asymptotically orbitally stable cycle of system (6.26) for $\mu \le \mu^*$, and the trajectory $x^*(t, \mu)$ is an unstable cycle of system (6.26) for $\mu > \mu^*$. In the latter case, the system (6.26) can have attractors in the form of other stable limit cycles of various periods or stable tori, and for larger values of parameter μ, the appearance of irregular attractors is possible, which indicates the chaotic dynamics of the system in this case.

The problem is to localize and stabilize the unstable cycle $x^*(t, \mu)$ of the system (6.26) with the help of small perturbations of the system parameter μ in the domain $\mu > \mu^*$ of chaotic behavior of trajectories of the system for the case in almost full absence of information about the cycle $x^*(t, \mu)$ itself.

The main idea of the method for solving this problem remains the same: construction of a dynamical system in the space of higher dimension, such that the unstable cycle $x^*(t,\mu)$ of the system (6.26) is the projection of some limit cycle of the new system; the latter cycle must be asymptotically orbitally stable in the domain $\mu > \mu^*$.

Let the cycle $x^*(t,\mu)$ of system (6.26) have the period $T = T(\mu)$. Note, that if $\mu > \mu^*$, then the period of the cycle $x^*(t,\mu)$ is unknown and cannot be found by analyzing solutions of system (6.26). Let us consider the $(m + k + 1)$-dimensional system

$$
\begin{aligned}
\dot{y} &= F(y,\mu) + Z(y,t,\mu)E(q - \mu e), \\
\dot{q} &= DQ(y,s,t,\mu) + \beta(q - \mu e), \\
\dot{s} &= C^{\mathrm{T}}Q(y,s,t,\mu),
\end{aligned}
\tag{6.27}
$$

where $s(t)$ is a scalar function, $y(t) \in \mathbb{R}^m$, $q(t) \in \mathbb{R}^k$ is a vector function, $D_{k \times m}$ and $E_{m \times k}$ are constant matrices, vector $e = (1,\ldots,1)^{\mathrm{T}}$, $\mu \in \mathbb{R}$, $C \in \mathbb{R}^m$ is a constant vector, $\beta \in \mathbb{R}$, $1 \leq k < m$. We define mappings $Q(y,s,t,\mu)$ and $Z(y,t,\mu)$ as follows:

$$
Q(y,s,t,\mu) = x(s,\mu) - x(0,\mu), \quad Z(y,t,\mu) = \frac{\partial y}{\partial x(0,\mu)},
\tag{6.28}
$$

where $x(\tau,\mu)$ is a solution of system (6.26) at time τ under the condition $x(t,\mu) = y$. Since the matrix $Z(y,t,\mu)$ in (6.28) is the derivative of solutions of system (6.26) with respect to initial conditions, then $Z(y,t,\mu) = Z\big(x(t,\mu)\big)$, where $Z\big(x(\tau,\mu)\big)$ is the solution of the non-autonomous matrix ordinary differential equation

$$
\dot{V}(\tau) = A\big(x(\tau,\mu)\big)V(\tau), \quad V(0) = I,
\tag{6.29}
$$

where $A\big(x(\tau,\mu)\big) = D_x F\big(x(\tau,\mu),\mu\big)$. Therefore, for any point $(y,t) \in \mathbb{R}^{m+1}$, the matrix $Z(y,t,\mu)$ occurring in (6.27) can be obtained as solution of the matrix non-autonomous Eq. (6.29) at time t, taken along the trajectory $x(\tau,\mu)$ of system (6.26) such that $x(t,\mu) = y$.

Obviously, $Q(x^*(t,\mu), T(\mu),t,\mu) = 0$ for all t and for all μ. Hence, for any μ (and, in particular, for $\mu > \mu^*$) vector $u^*(t,\mu) = \big(x^*(t,\mu),\mu e, T(\mu)\big)^{\mathrm{T}}$ is a periodic solution (a cycle) of the extended system (6.27). Now it is enough to choose in (6.27) matrices E and D of control parameters, vector C and scalar β so, that the cycle $u^*(t,\mu)$ is an asymptotically orbitally stable limit cycle of the system (6.27) in some neighborhood of the parameter value $\mu^* \leq \mu \leq \mu_1^*$. In this case, vector $y(t)$, representing the first m

coordinates of the solution $u(t, \mu)$ of system (6.27) with initial conditions

$$u(0, \mu) = \big(y(0, \mu), q(0, \mu), s(0, \mu)\big)^{\mathrm{T}} = \big(x^*(0, \mu^*), \mu^* e, T(\mu^*)\big)^{\mathrm{T}}$$

tends to the unstable limit cycle $x^*(t, \mu)$ of system (6.26) for all $\mu \in [\mu^*, \mu_1^*]$.

Theorem 6.5   *There exist constant matrices E and D of control parameters, vector C, scalar β, and also the value of system parameter $\mu_1^* > \mu^*$ such that the cycle $u^*(t, \mu)$ is an asymptotical orbitally stable limit cycle of the system (6.27) for all values of the parameter $\mu \in [\mu^*, \mu_1^*]$.*

Proof.   We linearize the system (6.27) in a neighborhood of the solution $u^*(t, \mu)$ and denote the linearization matrix by $K(t, \mu)$. Since

$$\left. \frac{\partial Q(y, s, t, \mu)}{\partial y} \right|_{u^*(t, \mu)} = \big[Z(x^*(T, \mu)) - I \big] Z^{-1}\big(x^*(t, \mu)\big),$$

$$\left. \frac{\partial Q(y, s, t, \mu)}{\partial s} \right|_{u^*(t, \mu)} = \dot{x}^*(T, \mu),$$

then $K(t, \mu) =$

$$= \begin{pmatrix} A\big(x^*(t, \mu)\big) & Z\big(x^*(t, \mu)\big) E & 0 \\ D\big(Z(x^*(T, \mu)) - I\big) Z^{-1}\big(x^*(t, \mu)\big) & \beta I & D\dot{x}^*(T, \mu) \\ C^{\mathrm{T}}\big(Z(x^*(T, \mu)) - I\big) Z^{-1}\big(x^*(t, \mu)\big) & 0 & C^{\mathrm{T}}\dot{x}^*(T, \mu) \end{pmatrix}.$$

Linear system of the first approximation has the form

$$\dot{w}(t) = K(t, \mu)w(t) \tag{6.30}$$

in a neighborhood of solution $u^*(t, \mu)$ of the system (6.27), where $w(t) = u(t, \mu) - u^*(t, \mu)$.

Since the matrix $Z\big(x^*(t, \mu)\big)$ is a fundamental matrix of the periodic linear system (6.29), it follows from the Floquet theory that it can be represented as $Z\big(x^*(t, \mu)\big) = R(t)e^{Bt}$, where $R(t)$ is a periodic matrix with the period $T = T(\mu)$ and $R(0) = R(T) = I$. We denote the Jordan form of the matrix B by $\Lambda = \Lambda(\mu)$, so $B = P\Lambda P^{-1}$. The matrix Λ has one zero eigenvalue for any value of the system parameter μ, lying in the neighborhood of the point μ^*. In addition, since the point μ^* lies on the boundary of stability domain of the cycle $x^*(t, \mu)$ of the system (6.26), it follows that the matrix Λ has at least one more eigenvalue λ, crossing the imaginary axis from left to right for $\mu = \mu^*$. Without loss of generality we assume that this condition is satisfied for the last k ($1 \le k < m$) eigenvalues of the matrix Λ. Hence, if $\mu = \mu^*$, then $\Re\{\lambda_i\} < 0$ $(i = 1, \ldots, m - k - 1)$, $\lambda_{m-k} = 0$, $\Re\{\lambda_i\} = 0$ $(i = m - k + 1, \ldots, m)$. The number k determines the order $m + k + 1$ of the system (6.27) and the linearized system (6.30).

In addition, suppose that the matrix $e^{\Lambda T}$ has a unique eigenvalue equal to unity for $\mu = \mu^*$ (that is, the multiplier of the cycle $x^*(t, \mu^*)$ of the system (6.26)), so that $\lambda_i \neq 0$, $i = m - k + 1, \ldots, m$. We represent the solution $w(t)$ of the linear system (6.30) in the form

$$
w(t) = \begin{pmatrix} R(t)P & 0 & 0 \\ 0 & e^{-\overline{\Lambda}t} & 0 \\ 0 & 0 & 1 \end{pmatrix} \omega(t) = G(t, \mu)\omega(t), \tag{6.31}
$$

where $\overline{\Lambda}$ is the matrix coinciding with the right bottom $k \times k$-block of the matrix Λ (the matrix $e^{\overline{\Lambda}T}$ has no eigenvalues equal to $+1$). By construction, the matrix $G(t, \mu)$ is periodic for $\mu = \mu^*$ and is bounded as $t \to \infty$ in some neighborhood $\mu > \mu^*$. Substituting (6.31) into (6.30), after simple transformations we obtain that $\dot{\omega}(t) = L(t, \mu)\omega(t)$, where $L(t, \mu)$ is

$$
\begin{pmatrix} P^{-1}R^{-1}(t)[A(t)R(t) - \dot{R}(t)]P & P^{-1}R^{-1}(t)Z(t)Ee^{-\overline{\Lambda}t} & 0 \\ e^{-\overline{\Lambda}t}D[Z(t) - I]Z^{-1}(t)R(t)P & \beta I + \overline{\Lambda} & e^{-\overline{\Lambda}t}D\dot{x}^*(T, \mu) \\ C^{\mathrm{T}}[Z(T) - I]Z^{-1}(t)R(t)P & 0 & C^{\mathrm{T}}\dot{x}^*(T, \mu) \end{pmatrix},
$$

$A(t) = A\big(x^*(t, \mu)\big)$ and $Z(t) = Z\big(x^*(t, \mu)\big)$. As

$$
Z(t) = R(t)Pe^{\Lambda t}P^{-1}, \quad Z(T) = Pe^{\Lambda T}P^{-1},
$$
$$
P^{-1}R^{-1}(t)[A(t)R(t)P - \dot{R}(t)P] = \Lambda, \quad Z^{-1}(t) = Pe^{-\Lambda t}P^{-1}R^{-1}(t),
$$

then

$$
\dot{\omega}(t) = \begin{pmatrix} \Lambda & e^{\Lambda t}P^{-1}Ee^{-\overline{\Lambda}t} & 0 \\ e^{\overline{\Lambda}t}DP[e^{\Lambda T} - I]e^{-\Lambda t} & \beta I + \overline{\Lambda} & e^{\overline{\Lambda}t}D\dot{x}^*(T, \mu) \\ C^{\mathrm{T}}P[e^{\Lambda T} - I]e^{-\Lambda t} & 0 & C^{\mathrm{T}}\dot{x}^*(T, \mu) \end{pmatrix} \omega(t). \tag{6.32}
$$

We choose matrices E, D and vector C in the following way:

$$
E = P\begin{pmatrix} 0 \\ F \end{pmatrix}, \quad D = (0\ H)\,P^{-1}, \quad C^{\mathrm{T}} = (0\ c\ 0)\,P^{-1},
$$

where $H_{k \times k}$ is an identity matrix, F is a matrix commuting with matrix $\overline{\Lambda}$ which is certain below, and vector C has an unique nonzero element c in

$(m - k)$-th column. Thus

$$e^{\Lambda t}P^{-1}Ee^{-\overline{\Lambda} t} = \begin{pmatrix} 0 \\ F \end{pmatrix}, \quad e^{\overline{\Lambda} t}DP[e^{\Lambda T} - I]e^{-\Lambda t} = \begin{pmatrix} 0 & e^{\overline{\Lambda} T} - I \end{pmatrix},$$
$$C^{\mathrm{T}}P[e^{\Lambda T} - I]e^{-\Lambda t} = 0.$$

Next, we note that vector $\dot{x}^*(t, \mu)$ is the solution of the linear system (6.29). Therefore, there exists vector $s = \dot{x}^*(0, \mu) = \dot{x}^*(T, \mu)$ such that $\dot{x}^*(t, \mu) = V(t)s = R(t)Pe^{\Lambda t}P^{-1}s$. Hence, $P^{-1}s = e^{\Lambda T}P^{-1}s$, whence follows, that the vector $P^{-1}s$ has a unique nonzero coordinate in the $(m - k)$-th position. We denote it by r. Therefore,

$$e^{\overline{\Lambda} t}D\dot{x}^*(T, \mu) = \begin{pmatrix} 0 & e^{\overline{\Lambda}(t+T)} \end{pmatrix} P^{-1}s = 0,$$
$$C\dot{x}^*(T, \mu) = \begin{pmatrix} 0 & c & 0 \end{pmatrix} P^{-1}s = cr.$$

Now we set $c = -\operatorname{sign}(r)$, $F = \operatorname{diag}(a_i)$, where the diagonal matrix F has identical diagonal elements in each block corresponding to the Jordan block of the matrix $\overline{\Lambda}$. Let us represent the matrix Λ in the form

$$\Lambda = \begin{pmatrix} \overline{A} & 0 \\ 0 & \overline{\Lambda} \end{pmatrix},$$

where the matrix $\overline{A}$ coincides with top left $(m - k) \times (m - k)$-block of the matrix Λ. Then the Eq. (6.32) acquires the form

$$\dot{\omega}(t) = \begin{pmatrix} \overline{A} & 0 & 0 & 0 \\ 0 & \overline{\Lambda} & F & 0 \\ 0 & e^{\overline{\Lambda} T} - I & \beta I + \overline{\Lambda} & 0 \\ 0 & 0 & 0 & -|r| \end{pmatrix} \omega(t) = N(\mu)\omega(t). \tag{6.33}$$

Let us show that diagonal elements a_i of the matrix F can always be chosen so that for $\mu = \mu^*$ the matrix $N(\mu)$ has a unique zero eigenvalue and the remaining eigenvalues of the matrix $N(\mu)$ have negative real parts. For this purpose it is enough to choose a_i such that all $2k$ eigenvalues of the matrix

$$U = \begin{pmatrix} \overline{\Lambda} & F \\ e^{\overline{\Lambda} T} - I & \beta I + \overline{\Lambda} \end{pmatrix}$$

have negative real parts. Let us consider the Jordan block $\widetilde{\Lambda}$ of the matrix $\overline{\Lambda}$ with multiplicity j, corresponding to some eigenvalue λ_i. Then the eigenvalues ν of the matrix U satisfy the equation

$$\det\left(U - \nu I\right) = \prod_i \det \begin{pmatrix} \widetilde{\Lambda} - \nu I & a_i I \\ e^{\widetilde{\Lambda}T} - I & \beta I + \widetilde{\Lambda} - \nu I \end{pmatrix} = 0. \tag{6.34}$$

In turn, each of the determinants occurring in the Eq. (6.34), vanishes if and only if

$$\left[(\lambda_i - \nu)(\beta + \lambda_i - \nu) - a_i(e^{\lambda_i T} - 1)\right]^j = 0.$$

The last equation has two j-multiple roots satisfying the quadratic equation

$$\nu^2 - \nu(\beta + 2\lambda_i) + (\beta + \lambda_i)\lambda_i - a_i(e^{\lambda_i T} - 1) = 0. \tag{6.35}$$

Let us choose parameters β and a_i as follows:

$$\beta = -2d, \quad a_i = \frac{-d^2}{e^{\lambda_i T} - 1}, \quad d > 0. \tag{6.36}$$

In this case, each of Eqs. (6.35) has two equal roots $\nu_i = -d + \lambda_i$ with real part in some neighborhood $\mu \in [\mu^*, \mu_1^*]$ of the value μ^* of the system parameter. Therefore, for all $\mu \in [\mu^*, \mu_1^*]$ the fundamental matrix solution $W(t, \mu)$ of the first-approximation linear system (6.30) can be represented in the form of $W(t, \mu) = G(t, \mu)e^{N(\mu)t}$, where the matrix $G(t, \mu)$ is bounded as $t \to \infty$, and the matrix $N(\mu)$ has exactly one zero eigenvalue corresponding to the unit multiplier of the periodic solution $u^*(t, \mu)$ of the system (6.27). All remaining $m + k$ eigenvalues of the matrix $N(\mu)$ have negative real parts for any $\mu \in [\mu^*, \mu_1^*]$, whence asymptotical orbital stability of the periodic solution $u^*(t, \mu)$ of the system (6.27) follows for all $\mu \in [\mu^*, \mu_1^*]$. The proof of the Theorem 6.5 is complete. $\qquad\square$

Example 6.5   As an example we consider the Rossler dynamical system (6.24). If $\mu > 1.88$, then the limit cycle of the system (6.24) loses stability, and an irregular singular attractor appears in the system for $\mu = 2.35$. We localized and stabilized the limit cycle of the system (6.24) with the help of the expanded system

$$\begin{aligned}
\dot{x} &= -(y + z) + (z_{11}\varepsilon_1 + z_{12}\varepsilon_2 + z_{13}\varepsilon_3)(q - \mu), \\
\dot{y} &= x + ay + (z_{21}\varepsilon_1 + z_{22}\varepsilon_2 + z_{23}\varepsilon_3)(q - \mu), \\
\dot{z} &= b + z(x - \mu) + (z_{31}\varepsilon_1 + z_{32}\varepsilon_2 + z_{33}\varepsilon_3)(q - \mu), \\
\dot{q} &= a_1\big(x(s) - x(0)\big) + a_2\big(y(s) - y(0)\big) + a_3\big(z(s) - z(0)\big) + \beta(q - \mu), \\
\dot{s} &= c_1 a_1\big(x(s) - x(0)\big) + c_2 a_2\big(y(s) - y(0)\big) + c_3 a_3\big(z(s) - z(0)\big),
\end{aligned} \tag{6.37}$$

where $\left(x(0), y(0), z(0)\right)^{\mathrm{T}}$ is a vector of initial data of the system (6.24) provided that its trajectory coincides in each moment of time with a projection of trajectory of the system (6.37) in $\mathbb{R}^3$.

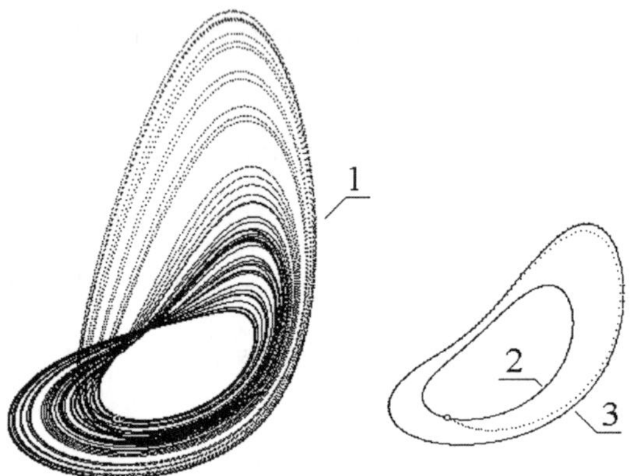

Fig. 6.3   Stabilization of an unstable limit cycle of the Rossler system.

Fig. 6.3 shows: 1 is a chaotic trajectory of the Rossler system for $\mu = 2.4$; 2 is a stable limit cycle of the Rossler system for the critical value $\mu^* = 1.85$; 3 is a projection of the limit cycle of the system (6.37) stabilizing the unstable limit cycle of the Rossler system (6.24) for $\mu = 2.4$.

6.2.4   *Chaos control in equations with delay argument*

Let us consider the nonlinear autonomous delay system

$$\dot{x}(t) = F(x(t), x(t - \tau_1), \ldots, x(t - \tau_m), \mu), \quad x \in \mathbb{R}^n, \quad \mu \in \mathbb{R}, \\ 0 \equiv \tau_0 \leq \tau_1 \leq \ldots \leq \tau_m, \tag{6.38}$$

where $F(x(t), x(t - \tau_1), \ldots, x(t - \tau_m), \mu)$ is a vector function smooth on a set of variables. Let $x^*(\mu)$ be a fixed point of system (6.38), depending on the scalar parameter μ, which, in particular, we can choose as one of delayed arguments. Just as above, we assume that there exists a critical value of the parameter μ^* such that the fixed point is stable for $\mu \leq \mu^*$, and the solution $x^*(\mu)$ becomes unstable for $\mu > \mu^*$. In this case, if the system (6.38) remains dissipative, then other attractors can appear in it, for example, stable limit cycles of various multiplicity and even irregular attractors, testifying about chaotic dynamics of the system for $\mu > \mu^*$.

Characteristic feature of equations with delay argument is the opportunity of presence of all specified types of attractors even in case when the system (6.38) contains only one equation as this equation has infinitely many degrees of freedom.

To solve the problem on localization and stabilization of the unstable fixed point $x^*(\mu)$ of the system (6.38) for $\mu > \mu^*$ let us consider the system

$$\begin{aligned}
\dot{x} &= F(x(t), x(t - \tau_1), \ldots, x(t - \tau_m), \mu) + \varepsilon(q(t) - \mu), \\
\dot{q} &= G(x(t), x(t - \tau_1), \ldots, x(t - \tau_m), \mu) + \beta(q(t) - \mu),
\end{aligned} \tag{6.39}$$

where scalar function $G(x(t), x(t - \tau_1), \ldots, x(t - \tau_m), \mu) = 0$ at the fixed point $(x^*(\mu), \mu)$, and $\varepsilon \in \mathbb{R}^n$ and $\beta \in \mathbb{R}$ are the control parameters of system (6.39).

Obviously, if point $x^*(\mu)$ is a fixed point of the system (6.38), then point $(x^*(\mu), \mu)$ is the fixed point of the system (6.39). Let us determine the values of control parameters ε and β in the system (6.39), at which point $(x^*(\mu), \mu)$ is a stable fixed point. We pass to new variables $u_i(t) = x_i(t) - x_i^*$, $(i = 1, 2, \ldots, n)$, $u_{n+1}(t) = q(t) - \mu$, and we extract the linear part of system (6.39) in a neighborhood of the fixed point $u^* = 0$

$$\dot{u}(t) = \sum_{l=0}^{m} A_l(\mu) u(t - \tau_l), \quad 0 \equiv \tau_0 \leq \tau_1 \leq \ldots \leq \tau_m, \tag{6.40}$$

where

$$A_0(\mu) = \begin{pmatrix} \dfrac{\partial F(\cdot)}{\partial u(t)} & \varepsilon \\ \dfrac{\partial G(\cdot)}{\partial u(t)} & \beta \end{pmatrix}_{u^*=0}, \quad A_l(\mu) = \begin{pmatrix} \dfrac{\partial F(\cdot)}{\partial u(t - \tau_l)} & 0 \\ \dfrac{\partial G(\cdot)}{\partial u(t - \tau_l)} & 0 \end{pmatrix}_{u^*=0}, \quad l = 1, \ldots, m.$$

According to the criterion of stability with respect to the first approximation, the fixed point of the system (6.39) is stable if all roots of the characteristic equation

$$\det\left(\sum_{l=0}^{m} A_l e^{-\lambda \tau_l} - \lambda E\right) = 0 \tag{6.41}$$

have negative real parts. Left-hand side of the Eq. (6.41) is the quasipolynomial

$$P(\lambda) = \sum_{k=0}^{n+1} \sum_{l=0}^{M} a_{kl} \lambda^k e^{-\lambda \tau_l}, \quad M = \max\{n, m\}, \tag{6.42}$$

where coefficients a_{kl} can be expressed via sums of all main minors of the matrices A_l. Usually, a quasipolynomial has infinitely many zeros; however, in case of delay equations (i.e. $a_{n+1,0} \neq 0$, $a_{n+1,l} = 0$, $l = 1,\ldots,M$) all roots of the quasipolynomial lie in the left half-plane, i.e. $\Re\{\lambda_i\} < d$. This circumstance allows to solve the problem of stabilizing an unstable stationary solution by a choice of control parameters so that to provide performance of condition $d < 0$. For delimitation of domain of stability of quasipolynomials we apply the following theorem.

Theorem 6.6 (the Rouche theorem [Shabat (1985)]). *Let $\varphi(z)$ and $\psi(z)$ be holomorphic in the field of D and continuous in $\overline{D} = D \cup \Gamma$ functions. If for each point $z \in \Gamma$ the inequality $|\varphi(z)| > |\psi(z)|$ takes place, then functions $\varphi(z)$ and $\varphi(z) + \psi(z)$ have in D the same number of roots.*

Let us present the quasipolynomial (6.42) in the form

$$P(\lambda) = \sum_{k=0}^{n+1} a_k \lambda^k + \sum_{k=0}^{n} \sum_{l=1}^{M} a_{kl} \lambda^k e^{-\lambda \tau_l} = \varphi(\lambda) + \psi(\lambda)$$

and choose a contour Γ consisting of the segment of imaginary axis $[-iR; iR]$ and the right semicircle of radius R centered at the origin. We apply the Rouche theorem to the functions

$$\varphi(\lambda) = \sum_{k=0}^{n+1} a_k \lambda^k, \ \ \psi(\lambda) = \sum_{k=0}^{n} \sum_{l=1}^{M} a_{kl} \lambda^k e^{-\lambda \tau_l},$$

Obviously, the inequality $|\varphi(\lambda)| > |\psi(\lambda)|$ is valid on the considered semi-circle for sufficiently large R. It is necessary to prove its validity

$$|\varphi(i\omega)| > |\psi(i\omega)| \tag{6.43}$$

on an imaginary axis for $\lambda = i\omega$. Since

$$|\psi(i\omega)| = \left| \sum_{k=0}^{n} \sum_{l=1}^{M} a_{kl}(i\omega)^k e^{-i\omega\tau_l} \right|$$

$$\leq \sum_{l=1}^{M} \left| e^{-i\omega\tau_l} \right| \left| \sum_{k=0}^{n} a_{kl}(i\omega)^k \right| = \sum_{l=1}^{M} \left| \sum_{k=0}^{n} a_{kl}(i\omega)^k \right|,$$

then the inequality

$$\left| \sum_{k=0}^{n+1} a_k(i\omega)^k \right| > \sum_{l=1}^{M} \left| \sum_{k=0}^{n} a_{kl}(i\omega)^k \right| \tag{6.44}$$

guarantees validity of the condition (6.43) on an imaginary axis. Inequality (6.44) together with the requirement that the polynomial $\varphi(\lambda)$ has no roots with negative real parts, determines the boundary of the domain of parameters ε and β for which the quasipolynomial $P(\lambda)$ is stable. Hence the fixed point of the expanded system (6.39) is asymptotically stable in this domain, that provides stabilization of unstable fixed point of the original system (6.38).

Example 6.6   As an example, we consider stabilization of unstable fixed point in the nonlinear Mackey–Glass equation (see Chapter 5)

$$\dot{x} = -ax(t) + \frac{\beta_0 \theta^n x(t-\tau)}{\theta^n + x^n(t-\tau)}, \tag{6.45}$$

where β_0, θ and n are positive constants such that $\beta_0 > a > 0$ and $nB > 2$, $6aB > \beta_0$, $B = (\beta_0 - a)/\beta_0$. The Eq. (6.45) has the unique stationary state $x^* = \theta \sqrt[n]{\dfrac{\beta_0 - a}{a}}$, which loses stability for

$$\tau > \tau^* = \frac{\arccos{(-a/b)}}{\sqrt{b^2 - a^2}},$$

where $b = a(nB - 1)$ [Hassard *et al.* (1981)]. In this case, a stable limit cycle appears first. For further growth of τ, a sequence of period doubling bifurcations takes place, and then chaotic dynamics is observed (see Sec. 5.3). For stabilization of unstable fixed point x^* of the Eq. (6.45) for $\tau > \tau^*$, we consider the system

$$\dot{x}(t) = -ax(t) + \frac{\beta_0 \theta^n x(t-\tau)}{\theta^n + x^n(t-\tau)} + \varepsilon(q(t) - \tau),$$
$$\dot{q}(t) = -a(x(t) - x^*) + \beta(q(t) - \tau). \tag{6.46}$$

Let us show that for any value of delay τ, values of control parameters ε and β in (6.46) can be chosen such that an unstable fixed point of the Eq. (6.45) will be stabilized. We consider the linear part of the Eq. (6.46) in a neighbourhood of the fixed point (x^*, τ)

$$\dot{u}_1(t) = -au_1(t) - bu_1(t-\tau) + \varepsilon u_2(t),$$
$$\dot{u}_2(t) = -au_1(t) + \beta u_2(t),$$

where $u_1(t) = x(t) - x^*$, $u_2(t) = q(t) - \tau$. The corresponding characteristic quasipolynomial has the form

$$P(\lambda) = \lambda^2 + \lambda(a - \beta) + a(\varepsilon - \beta) + be^{-\lambda\tau}(\lambda - \beta). \tag{6.47}$$

Let us set $\varphi(\lambda) = \lambda^2 + \lambda(a - \beta) + a(\varepsilon - \beta)$, $\quad \psi(\lambda) = be^{-\lambda\tau}(\lambda - \beta)$. For our special case $M = n = 1$ and accordingly to (6.44) we have

$$|-\omega^2 + a(\varepsilon - \beta) + i\omega(a - \beta)| > |b||i\omega - \beta|,$$

that is equivalent to the inequality

$$\omega^4 + \omega^2(a^2 - b^2 + \beta^2 - 2a\varepsilon) + a^2(\varepsilon - \beta)^2 - b^2\beta^2 > 0.$$

The necessary and sufficient conditions for validity of last inequality for the interval $\omega^2 \in \mathbb{R}_+$ are $a^2 - b^2 + \beta^2 - 2a\varepsilon > 0$, $\quad a|\varepsilon - \beta| > b|\beta|$. The last conditions together with conditions $a - \beta > 0$, $\quad \varepsilon - \beta > 0$, corresponding to absence of roots of function $\varphi(\lambda)$ in the right half-plane, give joint system of inequalities

$$\beta^2 - b^2 + a^2 - 2\varepsilon a > 0, \quad -\frac{a\varepsilon}{b - a} < \beta < \frac{a\varepsilon}{b + a}, \quad \beta < \varepsilon, \quad \beta < a \quad (6.48)$$

for finding a domain in the space of parameters ε and β, providing stability of the quasipolynomial (6.47). This implies an asymptotic stability of the stationary solution of the Eq. (6.46) for any τ.

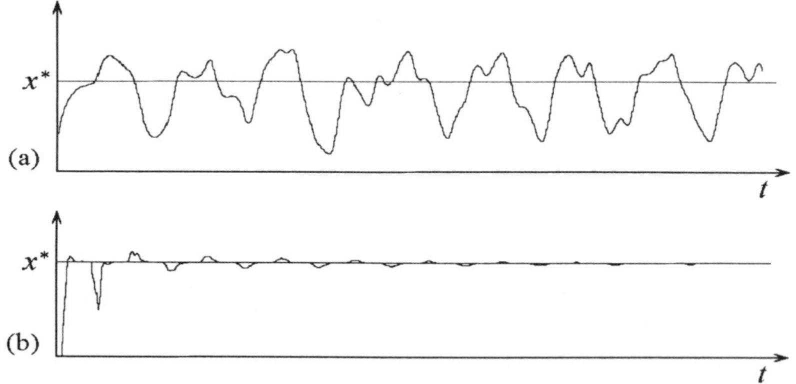

Fig. 6.4   Solutions of the Mackey–Glass Eq. (6.45) (a) and stabilizing the system (6.46) (b) for the value of bifurcation parameter $\tau = 5$, where x^* is the stationary solution.

The solution of the Eq. (6.45) is shown in Fig. 6.4a for values of parameters $a = 0.75$, $\beta_0 = 1.5$, $\theta = 2$, $n = 8$ and for the value of bifurcation parameter $\tau = 5$, corresponding to chaotic motion. The solution of the stabilizing system (6.46) is shown in Fig. 6.4b for the same values of parameters a, β_0, θ, n and for values of control parameters $\varepsilon = 12$, $\beta = -5$ taken from the domain (6.48).

Let us consider now the problem of localization and stabilization of an unstable periodic trajectory in the nonlinear equations with delay argument, having chaotic behaviour

$$\dot{x} = f(x(t), x(t - \tau), \mu), \ t \geq 0, \tag{6.49}$$

where $x(t)$, $f(\cdot)$ are scalar functions and $\tau > 0$ is a constant delay.

Let $x^*(t, \mu)$ be a periodic trajectory (a limit cycle) of the Eq. (6.49) having the period $T(\mu)$. As a rule, there exists a critical value of the parameter μ^* such that the cycle $x^*(t, \mu)$ is asymptotical orbitally stable for $\mu \leq \mu^*$ and unstable for $\mu > \mu^*$. The problem consists in localization and stabilization of this cycle for $\mu > \mu^*$, including parameter values μ for which the Eq. (6.49) has a chaotic behaviour.

We show, that the approach offered in Sec. 6.2.3 can be used for solution of this problem. Let $C[-\tau; 0]$ be the space of continuous real functions $\varphi(\cdot)$, setting initial conditions for the Eq. (6.49) on the interval $[-\tau; 0]$.

Let us designate $x_t(\theta) = x(t + \theta)$, $-\tau \leq \theta \leq 0$ and represent the Eq. (6.49) in the form of a finite-dimensional system of ordinary differential equations. For this purpose we divide the interval $[-\tau; 0]$ on m identical parts and designate

$$x_t(0) = y_0, \ x_t(-\tau) = y_m, \ x_t\left(\frac{-i\tau}{m}\right) = y_i, \quad i = 1, \ldots, m - 1. \tag{6.50}$$

Then, using the difference approximation of a derivative, we obtain

$$\dot{y}_0 = f(y_0, y_m, \mu), \ \dot{y}_i = \frac{m}{\tau}(y_{i-1} - y_i), \quad i = 1, \ldots, m. \tag{6.51}$$

Thus, the Eq. (6.49) is reduced to $(m + 1)$-dimensional system of ordinary differential equations

$$\dot{y} = F(y, \mu, \tau), \tag{6.52}$$

which vector of solutions $y = (y_0(t), y_1(t), \ldots, y_m(t))^{\mathrm{T}}$ determines a vector-function

$$\varphi_t^{[m]}(\vartheta) = y_i + \frac{(y_{i-1} - y_i)m}{\tau}\vartheta, \ \frac{-\tau i}{m} \leq \vartheta \leq \frac{-\tau(i - 1)}{m}, \quad i = 1, \ldots, m.$$

Each coordinate of this vector-function linearly approximates the function $x_t(\vartheta)$ on a segment of the length $h = \vartheta_{i-1} - \vartheta_i$ on two values of function in nodes y_i and y_{i-1} with an error, not exceeding $O(h^2)$. Obviously, the function $\varphi_t^{[m]}(\vartheta)$ is arbitrarily close to the function $x_t(\vartheta)$ on the interval $[-\tau; 0]$ for high enough order m. Thus values of the coordinate $y_0(t, \mu)$ of the system (6.52) correspond to the solution $x(t, \mu)$ of the Eq. (6.49), and trajectory of the system (6.52) in the phase space $\mathbb{R}^{m+1}$ corresponds to trajectory of the Eq. (6.49) in the expanded phase space $\mathbb{R} \times C[-\tau; 0]$. Hence, the problem of stabilization of the unstable periodic solution $x(t, \mu)$ of the Eq. (6.49) with delay is reduced to the problem of stabilization of a corresponding unstable periodic trajectory $y(t, \mu)$ of the system (6.52). So, the method stated in Sec. 6.2.3 can be applied to solution of the last problem. According to this method, stabilization of periodic solutions of the system (6.52) can be carried out by solutions of the expanded system (6.27).

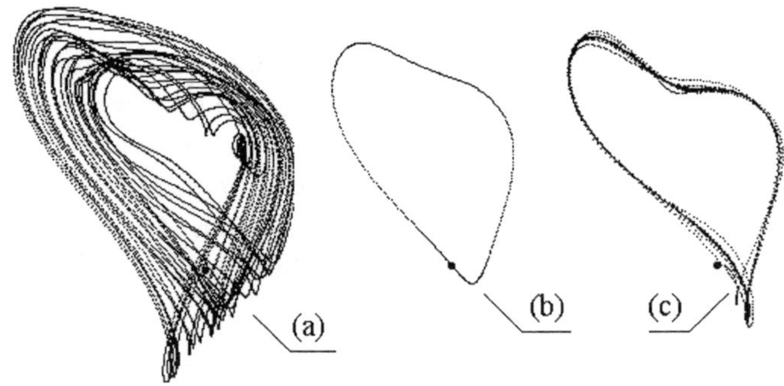

Fig. 6.5   Stabilization of an unstable periodic solution of the Mackey–Glass Eq. (6.45).

Example 6.7   As an example we consider the Mackey–Glass Eq. (6.45) in which the value of delay τ is a bifurcation parameter. For solution of the problem of localization and stabilization of its unstable periodic orbit for values τ, corresponding to chaotic oscillations, we transform the Eq. (6.45) to the finite-dimensional system

$$\dot{y}_0 = f_0(y_0, y_m, \tau), \quad \dot{y}_i = f_i(y_0, y_1, \ldots, y_m, \tau), \quad i = 1, \ldots, m, \qquad (6.53)$$

where function $f_0(y_0, y_m, \tau)$ coincides with the right part of the Eq. (6.45) according to accepted in (6.51) designations, and functions $f_i(y_0, y_1, \ldots, y_m)$ depend on the accepted way of approximation of the Eq. (6.49) by the system (6.53). For reduction of the order of approximating system we used interpolation of function $x_t(\vartheta)$ by a cubic spline on an interval $[-\tau; 0]$. Thus, as it was noted in Sec. 5.3, the order of the system of ordinary differential Eqs.(6.53), approximating the Eq. (6.45), can be accepted equal to $m = 20$. Results of numerical experiment on stabilization of unstable periodic solution of the system (6.53) for the parameter value $\tau = 1.8$, corresponding to chaotic oscillations, are shown in Fig. 6.5: (a) is a projection of a phase portrait of irregular singular attractor on the plane (y_0, y_m) for the system (6.53) or, accordingly, on the plane $(x(0), x(-\tau))$ for the Eq. (6.45) for $\tau = 1.8$; (b) is a similar projection of a phase portrait of a limit cycle of the system (6.53) (and, accordingly, the Eq. (6.45) for $\tau = 1.2$; (c) is a projection of a phase portrait of an unstable limit cycle stabilized by the system (6.27) on the plane (y_0, y_m) (in case of approximating system (6.53)) and, accordingly, on the plane $(x(0), x(-\tau))$ (in case of the original Eq. (6.45)) for $\tau = 1.8$.

6.2.5 *Stabilization of a thermodynamic branch in reaction–diffusion systems of equations*

In Sec. 5.1 it has been noted, that the system of equations of the reaction–diffusion type (5.1) is one of the most important and widely used models of different processes and phenomena in physics, chemistry, biology, ecology and in many other fields of science. Depending on parameter values μ, and also on coefficients of diffusion D_1 and D_2, and on the form and the size of spatial area and boundary conditions, the system (5.1) has a plenty of qualitatively various solutions after the loss of stability of a thermodynamic branch including periodic solutions, spatially homogeneous and inhomogeneous dissipative structures, and also inhomogeneous nonperiodic solutions which were called diffusion or spatio-temporal chaos (see Chapter 5). In this connection, *the control problem for diffusion chaos* in a broad sense can be formulated as follows: localize (find out) and stabilize an unstable periodic or dissipative structure that is the solution of the system (5.1) for those values μ such that the system (5.1) has spatially inhomogeneous nonperiodic solutions. In the restricted sense, we treat the control problem for diffusion chaos as the problem of stabilization the unstable thermodynamic branch of the system (5.1) for the case in which this system has spatially inhomogeneous nonperiodic solutions. Here we consider a problem of controlling diffusing chaos exclusively in the restricted sense. As it was shown in Chapter 5, this problem can be reduced to the stabilization problem of the absolutely unstable zero solution of the Kuramoto–Tsuzuki equation in the case of when it possesses nonperiodic spatially inhomogeneous solutions. In the case of boundary conditions of the second kind, the problem of diffusion chaos control in the restricted sense is reduced to stabilization of the zero solution of the boundary value problem

$$W_t = W + (1 + ic_1)W_{xx} - (1 + ic_2)W|W|^2,$$
$$0 \le x \le l, \quad 0 \le t < \infty, \tag{6.54}$$
$$W(x,0) = W_0(x), \quad W_x(0,t) = W_x(l,t) = 0.$$

Since $W(x,t) = u(x,t) + iv(x,t)$, we can rewrite the boundary value problem (6.54) in the form

$$u_t = u + u_{xx} - c_1 v_{xx} - (u^2 + v^2)(u - c_2 v),$$
$$v_t = v + c_1 u_{xx} + v_{xx} - (u^2 + v^2)(u + c_2 v),$$
$$0 \le x \le l, \quad 0 \le t < \infty, \tag{6.55}$$
$$u(x,0) = u_0(x), \quad v(x,0) = v_0(x),$$
$$u_x(0,t) = v_x(0,t) = u_x(l,t) = v_x(l,t) = 0.$$

It follows from the Eq. (6.55) that

$$\frac{d}{dt}\begin{pmatrix} u \\ v \end{pmatrix} = L\begin{pmatrix} u \\ v \end{pmatrix} + G\begin{pmatrix} u \\ v \end{pmatrix}. \tag{6.56}$$

Here the linear operator L can be represented in the form

$$L = \begin{pmatrix} 1 + \dfrac{\partial^2}{\partial x^2} & -c_1\dfrac{\partial^2}{\partial x^2} \\ c_1\dfrac{\partial^2}{\partial x^2} & 1 + \dfrac{\partial^2}{\partial x^2} \end{pmatrix} = I + C\Delta = \begin{pmatrix} 1 & 0 \\ 0 & 1 \end{pmatrix} + \begin{pmatrix} 1 & -c_1 \\ c_1 & 1 \end{pmatrix}\dfrac{\partial^2}{\partial x^2}, \tag{6.57}$$

where I is the identity matrix, and eigenvalues of the operator Δ on the interval $[0, l]$ with boundary conditions of the second kind are equal to $-\pi^2 n^2/l^2$ $(n = 0, 1, \ldots)$ with the corresponding eigenfunctions $\cos\dfrac{\pi n}{l}x$.

Let us calculate the eigenvalues of the operator L. Let λ be such an eigenvalue. The corresponding eigenfunction is $\varphi = \sum\limits_{n=0}^{\infty} \begin{pmatrix} a_n \\ b_n \end{pmatrix} \cos\dfrac{\pi n}{l}x$.

Then

$$L\varphi = (I + C\Delta)\sum_{n=0}^{\infty} \begin{pmatrix} a_n \\ b_n \end{pmatrix} \cos\frac{\pi n}{l}x$$
$$= \sum_{n=0}^{\infty}\left[\begin{pmatrix} a_n \\ b_n \end{pmatrix} - C\frac{\pi^2 n^2}{l^2}\begin{pmatrix} a_n \\ b_n \end{pmatrix}\right]\cos\frac{\pi n}{l}x = \lambda\sum_{n=0}^{\infty}\begin{pmatrix} a_n \\ b_n \end{pmatrix}\cos\frac{\pi n}{l}x.$$

Hence,

$$\left(I - \frac{\pi^2 n^2}{l^2} C\right)\begin{pmatrix} a_n \\ b_n \end{pmatrix} = \lambda \begin{pmatrix} a_n \\ b_n \end{pmatrix}, \quad n = 0, 1, \ldots$$

Thus, eigenvalues of the operator L are just the eigenvalues $1 - \pi^2 n^2/l^2 \pm ic_1(\pi^2 n^2/l^2)$ of the matrices

$$B_n = I - \frac{\pi^2 n^2}{l^2} C.$$

We see that if the domain is small ($l < \pi$), then only two eigenvalues of the operator L, namely, those corresponding to the zero-order harmonics, lie in the right half-plane. This results in existence of spatially homogeneous self-oscillations ($W = e^{-ic_2(t+\theta)}$) for Eq. (6.54). For larger l, the operator L has more eigenvalues in the right half-plane, which results in appearance of spatially inhomogeneous self-oscillations in the Eq.(6.54) and more complex dissipative structures.

The idea of a method of stabilization of the zero solution of the second boundary value problem (6.54) consists in construction of such expanded system for which all eigenvalues of the operator of the problem linearized around the zero solution lie in the left half-plane regardless of values of the parameters c_1, c_2 in the original Eq. (6.54) and the instability degree of its zero solution.

We note, that for $n = 0$ the matrix B_0 coincides with an identity matrix I, i.e. it has a diagonal form with multiple eigenvalues. We show, that in this case for stabilization of the zero solution of the boundary value problem (6.54) the system must be supplemented by two new equations. So, we consider the boundary value problem

$$\begin{aligned}
u_t &= u + u_{xx} - c_1 v_{xx} - (u^2 + v^2)(u - c_2 v) + \varepsilon\varphi, \\
v_t &= v + c_1 u_{xx} + v_{xx} - (u^2 + v^2)(u + c_2 v) + \varepsilon\psi, \\
\varphi_t &= u + \beta\varphi, \quad \psi_t = v + \beta\psi, \\
&\quad 0 \leq x \leq l, \quad 0 \leq t < \infty, \\
u(x,0) &= u_0(x), \quad v(x,0) = v_0(x), \\
\varphi(x,0) &= \varphi_0(x), \quad \psi(x,0) = \psi_0(x), \\
u_x(0,t) &= u_x(l,t) = v_x(0,t) = v_x(l,t) = 0, \\
\varphi_x(0,t) &= \varphi_x(l,t) = \psi_x(0,t) = \psi_x(l,t) = 0.
\end{aligned} \quad (6.58)$$

Theorem 6.7 *For any values of parameters c_1, c_2, l of problem (6.54), there exist values of control parameters ε, β such that the zero solution of the boundary value problem (6.58) is uniformly and asymptotically stable.*

Proof. Let us rewrite equations of the boundary value problem (6.58) in the form

$$\frac{dz}{dt} = Lz + F(z,t),\tag{6.59}$$

where $z(x,t) = \big(u(x,t), v(x,t), \varphi(x,t), \psi(x,t)\big)^{\mathrm{T}}$, and the linear operator L has the kind

$$L = \begin{pmatrix} 1 + \dfrac{\partial^2}{\partial x^2} & -c_1 \dfrac{\partial^2}{\partial x^2} & \varepsilon & 0 \\[2mm] c_1 \dfrac{\partial^2}{\partial x^2} & 1 + \dfrac{\partial^2}{\partial x^2} & 0 & \varepsilon \\[2mm] 1 & 0 & \beta & 0 \\[2mm] 0 & 1 & 0 & \beta \end{pmatrix}$$

with conditions at endpoints of the interval being determined by the second boundary value problem. To prove the theorem, it suffices to choose values of control parameters ε, β such that the spectrum of operator L lies in the left half-plane [Daletskii and Krein (1974)]. We can readily show that, just as above for the boundary value problem (6.55), the spectrum of the operator L is discrete and consists of the eigenvalues of matrices

$$B_n = \begin{pmatrix} 1 - k_n & c_1 k_n & \varepsilon & 0 \\ -c_1 k_n & 1 - k_n & 0 & \varepsilon \\ 1 & 0 & \beta & 0 \\ 0 & 1 & 0 & \beta \end{pmatrix}, \quad k_n = \frac{\pi^2 n^2}{l^2}, \quad n = 0, 1, \ldots$$

The characteristic equation for the matrix B_n has the form

$$\big[(\beta - \lambda)(1 - k_n - \lambda) - \varepsilon\big]^2 + (\beta - \lambda)^2 c_1^2 k_n^2 = 0.$$

Roots of this equation can be found from the two quadratic equations

$$\lambda^2 - \lambda(\beta + 1 - k_n \mp ic_1 k_n) + \beta(1 - k_n \mp ic_1 k_n) - \varepsilon = 0.$$

From corresponding expressions for complex roots of the last equation, we can easily derive sufficient conditions for the negativeness of their real parts in the form

$$\begin{aligned} \beta &< -1, \\ \beta c_1^2 k_n^2 (k_n - 1) &+ (\beta + 1 - k_n)^2 (\beta k_n + \varepsilon - \beta) < 0. \end{aligned}\tag{6.60}$$

If the domain is small ($l \leq \pi$), then we have $k_n \geq 1$ for all $n \geq 1$. Therefore, conditions (6.60) are satisfied for all n when $\varepsilon < \beta < -1$. If $l > \pi$, then

$k_n < 1$ for some $n = 1, \ldots, n_0$; and hence, the choice of ε depends on the value of parameter c_1:

$$\varepsilon < \min_{0 \le n \le n_0} \left\{ \frac{\beta(1 - k_n)[c_1^2 k_n^2 + (\beta + 1 - k_n)^2]}{(\beta + 1 - k_n)^2} \right\}.$$

In any case, control parameters ε, β can be chosen so as to ensure that the spectrum of operator L lies in the left half-plane. The proof of the Theorem 6.7 is complete. $\qquad\square$

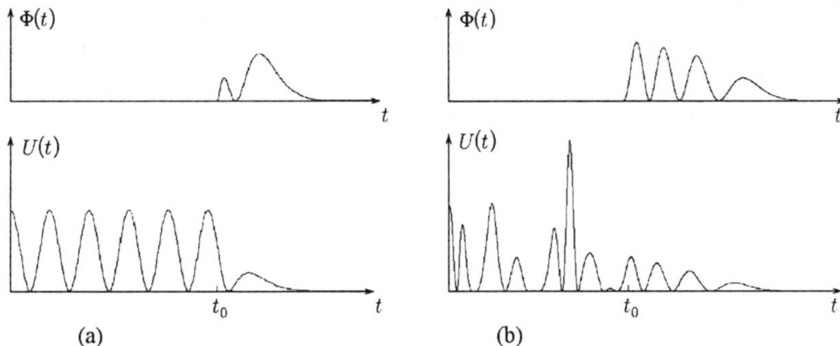

Fig. 6.6   Stabilization of the zero solution of the problem (6.55) in a case when its solution is: a) stable spatially homogeneous self-oscillating mode ($l = \pi$, $c_1 = 5$, $c_2 = -2$, $\varepsilon = -4$, $\beta = -3$) and b) spatially inhomogeneous nonperiodic mode (diffusion chaos) ($l = \pi$, $c_1 = 5$, $c_2 = -5$, $\varepsilon = -4$, $\beta = -3$).

To visualize the behaviour of solutions of problems (6.55) and (6.58), we use the variables $U(t) = \int_0^l u^2(x,t)\, dx$ and $\Phi(t) = \int_0^l \varphi^2(x,t)\, dx$. The dependence of functions $U(t)$ and $\Phi(t)$ on time is shown in Fig. 6.6 for the following cases: a) $c_1 = 5$, $c_2 = -2$, $l = \pi$ (the problem (6.55) has spatially homogeneous stable self-oscillating mode); b) $c_1 = 5$, $c_2 = -5$, $l = \pi$ (the problem (6.55) has spatially inhomogeneous nonperiodic mode (diffusion chaos)). These figures show the solution of original problem (6.55) until the moment $t = t_0$, and the solution of problem (6.58) stabilizing the zero solution of original problem for the control parameters $\varepsilon = -4$, $\beta = -3$ and for $t > t_0$. In both cases the zero solution of problem (6.55) is stabilized by the first two coordinates u and v of the solution of problem (6.58).

In the case of boundary conditions of the first kind, the restricted control problem for diffusion chaos can be reduced to stabilization of the zero solution of the first boundary value problem for the Kuramoto–Tsuzuki equation

$$W_t = W + (1 + ic_1)W_{xx} - (1 + ic_2)W|W|^2,$$
$$0 \le x \le l, \quad 0 \le t < \infty, \tag{6.61}$$
$$W(x,0) = W_0(x), \quad W(0,t) = W(l,t) = 0.$$

This problem can be written in the form (6.55) or (6.56) with zero boundary conditions at endpoints of the interval. In this case, operator Δ has the eigenvalues $-\pi^2 n^2/l^2$, $n = 1, 2, \ldots$, corresponding to the eigenfunctions $\sin \dfrac{\pi n}{l} x$. Furthermore, operator L has only distinct eigenvalues $1 - \pi^2 n^2/l^2 \pm ic_1(\pi^2 n^2/l^2)$, $n = 1, 2, \ldots$. For a small domain ($l < \pi$), all eigenvalues of the operator L have negative real parts. In this case the zero solution of the first boundary value problem (6.61) is uniformly and asymptotically stable. If $l \ge \pi$, then to stabilize the zero solution of the boundary value problem (6.61) one can use a system of four equations similar to the system (6.58). However, whereas operator L in this case has only distinct eigenvalues, result of the Theorem 6.7 can be substantially strengthened. Namely, let us consider an extended system of the following kind

$$u_t = u + u_{xx} - c_1 v_{xx} - (u^2 + v^2)(u - c_2 v) + \varepsilon_1 \varphi,$$
$$v_t = v + c_1 u_{xx} + v_{xx} - (u^2 + v^2)(u + c_2 v) + \varepsilon_2 \varphi,$$
$$\varphi_t = u + v + \beta \varphi,$$
$$0 \le x \le l, \quad 0 \le t < \infty, \tag{6.62}$$
$$u(x,0) = u_0(x), \quad v(x,0) = v_0(x), \quad \varphi(x,0) = \varphi_0(x),$$
$$u(0,t) = u(l,t) = v(0,t) = v(l,t) = \varphi(0,t) = \varphi(l,t) = 0.$$

Theorem 6.8   *For any values of the system parameters c_1, c_2, and l of problem (6.61), there exist values of control parameters ε_1, ε_2, β such that the zero solution of the problem (6.62) is uniformly and asymptotically stable.*

Proof.   We rewrite equations of the boundary value problem (6.62) in the form (6.59), where $z(x,t) = \big(u(x,t), v(x,t), \varphi(x,t)\big)^{\mathrm{T}}$, and the linear operator L has the form

$$L = \begin{pmatrix} 1 + \dfrac{\partial^2}{\partial x^2} & -c_1 \dfrac{\partial^2}{\partial x^2} & \varepsilon_1 \\ c_1 \dfrac{\partial^2}{\partial x^2} & 1 + \dfrac{\partial^2}{\partial x^2} & \varepsilon_2 \\ 1 & 1 & \beta \end{pmatrix} \tag{6.63}$$

with conditions at endpoints of the interval being determined by the first boundary value problem. To prove the theorem, it suffices to choose the values of control parameters ε_1, ε_2, β such that the spectrum of operator L lies in the left half-plane. Obviously, operator L has a discrete spectrum consisting of eigenvalues of the matrices

$$B_n = \begin{pmatrix} 1 - k_n & c_1 k_n & \varepsilon_1 \\ -c_1 k_n & 1 - k_n & \varepsilon_2 \\ 1 & 1 & \beta \end{pmatrix}, \quad k_n = \frac{\pi^2 n^2}{l^2}, \quad k_1 \leq 1, \quad n = 1, 2, \ldots . \quad (6.64)$$

The characteristic equation for the matrix B_n has a kind of

$$\lambda^3 - \lambda^2(\beta + 2 - 2k_n) + \lambda(2\beta(1 - k_n) + (1 - k_n)^2 + c_1^2 k_n^2 - \varepsilon_1 - \varepsilon_2)$$
$$- \left[\beta\left((1 - k_n)^2 + c_1^2 k_n^2\right) - (1 - k_n)(\varepsilon_1 + \varepsilon_2) + c_1 k_n(\varepsilon_2 - \varepsilon_1)\right] = 0. \quad (6.65)$$

According to the Routh–Hurwitz criterion, necessary and sufficient conditions of negativity of real parts of roots of the characteristic Eq. (6.65) are the follows

$$\begin{aligned}
-(\beta + 2 - 2k_n) &> 0, \\
0 < -[\beta d_n - (1 - k_n)\gamma_1 + c_1 k_n \gamma_2] & \\
< -(\beta + 2 - 2k_n)[2\beta(1 - k_n) + d_n - \gamma_1] &,
\end{aligned} \quad (6.66)$$

where $d_n = (1 - k_n)^2 + c_1^2 k_n^2$, $\gamma_1 = \varepsilon_1 + \varepsilon_2$, $\gamma_2 = \varepsilon_2 - \varepsilon_1$. Let us find conditions sufficient for the validity of the system of inequalities (6.66) for all $n \geq 1$. We seek parameters β, γ_1, γ_2 in the form $\beta < -2(1 - k_1) < 0$; $\gamma_2 < 0$; $\gamma_1 = 2\beta(1 - k_1) - d_1 < 0$. We suppose also, that $k_1(1 + c_1^2) \geq 1$. In this case $d_{n+1} > d_n$ for all $n \geq 1$. Then

$$\begin{aligned}
-\beta d_n + (1 - k_n)\gamma_1 - c_1 k_n \gamma_2 &\geq -\beta d_1 + (1 - k_1)\gamma_1 - c_1 k_1 \gamma_2 \\
&= -\beta d_1 + 2\beta(1 - k_1)^2 - (1 - k_1)d_1 - c_1 k_1 \gamma_2.
\end{aligned}$$

Hence, the first inequality in (6.66) is valid for all $n \geq 1$, and the inequality

$$d_1(\beta + 1 - k_1) < 2\beta(1 - k_1)^2 - c_1 k_1 \gamma_2$$

is a sufficient condition for validity of the left part of the second inequality of (6.66) for all $n \geq 1$. If $\gamma_2 \leq (2\beta(1 - k_1)^2)/(c_1 k_1)$, then the last condition is executed for any $\beta < -2(1 - k_1)$. Thus, the right part of the second inequality of system (6.66) acquires the form

$$\beta d_n - 2\beta(1-k_n)(1-k_1) + (1-k_n)d_1 + \frac{2\beta k_n(1-k_1)^2}{k_1}$$
$$> \left[\beta + 2(1-k_n)\right]\left[2\beta(k_1-k_n) + d_1 + d_n\right]$$
$$= 2\beta^2(k_1-k_n) + 4\beta(k_1-k_n)(1-k_n) + \beta(d_1+d_n) + 2(1-k_n)(d_1+d_n),$$

which is equivalent to the inequality

$$2\beta^2(k_n-k_1) + 4\beta(k_n-k_1)\left(1-k_n + \frac{1-k_1}{2k_1}\right)$$
$$-\beta d_1 - (1-k_n)(d_1 + 2d_n) > 0. \qquad (6.67)$$

It follows from the last relation with $n = 1$, that $\beta < -3(1-k_1)$. Let $k_n \geq 1$ for $n \geq n_0$. Then the inequality $\beta - 1 + 1/k_1 < 0$ is a sufficient condition for validity of (6.67) for all $n \geq n_0$. It remains to find a condition for validity of finitely many inequalities (6.67) for $2 \leq n < n_0$, $n_0 > 2$. A sufficient condition for validity of all these inequalities is

$$\beta^2 + 2\beta\left(1 - k_2 + \frac{1-k_1}{2k_1}\right) - \frac{3(1-k_1)c_1^2}{k_2 - k_1} > 0,$$

which is, obviously, carried out for all

$$\beta < -\left(1 - k_2 + \frac{1-k_1}{2k_1}\right) - \sqrt{\left(1 - k_2 + \frac{1-k_1}{2k_1}\right)^2 + \frac{3(1-k_1)c_1^2}{k_2 - k_1}} < -\frac{1-k_1}{k_1}.$$

The proof of the Theorem 6.8 is complete. $\square$

In many mathematical models, it is of great interest to solve the system of equations of the reaction–diffusion type (5.1) in a two-dimensional case. In a neighbourhood of the first bifurcation point, such a system can have as spatially homogeneous self-oscillating solutions as well as more complicated dissipative structures like leading centers, spiral waves, two-frequency self-similar solutions and diffusion chaos [Akhromeeva *et al.* (1992)]. The problem of stabilizing the thermodynamic branch of such a system in rectangular domain with second-kind boundary conditions is reduced to stabilizing a zero solution of the two-dimensional Kuramoto–Tsuzuki equation

$$W_t = W + (1 + ic_1)(W_{xx} + W_{yy}) - (1 + ic_2)W|W|^2,$$
$$0 \leq x \leq l_1, \quad 0 \leq y \leq l_2, \quad 0 \leq t < \infty,$$
$$W(x,y,0) = W_0(x,y), \qquad (6.68)$$
$$W_x(0,y,t) = W_x(l_1,y,t) = W_y(x,0,t) = W_y(x,l_2,t) = 0.$$

Similarly previous, we represent this boundary value problem (6.68) in the form (6.56), where the linear operator L looks like

$$L = I + C\Delta = \begin{pmatrix} 1 & 0 \\ 0 & 1 \end{pmatrix} + \begin{pmatrix} 1 & -c_1 \\ c_1 & 1 \end{pmatrix} \left(\frac{\partial^2}{\partial x^2} + \frac{\partial^2}{\partial y^2} \right).$$

Operator Δ in the rectangle $[0, l_1] \times [0, l_2]$ with second-kind boundary conditions has eigenvalues

$$k_{nm} = -(\pi^2 n^2 / l_1^2 + \pi^2 m^2 / l_2^2), \quad n, m = 0, 1, \ldots,$$

corresponding to eigenfunctions $\cos\left(\dfrac{\pi n}{l_1} x\right) \cos\left(\dfrac{\pi m}{l_2} y\right)$. Consequently, eigenvalues of the operator L are just the eigenvalues $1 - k_{nm} \pm i c_1 k_{nm}$ of the matrices

$$B_{nm} = I - k_{nm} C. \tag{6.69}$$

It is visible from (6.69), that for $l_1 > \pi$ and $l_2 > \pi$ the number of eigenvalues of operator L in the right half-plane substantially exceeds that in the one-dimensional case, which results in appearance in the Eq. (6.68) of complicated dissipative structures, including diffusion chaos, even if the rectangular domain is small.

For stabilization of the zero solution of the boundary value problem (6.68), we shall consider the boundary value problem

$$
\begin{aligned}
u_t &= u + u_{xx} + u_{yy} - c_1(v_{xx} + v_{yy}) - (u^2 + v^2)(u - c_2 v) + \varepsilon\varphi, \\
v_t &= v + c_1(u_{xx} + u_{yy}) + v_{xx} + v_{yy} - (u^2 + v^2)(u + c_2 v) + \varepsilon\psi, \\
\varphi_t &= u + \beta\varphi, \qquad \psi_t = v + \beta\psi, \\
0 &\leq x \leq l_1, \qquad 0 \leq y \leq l_2, \qquad 0 \leq t < \infty,
\end{aligned}
\tag{6.70}
$$

with the Neumann conditions on the boundary of the rectangle and with arbitrary initial conditions.

Theorem 6.9 *For any values of parameters c_1, c_2, l_1, l_2 of the problems (6.68), there exist values of control parameters ε, β such that zero solution of the boundary value problem (6.70) is uniformly and asymptotically stable.*

Proof. The proof of this theorem is similar to that of the Theorem 6.7. Equations of the boundary value problem (6.70) can be represented in the form (6.59) with operator L that has the discrete spectrum consisting of eigenvalues of the matrices

$$
B_{nm} = \begin{pmatrix} 1 - k_{nm} & c_1 k_{nm} & \varepsilon & 0 \\ -c_1 k_{nm} & 1 - k_{nm} & 0 & \varepsilon \\ 1 & 0 & \beta & 0 \\ 0 & 1 & 0 & \beta \end{pmatrix}, \quad n, m = 0, 1, \ldots
$$

Sufficient conditions of negativity of real parts of eigenvalues of all matrices B_{nm} are the conditions similar to conditions (6.60)

$$
\beta < -1,
$$

$$
\beta c_1^2 k_{nm}^2 (k_{nm} - 1) + (\beta + 1 - k_{nm})^2 (\beta k_{nm} + \varepsilon - \beta) < 0.
$$

Clearly, that all these conditions are satisfied for any n and m such that $k_{nm} \geq 1$ for $\varepsilon < \beta < -1$. For a finite set of n and m such that $k_{nm} < 1$, the choice of ε depends only on the value of parameter c_1 similar to how it takes place in the Theorem 6.7. Theorem 6.9 is proved. $\qquad\square$

Remark 6.2 *Zero solution of the first boundary value problem for the two-dimensional Kuramoto–Tsuzuki equation can be stabilized either by system (6.70) with the first-kind boundary conditions or by a two-dimensional boundary value problem of the form (6.62).*

Presence of chaos in many cases can be considered as the advantage of dynamical system allowing to change qualitatively its dynamics by small perturbations of system parameters. Thus for systems with concentrated parameters, there is a necessity for localization and stabilization of their unstable, especially periodic trajectories twisted in a web of singular attractor. For systems with distributed parameters, having spatially inhomogeneous nonperiodic solutions (diffusion or spatio-temporal chaos), a necessity arises for localization and stabilization of their various unstable dissipative structures. In the present chapter, the method is offered allowing to carry out these procedures successfully both for chaotic mappings and for finite-dimensional and infinite-dimensional concentrated and distributed chaotic dynamical systems.

6.3 Reconstruction of Dynamical System on Trajectory of Irregular Attractor

In some cases, including a chaos control problem, it is necessary to solve *the problem of reconstruction of system of differential equations* proceeding from the given set of points in the phase space, belonging to irregular attractor of the system. For solution of this problem, both methods of synchronization and direct methods were used (see, for example, [Brown *et al.* (1994); Baker *et al.* (1996)]). Methods of synchronization are based on the use of unidirectional connection of two chaotic systems. For strong connection, amplitudes of oscillations of the connected systems are identical and change equally chaotically. Parameters of the synchronizing system are selected such that its oscillations coincide with oscillations of the unknown system which solutions are given by the original set of points. Direct methods are based on approximation of derivatives calculated approximately on the given set of points, by some functions, more often by polynomials. As follows from papers [Rulkov *et al.* (1995); Parlitz (1996)], these methods yield good enough results if the kind of the right part of system is known and the problem is linear on unknown parameters. If the right part of system of differential equations is unknown, then such approach as our research had shown, does not allow to solve the problem as solutions of the system determined by this way do not coincide with solutions for the given set of points. Here we represent a direct method of solution of the specified problem, based on the Newton–Gauss algorithm.

The set of points determining an interval of a trajectory of an irregular attractor in the space $\mathbb{R}^m$, defines some curve which can be described, for example, parametrically $x_i(t_k)$, $i = 1, 2, \ldots, m$, $k = 1, 2, \ldots, n$. Here n is a number of values of an independent variable t (or number of points in the given set). The set $x_i(t_k)$ actually defines values x_{ik} of the net functions given on a grid $\tau = \{t_k : k = 1, 2, \ldots, n\}$. The last set defines some net trajectory $x(t_k)$ in the phase space $\mathbb{R}^m$.

We approximate the set of values x_{ik}, $i = 1, \ldots, m$, $k = 1, \ldots, n$ of the net function $x(t_k)$ in phase space $\mathbb{R}^m$ by the solution $x(t, \theta) = (x_1(t, \theta), \ldots, x_m(t, \theta))^{\mathrm{T}}$ of the system of differential equations

$$\dot{x} = F(x, \theta), \quad x \in \mathbb{R}^m, \ \theta \in \mathbb{R}^p \tag{6.71}$$

on an interval $t \in [0, t_n]$ with the initial condition $x(0) = x_0$, where $\theta = (\vartheta_1, \vartheta_2, \ldots, \vartheta_p)^{\mathrm{T}}$ is a vector of parameters and p is a dimension of a space of parameters. It is necessary to choose a solution $x(t, \theta)$ from the set of solutions of the system (6.71) such that its trajectory is closest to the net function $x(t_k)$ in the phase space $\mathbb{R}^m$. For estimation of closeness, we use the functional

$$\Phi(x,\theta) = \sum_{i=1}^{m} \sum_{k=1}^{n} \left(x_{ik} - x_i(t_k,\theta)\right)^2. \qquad (6.72)$$

The problem is to find such value of a vector of parameters θ^* for which the functional (6.72) has the least value. The necessary condition of extremum of the functional

$$\frac{\partial \Phi(x,\theta)}{\partial \vartheta_j} = 0, \quad j = 1, \ldots, p,$$

is equivalent to the system of p equations

$$\sum_{i=1}^{m} \sum_{k=1}^{n} \left(x_{ik} - x_i(t_k,\theta)\right) \cdot \frac{\partial x_i(t_k,\theta)}{\partial \vartheta_j} = 0, \quad j = 1, \ldots, p, \qquad (6.73)$$

which are nonlinear equations concerning an unknown vector of parameters $\theta = (\vartheta_1, \vartheta_2, \ldots, \vartheta_p)^{\mathrm{T}}$. For numerical solution of the system (6.73) with the help of iterative process, we expand the solution $x(t,\theta)$ of the system (6.71) in the Taylor series at a point θ^0

$$x(t_k,\theta) = x(t_k,\theta^0) + \sum_{l=1}^{p} \left.\frac{\partial x(t_k,\theta)}{\partial \vartheta_l}\right|_{\theta=\theta^0} \cdot (\theta_l - \theta_l^0) + O(|\theta - \theta^0|^2) \quad (6.74)$$

and reject all values of the second order of smallness. Then, substituting (6.74) into (6.73), we obtain a system of algebraic equations

$$\sum_{i=1}^{m} \sum_{k=1}^{n} \sum_{l=1}^{p} \left[\frac{\partial x_i(t_k,\theta)}{\partial \vartheta_j} \cdot \frac{\partial x_i(t_k,\theta)}{\partial \vartheta_l}\right]_{\theta=\theta^0} (\theta_l - \theta_l^0)$$

$$= \sum_{i=1}^{m} \sum_{k=1}^{n} \left(x_{ik} - x_i(t_k,\theta^0)\right) \cdot \left.\frac{\partial x_i(t_k,\theta)}{\partial \vartheta_j}\right|_{\theta=\theta^0}, \quad j = 1, \ldots, p,$$

which is a linear system concerning to increments $\delta_l^0 = \theta_l - \theta_l^0$. The last system has the vector form

$$\sum_{k=1}^{n} U^{\mathrm{T}}(t_k,\theta^0) U(t_k,\theta^0) \delta^0 = \sum_{k=1}^{n} U^{\mathrm{T}}(t_k,\theta^0)\left(x_k - x(t_k,\theta^0)\right), \qquad (6.75)$$

where $U_{m \times p}(t_k,\theta^0)$ is a matrix with elements $u_{ij} = \partial x_i(t_k,\theta)/\partial \theta_j$ for the value $\theta = \theta^0$, $\delta^0 = \theta - \theta^0$, x_k is a vector $(x_{1k}, x_{2k}, \ldots, x_{mk})^{\mathrm{T}}$ of the net function $x(t_k)$ at the point with number k in the phase space $\mathbb{R}^m$, $x(t_k, \theta^0)$ is the solution of the Eq. (6.71) at the moment of time $t = t_k$ and for the value $\theta = \theta^0$. Matrix U is the solution of the linear matrix inhomogeneous differential equation

$$\dot{U} = PU + Q$$

with the initial condition $U(0) = O_{m \times p}$, where $O_{m \times p}$ is the zero matrix, and matrices $P_{m \times m}$ and $Q_{m \times p}$ are the derivatives of the right part of the system (6.71) accordingly on variables x and θ

$$P = D_x F(x, \theta), \quad Q = D_\theta F(x, \theta).$$

Thus, all necessary components for solving the system (6.75) of linear algebraic equations can be obtained from the solution of the following system of differential equations

$$
\begin{aligned}
\dot{x} &= F(x, \theta^0), \\
\dot{U} &= P^0 U(t, \theta^0) + Q^0(x).
\end{aligned}
\tag{6.76}
$$

on an interval $t \in [0, t_n]$ under the initial conditions, mentioned above. Matrices P^0 and Q^0 in (6.76) are calculated for the vector value $\theta = \theta^0$ and for $x = x(t, \theta^0)$. Value δ^0 obtained at solving the system (6.75) minimizes the sum of squares

$$\Phi^0(x, \theta) = \sum_{k=1}^{n} |x_k - x(t_k, \theta^0) - U(t_k, \theta^0)\delta^0|^2$$

and this is the least squares estimation for vector $\delta = \theta - \theta^0$ in the m-dimensional phase space. Therefore, the value $\theta^1 = \theta^0 + \delta^0$ will be a specified estimation of a vector of parameters θ in relation to θ^0. It can be used further for subsequent improvement of the solution, if to use the value θ^1 instead of the value θ^0 in the above mentioned algorithm. Thus, starting with value θ^0 we find a sequence of vectors $\{\theta^\nu\}$

$$\theta^\nu = \theta^{\nu-1} + \delta^{\nu-1}, \quad \nu = 1, 2 \ldots,$$

where ν is the number of iteration. The limit of this sequence for $\nu \to \infty$ is the solution θ^* of our problem.

Example 6.8 As an example we considered a problem of reconstruction of the dynamical Rossler system on the given interval of values of a trajectory created as a result of the numerical solution of the system (6.24) for the following values of parameters: $a = 0.5$, $b = 0.75$, $\mu = 2.4$. For approximation of the given set of values, we used the system of differential equations

$$\dot{x}_i = a_i + \sum_{j=1}^{3} a_{ij} x_j + \sum_{j=1}^{3} \sum_{k=1}^{3} a_{ijk} x_j x_k, \quad i = 1, 2, 3,$$

with a quadratic polynomial in the right part, containing 30 unknown parameters a_i, a_{ij}, a_{ijk}. The developed method has allowed to solve this problem for zero approach of an initial vector θ^0. We established, that accuracy of calculation of parameters depends, mainly, on the error with which points of the initial set are given. In our case, this accuracy was defined by errors of numerical integration of the original system by the Runge–Kutta method and by mistakes of rounding. Therefore, the accuracy of calculation of values of parameters appeared so small, that the obtained nonzero values: $a_{12} = a_{13} = -1$, $a_{21} = 1$, $a_{22} = 0.5$, $a_3 = 0.75$, $a_{33} = -2.4$, $a_{313} = 1$ have completely coincided with parameters of the system (6.24) for not reduced value of the functional $\Phi \sim 10^{-10}$.

Bibliography

Akhromeeva, T.S., Kurdumov, S.P., Malinetskii, G.G., Samarskii, A.A. (1992). Nonstationary dissipative structures and diffusion chaos. — Moscow: Nauka, 541p. (in Russian).

Andrievskii, B.R., Fradkov A.L. (2003). Control of chaos: methods and applications. *I. Methods. Aut. Rem. Control*, **64**, 5, pp. 673–713 (in Russian).

Andronov, A.A., Pontryagin, L.S. (1937). Systemes grossiers. *Dokl. Akad. Nauk USSR*, **14**, pp. 247–251.

Andronov, A.A., Vitt, A.A., Khaikin, S.E. (1959). The theory of oscillations. — Moscow: Fizmatgiz, 916p. (in Russian).

Anishchenko, V.S. (2002). Nonlinear dynamics of chaotic and stochastic systems. Springer Verlag, New York, 386p.

Anishchenko, V.S., Vadivasova, T.A., Astakhov, V.V. (1999). Non-linear dynamics of chaotic and stochastic systems. Fundamental basis and selected problems. Edited by V.S. Anishchenko. — Saratov: Published by Saratov University, 368p. (in Russian).

Anosov, D.V. (1963). Ergodic properties of geodesic flows on closed Riemannian manifolds of negative curvature. *Dokl. Akad. Nauk USSR*, **151**, pp. 1250–1252. English translation in *Soviet Math. Dokl.* **4**, pp. 1153–1156.

Anosov, D.V. (1985). Dynamical systems. Contemporary mathematical problems. — Moscow: VINITI, 1, (in Russian).

Arnold, V.I. (1978a). Additional chapters of the theory of ordinary differential equations. — Moscow: Nauka, 304p. (in Russian).

Arnold, V.I. (1978b). Mathematical methods of classical mechanics. Springer Verlag, New York.

Arnold, V.I. (1983). Geometrical methods in the theory of ordinary differential equations. Springer Verlag, New York.

Arnold, V.I. (1990). Catastrophe theory. Third edition, extended. — Moscow: Nauka, 128p. (in Russian)

Arnold, V.I., Kozlov, V.V. and Neishtadt, A.I. (1988). Dynamical systems. III. Springer Verlag, Berlin, 291p.

Arnold, V.I., Afraimovich, V.S., Il'yashenko, Yu.S. and Shil'nikov, L.P. (1999). Bifurcation theory and catastrophe theory. Springer Verlag.

Baker, C.L., Collub J.P. and Blackburn J.A. (1996). Inverting chaos: extracting system parameters from experimental data. *Chaos*, 4, pp. 528–533.

Bautin, N.N. and Leontovich, E.A. (1990). Methods and techniques for quality analysis of planar dynamic systems. — Moscow: Nauka, 488p. (in Russian).

Berger, P., Pomeau, Y. and Vidal, C. (1984). L'ordre dans le chaos: vers une approche deterministe de la turbulence. Paris: Hermann.

Bilov, B.F., Vinograd, R.E., Grobman, D.M. and Nemytskii, V.V. (1966). Theory of Lyapunov exponents. — Moscow: Nauka, 576p. (in Russian).

Brown, R., Rulkov, N.F. and Tracy, E.R. (1994). Modelling and synchronizing chaotic systems from time-series data. *Phys. Rev.* E, **49**, pp. 3784–3800.

Chen, G. and Dong, X. (1998). From chaos to order: perspectives, methodologies and applications. World Scientific, Singapore, 743p.

Chen, X. (1996). Lorenz equations. Part I: Existence and nonexistence of homoclinic orbits. *SIAM J. Math. Anal.*, **27**, 4, pp. 1057–1069.

Chua, L.O., Komuro, M. and Matsumoto, T. (1986). The double scroll family. *IEEE Trans. Circuits and Syst.* CAS-33, pt. 1, 2, pp. 1073–1118.

Chua's circuit: a paradigm for chaos. (1993). Edited by Madan, R.N. World scientific, Singapore.

Coddington, E.A. and Levinson N. (1955). Theory of ordinary differential equations. McGraw-Hill, New York, 429p.

Collet, P., Eckmann, J.P. and Lanford, O.E. (1980). Universal properties of maps of an interval. *Comm. Math. Phys.* **76**, pp. 211–254.

Cook, A.E. and Roberts, P.H. (1970). The Rikitaki two-disc dynamo system. *Proc. Camb. Phil. Soc.*, **68**, pp. 547–569.

Daletskii, Y.L. and Krein, M.G. (1974). Stability of solutions of differential equations in Banach space. Amer. Math. Soc. 534p. (translated from Russian).

Dernov, A.V. (2001a). Regular dynamics and diffusion chaos in the Brusselator model. *Differential equations*, **37**, 11, pp. 1631–1633.

Dernov, A.V. (2001b). Stabilization of unstable periodic orbits of one-dimensional chaotic mappings. *Nonlinear Dynamics and Control*, **1**, Eds. S.V. Emelyanov and S.K. Korovin. — Moscow: Fizmatlit, pp. 247–252 (in Russian).

Dernov, A.V. (2002a). Diffusion of the capital and demand in the model of self-developing market economy. *Nonlinear Dynamics and Control*, **2**, Eds. S.V. Emelyanov, S.K. Korovin. — Moscow: Fizmatlit, pp. 233–242 (in Russian).

Dernov, A.V. (2002b). On new approaches to a problem of chaos control. *Proc. of Moscow State University*, **1**, p. 31 (in Russian).

Dernov, A.V. and Magnitskii, N.A. (2005). Transition to chaos in a nonclassical reaction–diffusion system. *Differential equations*, **41**, 12, pp. 1751–1756.

Eckmann, J.P. and Ruelle, D. (1985). Ergodic theory of chaos and strange attractors. *Rev. Mod. Phys.*, **57**, 3, pp. 617–656.

Ershov, S.V., Malinetskii, G.G. and Rusmaikin, A.A. (1989). A generalised two-disk dynamo model. *Geophys. Astrophys. Fluid. Dyn.*, **47**, pp 251–277.

Feigenbaum, M.J. (1978). Quantitative universality for a class of nonlinear transformations. *J. Stat. Phys.*, **19**, pp. 25–52.

Feigenbaum, M.J. (1980). Universal behavior in nonlinear systems. *Los Alamos. Sci.*, **1**, 1, pp. 4–27.

Gibbon, J.D. (1981). Dispersive instabilities in nonlinear systems: the real and complex Lorenz equations. *Proc. Int. Symp. on Synergetics*. Ed. H. Haken, pp. 92–101. Springer Verlag, New York.

Gilmor, R. (1981). Catastrophe theory for scientists and engineers. Wiley and Sons, New York.

Gribov, A.F. and Krischenko, A.P. (2001). Analytical conditions of existence of homoclinic loop in the Chua circuits. *Nonlinear Dynamics and Control*, **1**, Eds. S.V. Emelyanov, S.K. Korovin. — Moscow: Fizmatlit, pp. 263–268 (in Russian).

Guckenheimer, J.A. (1976). Strange, strange attractor. In *The Hopf Bifurcation and Its Applications*, Eds. J.E. Marsden and M. McCracken. Springer Verlag, New York.

Guckenheimer, J. and Holmes, P. (1983). Nonlinear oscillations, dynamical systems and bifurcations of vector fields. Springer Verlag, New York, 453p.

Guckenheimer, J. and Williams, R.F. (1979). Structural stability of Lorenz attractors. *Publ. Math. IHES*, **50**, pp. 59–72.

Gurel, D., and Gurel, O. (1983). Topics in current chemistry: Oscillating chemical reactions. Springer Verlag, Heidelberg.

Haken, H. (1983). Synergetics, an introduction. Springer Verlag, Berlin, Third edition.

Hartman, P. (1964). Ordinary differential equations. Wiley and Sons, New York.

Hassard, B.D., Kazarinoff, N.D. and Wan, Y.H. (1981). Theory and applications of Hopf bifurcation. Cambridge Univ. Press, Cambridge, 311p.

Hasselblatt, B. and Katok, F. (2003). A first course in dynamics. Cambridge Univ. Press, 436p.

Henon, M. (1976). A two-dimensional mapping with a strange attractor. *Comm. Math. Phys.*, **50**, p. 69.

Hirsch, M. and Smale, S. (1974). Differential equations, dynamical systems and linear algebra. Academic Press, New York, 358p.

Hirsch, M.W., Smale, S. and Devaney, R.L. (2004). Differential equations, dynamical systems and an introduction to chaos. Elsevier, Amsterdam, 417p.

Jakobson, M.V. (1981). Absolutely continuous measures for one parameter families of one-dimensional maps. *Comm. Math. Phys.*, **81**, pp. 39–88.

Kaloshin, D.A. (2001). Search and stabilization of unstable saddle cycles in Lorenz system. *Differential equations*, **37**, 11, pp. 1636–1639.

Kaloshin, D.A. (2003). On construction of bifurcation surface of homoclinic butterfly in the Lorenz system. *Differential equations*, **39**, 11, pp. 1648–1650.

Kaloshin, D.A., Magnitskii, N.A. and Sidorov, S.V. (2003). On some features of transition to chaos in the system of Lorenz equations. *Nonlinear Dynamics and Control*, **3**, Eds. S.V. Emelyanov, S.K. Korovin. — Moscow: Fizmatlit, pp. 99–106 (in Russian).

Komuro, M., Matsumoto, T., Chua, L.O. *et al.* (1991). Global bifurcation analysis of the double scroll circuit. *Intern. J. Bifurcation and chaos*, **1**, 1, pp. 139–182.

Knyazev, E.A., Magnitskii, N.A. and Sidorov, S.V. (1999). Stabilization of unstable stationary points in the equations with delay argument. *Nonlinear dynamics. Trans. ISA RAS.* — Moscow: Editorial URSS, pp. 133–141.

Kuramoto, Y. and Tsuzuki, T. (1975). On the formation of dissipative structures in reaction–diffusion systems. *Progr. Theor. Phys.*, **54**, 3, pp. 687–699.

Kuznetzov, S.P. (2001). Dynamical chaos. — Moscow: Fizmatlit, 296p. (in Russian).

Landa, P.S. (1996). Nonlinear oscillations and waves in dynamical systems. Kluwer Acad. Publ., Boston.

Lefever, R. and Prigogine, I. (1968). Symmetry-breaking instabilities in dissipative systems. *J. Chem. Phys.*, **48**, pp. 1695–1700.

Leonov, G.A. (1988). An estimate of the bifurcation parameters of the separatrix loop of a saddle point of the Lorenz system. *Differential equations*, **24**, 6, pp. 634–638.

Li, T.Y. and Yorke, J.A. (1975). Period three implies chaos. *Amer. Math. Monthly*, **82**, 10, pp. 982–985.

Lorenz, E.N. (1963). Deterministic nonperiodic flow. *J. Atmos. Sci.*, **20**, pp. 130–141.

Loskutov, A.Y. (2001). Chaos and control in dynamical systems. *Nonlinear Dynamics and Control*, **1**, Eds. S.V. Emelyanov, S.K. Korovin. — Moscow: Fizmatlit, pp. 163–216 (in Russian).

Loskutov, A.Y. and Mikhailov, A.S. (1990). Introduction into synergetics. — Moscow: Nauka, 272p. (in Russian).

Machta, J. (1983). Power low decay of correlation in a billiard problem. *J. of Stat. Phys.*, **32**, 3, pp. 555–564.

Mackey, M. and Glass, L. (1977). Oscillations and chaos in physiological control systems. *Science*, **197**, pp. 287–289.

Magnitskii, N.A. (1991). Mathematical model of self-developing market economy. *Trans. ISR AS USSR*, pp. 16–22. (in Russian).

Magnitskii, N.A. (1992). Asymptotic methods of analysis of nonstationary control systems. — Moscow: Nauka, 160p. (in Russian).

Magnitskii, N.A. (1995). Hopf bifurcation in the Rossler system. *Differential equations*, **31**, 3, pp. 538–541.

Magnitskii, N.A. (1996). On stabilization of fixed points of chaotic mappings. *Dokl. Akad. Nauk USSR*, **351**, 2, pp. 175–177.

Magnitskii, N.A. (1997a). On stabilization of fixed points of chaotic dynamical systems. *Dokl. Akad. Nauk USSR*, **352**, 5, pp. 610–612.

Magnitskii, N.A. (1997b). On stabilization of unstable cycles of chaotic mappings. *Dokl. Akad. Nauk USSR*, **355**, 6, pp. 747–749.

Magnitskii, N.A. (1997c). Stabilization of unstable periodic orbits of chaotic maps. *Computers Math. Applic.*, **34**, 2–4, 1997, pp. 369–372.

Magnitskii, N.A. (2004). On nature of chaotic attractors. *Nonlinear Dynamics and Control*, **4**, Eds. S.V. Emelyanov, S.K. Korovin. — Moscow: Fizmatlit, pp. 37–58 (in Russian).

Magnitskii, N.A. (2005). On singular attractors of dissipative systems of nonlinear ordinary differential equations. *Proc. of ENOC-2005 Int. Conf.*, Univ.

Technol., Eindhoven, pp. 1285–1294.

Magnitskii, N.A. and Sidorov, S.V. (1998). Control of chaos in nonlinear dynamical systems. *Differential equations*, **34**, 11, pp. 1051–1509 (in Russian).

Magnitskii, N.A. and Sidorov, S.V. (1999). Some approaches to the control problem for diffusion chaos. *Differential equations*, **35**, 5, pp. 669–674.

Magnitskii, N.A. and Sidorov, S.V. (2000). Stabilization of unstable periodic solutions of delay equations. *Differential equations*, **36**, 11, pp. 1634–1638.

Magnitskii, N.A. and Sidorov, S.V. (2001a). Nonlinear dynamics on the life and social science. *IOS Press, NATO Science Series. Ser. A. Life Science*, **320**, pp. 33–44.

Magnitskii, N.A. and Sidorov, S.V. (2001b). Localization and stabilization of unstable solutions of chaotic dynamical systems. *Nonlinear Dynamics and Control*, **1**, Eds. S.V. Emelyanov, S.K. Korovin. — Moscow: Fizmatlit, pp. 217–246 (in Russian).

Magnitskii, N.A. and Sidorov, S.V. (2001c). A new view of the Lorenz attractor. *Differential equations*, **37**, 11, pp. 1568–1579.

Magnitskii, N.A. and Sidorov, S.V. (2002). Transition to chaos in nonlinear dynamical systems through a subharmonic cascade of bifurcations of two-dimensional tori. *Differential equations*, **38**, 12, pp. 1703–1708.

Magnitskii, N.A. and Sidorov, S.V. (2003a). On scenarios of transition to chaos in nonlinear dynamical systems described by ordinary differential equations. *Nonlinear Dynamics and Control*, **3**, Eds. S.V. Emelyanov, S.K. Korovin. — Moscow: Fizmatlit, pp. 73–98 (in Russian).

Magnitskii, N.A. and Sidorov, S.V. (2003b). Finding homoclinic and heteroclinic contours of singular points of nonlinear systems of ordinary differential equations. *Differential equations*, **39**, 11, pp. 1593–1602.

Magnitskii, N.A. and Sidorov, S.V. (2004a). Actual problems of chaotic dynamics in dissipative systems of nonlinear ordinary differential equations. *Dynamics of non-homogeneous systems*, **7**, pp. 5–43.

Magnitskii, N.A. and Sidorov, S.V. (2004b). New methods for chaotic dynamics. — Moscow: Editorial URSS, 320p. (in Russian).

Magnitskii, N.A. and Sidorov, S.V. (2004c). Rotor type singular points of non-autonomous systems of differential equations and their role in generation of singular attractors of nonlinear autonomous systems. *Differential Equations*, **40**, 11, pp. 1579–1593.

Magnitskii, N.A. and Sidorov, S.V. (2005a). Transition to chaos in the Lorenz system via a compete double homoclinic bifurcation cascade. *Computational Mathematics and Modelling*, **16**, 1, pp. 26–37.

Magnitskii, N.A. and Sidorov, S.V. (2005b). Distributed model of a self-developing market economy. *Computational Mathematics and Modelling*, **16**, 1, pp. 83–97.

Magnitskii, N.A. and Sidorov, S.V. (2005c). On transition to diffusion chaos through a subharmonic cascade of bifurcations of two-dimensional tori. *Differential Equations*, **41**, 11, pp. 1550–1558.

Malinetskii, G.G. and Potapov, A.B. (2000). Modern problems of nonlinear dynamics. Second edition. — Moscow: Editorial URSS, 335p. (in Russian).

Malkin, I.G. (1966). Theory of stability of motion. — Moscow: Nauka, 530p. (in Russian).

Mandelbrot, B. (1982). The fractal geometry of nature. Freeman, San Francisco.

Manneville, P. and Pomeau, Y. (1980). Different ways to turbulence in dissipative dynamical systems. *Physica*, **1D**, p. 219.

Marsden, J.E. and McCracken, M. (1976). The Hopf bifurcation and its applications. Springer Verlag, New York.

Moser, J. (2001). Stable and random motions in dynamical systems. *UK Nonlinear News*, **25**. Princeton Univ. Press, 210p.

Neimark, Yu.I. and Landa, P.S. (1992). Stochastic and chaotic oscillations. Kluwer, Dordrecht.

Newhouse, S., Ruelle, D. and Takens, F. (1979). Occurrence of strange axiom A attractors near quasi-periodic flows on T^m, $m \geq 3$. *Comm. Math. Phys.*, **64**, 1, pp. 35–40.

Nicolis, G. and Prigogine, I. (1977). Self-organization in non-equilibrium systems. Wiley and Sons, New York.

Nitecki, Z. (1971). Differentiable dynamics: an introduction to the orbit structure of diffeomorphisms. MIT Press, Cambridge, Mass., 282p.

Novikov, M.D. and Pavlov, B.M. (2000). On one nonlinear model with complex dynamics. *Vestnik MSU*, **2**, pp. 3–7 (in Russian).

Ott, E., Grebogi, C. and Yorke, J.A. (1990). Controlling chaos. *Phys. Rev. Lett.*, **4**, pp. 1196–1199.

Palis, J. and De Melo, W. (1982). Geometrical theory of dynamical systems: an introduction. Springer Verlag, New York.

Palis, J. and Takens, F. (1993). Hyperbolicity and sensitive chaotic dynamics at homoclinic bifurcations. Univ. Press, Cambridge.

Parlitz, U. (1996). Estimating model parameters from times series by autosynchronization. *Phys. Rev. Lett.*, **76**, pp. 1232–1235.

Pontryagin, L.S. (1982). Ordinary differential equations. — Moscow: Nauka, 331p. (in Russian).

Popov, V.V. (1989). Business cycle and rate of return in the USA. — Moscow: Nauka, 176p. (in Russian).

Poston, T. and Stewart, I. (1972). Catastrophe theory and its applications. Pitman, London, 608p.

Prigogine, I. and Stengers, I. (1984). Order out of chaos. Bentam Books, New-York.

Pyragas, K. (1992). Continuous control of chaos by self-controlling feedback. *Phys. Lett. A.*, **170**, pp. 421–428.

Rabinovich, M.I. and Fabrikant, A.L. (1979). Stochastic self-modulation of waves in non equilibrium environments. *Zh. Eksp. Teor. Fiz.*, **77**, pp. 617–629.

Robinson, C. (1995). Dynamical systems. Second edition. CRC Press, New York.

Rossler, O.E. (1976). An equation for continuous chaos. *Phys. Lett. A*, **57**, 5, pp. 397–398.

Ruelle, D. and Takens, F. (1971). On the nature of turbulence. *Commun. Math. Phys.*, **20**, 3, pp. 167–192.

Rulkov, N.F., Sushchik, M.M., Tsimring, L.S. and Abarbanel, H.D. (1995). Generalized synchronization of chaos in directional coupled chaotic systems. *Phys. Rev. E*, **51**, 2, pp. 980–994.

Saltzman, B. (1981). Finite amplitude free convection as an initial value problem. *J. Atmos. Sci.*, **19**, p. 329.

Samarskii, A.A. and Gulin, A.V. (1989). Computational methods. — Moscow: Nauka, 430p.

Schuster, H.G. (1984). Deterministic chaos. Physik-Verlag, Weinheim, 240p.

Sepulchre, J. and Babloyantz, A. (1994). Controlling chaos in a network of oscillators. *Phys. Rev.*, **E48**, 2, pp. 119–125.

Shabat, B.V. (1985). Introduction into complex analysis. Part 1. — Moscow: Nauka, 336p.

Sharkovskii, A.N. (1964). Cycles coexistence of continuous transformation of line in itself. *Ukr. Math. Journal*, **26**, 1, pp. 61–71.

Sharkovskii, A.N, Maistrenko, Y.L. and Romanenko, E.Y. (1993). Difference equations and their applications. Kluwer, Dordrecht.

Shil'nikov, L.P. (1970). A contribution to the problem of the structure of an extended neighborhood of a rough equilibrium state of saddle–focus type. *Math. USSR Sb.*, **81(123)**, 1, pp. 92–103.

Shil'nikov, L.P. (1980). Bifurcation theory and Lorenz model. In *Marsden J.E. and McCracken M., The Hopf Bifurcation and its Applications.* — Moscow: Mir, pp. 317–335. (Addition II to Russian translation).

Shil'nikov, L.P. (1994). Chua's circuit: rigorous results and future problems. *Int. J. Bifurcation and Chaos*, **4**, 3, pp. 488–519.

Shil'nikov, A.L., Shil'nikov, L.P. and Turaev, D.V. (1993). Normal forms and Lorenz attractors. *Int. J. Bifurcation and Chaos*, **3**, 5, pp. 1123–1139.

Shinbort, T., Grebogi, C. and Yorke, J.A. (1993). Using small perturbations to control chaos. *Nature*, **363**, 3, 1993, pp. 411–417.

Sinai, J.G. (1970). Dynamical systems with elastic mappings. Ergodic properties of scattering billiards. *Russ. Math. Surv.*, **25**, 2, pp. 141–192 (in Russian).

Sinai, J.G. (1995). Modern problems of ergodic theory. **1**. — Moscow: Fizmatgiz, 243p. (in Russian).

Smale, S. (1966). Structurally stable systems are not dense. *Amer. J. Math.*, **88**, pp. 491–496.

Smale, S. (1967). Differentiable dynamical systems. *Bull. Amer. Math. Soc.*, **73**, pp. 747–817.

Smale, S. (1998). Mathematical problems for the next century. *Math. Intelligencer*, **20**, 2, pp. 7–15.

Sparrou, C. (1982). The Lorenz equations: bifurcations, chaos and strange attractors. Springer Verlag, New York.

Strogatz, S.H. (2001). Nonlinear dynamics and chaos: with applications to physics, biology, chemistry and engineering. Second edition. Rerseus Books Group, Cambridge, 478p.

Sviregev, Y.M. (1987). Nonlinear waves, dissipative structures and catastrophes in ecology. — Moscow: Nauka, 368p. (in Russian).

Thompson, J.M.T. and Stewart, H.B. (2002). Nonlinear dynamics and chaos: geometrical methods for engineers and scientists. Wiley and Sons, New

York, 460p.

Tikhonov, A.N. and Arsenin, V.Y. (1977). Solutions of ill-posed problems. Wiley and Sons, New York.

Tucker, W. (2002). A rigorous ODE solver and Smale's 14th problem. *Found. Comput. Math.*, pp. 53–117.

Turing, A. (1952). On the chemical basic of morphogenesis. *Phil. Trans. Roy. Soc. London, Ser. A*, **237**, pp. 37–52.

Vallis, G.K. (1986). A chaotic dynamical system. *Science*, **232**, pp. 243–245.

Vallis, G.K. (1988). Conceptual models of El Nino. *J. Geophys. Res.*, **93**, pp. 13979–13991.

Wiggins, S. (2003). Introduction to applied nonlinear dynamical systems and chaos. Second edition. Springer Verlag, New-York.

Williams, R.F. (1979). The structure of the Lorenz attractors. *Publ. Math. IHES*, **50**, pp. 321–347.

Yorke, J.A. and Yorke, E.O. (1979). Metastable chaos: the transition to sustained chaos oscillations in a model of Lorenz. *J. Stat. Phys.*, **21**, pp. 263–277.

Zaslavskii, G.M. (1984). Stochasticity of dynamical systems. — Moscow: Nauka, 272p. (in Russian).

Zhou, T. and Chen, G. (2005). Classification of chaos in 3-d autonomous quadratic systems. **I**: Basic framework and methods. *Int. J. of Bifur. Chaos*, accepted.

Index